Switching Networks: Recent Advances

Network Theory and Applications

Volume 5

Switching Networks: Recent Advances

Edited by

Ding-Zhu-Du

and

Hung Q. Ngo

Department of Computer Science and Engineering,
University of Minnesota,
Minneapolis, U.S.A.

KLUWER ACADEMIC PUBLISHERS

DORDRECHT / BOSTON / LONDON

A C.I.P. Catalogue record for this book is available from the Library of Congress.

ISBN-13:978-1-4613-7976-8 e-ISBN-13:978-1-4613-0281-0
DOI: 10.1007/978-1-4613-0281-0

Published by Kluwer Academic Publishers,
P.O. Box 17, 3300 AA Dordrecht, The Netherlands.

Sold and distributed in North, Central and South America
by Kluwer Academic Publishers,
101 Philip Drive, Norwell, MA 02061, U.S.A.

In all other countries, sold and distributed
by Kluwer Academic Publishers,
P.O. Box 322, 3300 AH Dordrecht, The Netherlands.

Printed on acid-free paper

Contents

Foreword

The switching network is an important classic research area in telecommunication and computer networks. Its importance stems from both theory and practice. In fact, some open problems, such as Beneš conjecture on shuffle-exchange networks and Chung-Ross conjecture on multirate rearrangeability, still attract many researchers and the further development in optical networks requires advanced technology in optical switching networks.

In 1997, we had a workshop in switching networks held in NSF Science and Technology Center in Discrete Mathematics and Theoretical Computer Science (DIMACS), at Princeton University. This workshop was very successful. Many participants wished to have a similar activity every two or three years. This book is a result of such a wish. We are putting together some important developments in this area during last several years, including articles in fault-tolerance, rearrangeability, nonblocking, optical networks, random permutation generation, and layout complexity. Some of those articles are research papers and some are surveys. All articles were reviewed. We would like to mention two special problems studied in those articles.

The first one is the wide-sense nonblocking in symmetric Clos networks $C(n, m, r)$ where n, m, are r are the number of inlets in each input crossbar switches, the number of center switches, are the number of input switches, respectively. The paper of Du/Fishburn/Gao/Hwang (page 89-100) presents a very interesting negative result that a well-known routing algorithm Packing which saves $n/2$ center switches for $r = 2$ does not save anything for $r = 3$. Motivated from this result, Tsai, Wang, and Hwang (see reference [11] in page 116) showed that for r sufficiently large, no routing algorithm can save anything. Now, the paper of Kwork/Li (page 101-116) presents a routing algorithm which does save something for $r = 3$. Thus, the study of wide-sense nonblocking is still requiring a lot of efforts.

The second one is Beneš conjecture. This conjecture is very well-known in switching networks since it had been solved twice in past twenty years and both proofs were pointed out to be wrong later. In 1997, Çam claimed that he has a correct new proof. However, during the last three years, his proof has not been confirmed officially by the journal which processes his paper. Whether his proof is correct or not is still a mystery. We are including a survey article of Çam (page 215-257) in which some of his idea are explained. We hope that this may speed up the process to solve this mystery.

We wish to thank all who contributed their articles to this book and reviewed articles for this book. We hope this collection of articles would serve as a useful reference in study of switching networks.

Ding-Zhu Du
Hung Quang Ngo

SWITCHING NETWORKS: RECENT ADVANCES
D.-Z. Du and H.Q. Ngo (Eds.) pp. 1-23

Fault-tolerance for Switching Networks

John D. Carpinelli
Department of Electrical and Computer Engineering
New Jersey Institute of Technology, Newark, NJ 07102
E-mail: `carpinelli@njit.edu`

Hamed Nassar
Department of Engineering Sciences
Faculty of Petroleum Engineering
Suez Canal University, Suez - Egypt
E-mail: `nassar@hotmail.com`

Contents

1 Introduction

Switching networks (SNs) are typically placed at the center of the systems they serve. Consequently, if an SN fails, system performance is bound to deteriorate or, in the worst case, the system may collapse totally. To limit these troubling consequences, the SN has to be *fault tolerant* [11]. Loosely defined, a fault tolerant SN is a SN that can still work in the presence of faults. Unfortunately, there is no universal, unique definition which can be applied to a network to see whether it is fault tolerant or not. To a great extent, fault tolerance is "in the eyes of the designer!" Specifically, a designer arbitrarily sets for the network in question a *fault tolerance model* [3, 4], defining what a fault is, how many such faults the network can (or should) tolerate, and exactly what is meant by "tolerate."

The basic idea of fault tolerance is to have in the SN some redundant paths [12, 22, 24], since, when a fault occurs, one or more of the original network paths may be disabled. Most fault tolerance techniques work at the *SN level*, by adding extra switches and links, preserving the basic design of the original (nonfault tolerant) SN switch [8, 26]. Some techniques, however, work also at the *switch level*, by introducing modifications to the switch design [14, 16].

Whatever technique is used, a fault tolerant SN must be equipped with a mechanism to automatically switch a normal path disrupted by a fault to a suitable redundant path. This switching mechanism may depend on multiplexers (1×2 switches) and demultiplexers (2×1 switches), placed at the SN terminals. Otherwise, the mechanism may depend on dual ports [2] built into the components using the SN.

An obvious and easy technique to achieve fault tolerance is to duplicate the SN [13]. In this approach, two copies of the same SN are configured in parallel, one active and the other standby. If a fault occurs in the active copy, the paths affected are established on the standby copy. The advantage of this technique is that performance remains the same under faulty conditions as under normal (no fault) conditions. The disadvantage of duplication is increased cost and size.

This duplication technique gives an immediate idea of the relationship between the amount of the extra hardware and performance. Adding extensive extra hardware guarantees good performance under faulty conditions, but increases cost and size. Adding little hardware, on the other hand, decreases cost and size but presents performance degradation under faulty conditions. As a consequence, a compromise must be sought by the fault tolerance designer to achieve minimal extra hardware and maximal performance under faulty conditions. Needless to say, any fault tolerance technique should cause no performance degradation under normal conditions.

This chapter is organized as follows. Section 2 introduces some basic terminology and notation. Section 3 presents the fault tolerance model, which in effect is the definition of fault tolerance for a specific SN. Sections 4, 5, 6 and 7 give four fault tolerance techniques with illustrative applications. Section 8 gives an idea of the complementary topic: fault detection and location. Finally, Section 9 gives the concluding remarks.

2 Terminology and Notation

Most of the terminology and notation used in this chapter will be introduced as they arise. However, the highly recurring few will be given here to ensure easy reading.

An SN is composed of a number of stages, each basically a column of (crossbar) switches. The stages are numbered from left to right starting with 0. In each stage, the switches are numbered from top to bottom starting also with 0. Switch i in stage j is denoted by $X(i,j)$.

The input terminals and output terminals of the SN are called *inlets* and *outlets*, respectively. This is done to differentiate them from the inputs and outputs of the individual switches or of the stages. The inlets and outlets, as well as the inputs and outputs of the switches and of the stages, are numbered from top to bottom starting with 0.

An SN with M inlets and N outlets, is said to be of *size* $M \times N$. Without loss of generality, this chapter deals with square SNs, i.e. SNs having $M = N$. For simplicity, then, an $N \times N$ SN will be called an SN of size N. To avoid trivialities, the networks of this chapter are assumed to have $N \geq 8$.

The word *fault* is pivotal in the chapter, and therefore its meaning must be made clear. Faults in SNs can be either in switches or in links. A switch fault can be manifested in many ways, such as the inability to execute some mappings properly, or the inability to transmit signals from one side to the other faithfully. For instance, the switch can be stuck in one of its legal states, such as connecting each input to the output with the same number, giving a proper connection only if that state happens to be the desired state. Also, a switch can be stuck in a partially legal state, such as connecting only one input to one output. This permanent state again may happen to be the desired connection to establish a given path. A switch can be stuck in an illegal state, such as connecting all its inputs to each other, being in this case a useless switch. All these faults will be called switch faults, and counted by the number of paths they disrupt. For example, if a switch fault disrupts *three* paths, we say that we have *three* faults.

A link fault is manifested in an open circuit, stopping signals from propagating down the link. Clearly, a link fault gives rise always to only one fault.

3 Fault Tolerance Model

Given an SN, one cannot decide whether it is fault tolerant or not. Only when one is given a fault tolerance model [3, 14] can this decision be made. Actually, a fault tolerant SN is developed to meet a predefined fault tolerance model. This model contains three elements: the *fault model*, the *fault tolerance criterion*, and the *fault tolerance size.*

The first element, the fault model, lists the faults that the network is assumed to encounter, and those the network is assumed to be immune from. In effect, it lists the faults the network can tolerate and those it cannot. A typical fault model (which is also adopted in all the fault tolerant SNs of this chapter) is as follows.

1. *Any component can fail.* The components meant here are those of the original network, i.e. the switches and links, before adding the extra hardware to supply fault tolerance. This assumption in effect states that the extra hardware is installed to tolerate faults in these components.

2. *The extra hardware added to provide fault tolerance cannot fail.* This assumption is clearly unrealistic, but has to be made for if it is not made it would be impossible to propose any fault tolerance SN. From a practical point of view, this assumption can be justified because the extra hardware usually remains idle under normal conditions and thus remains in good shape all the time.

The second element of the fault tolerance model is the fault tolerance criterion. This criterion states basically what is meant by 'fault tolerance' for the SN under consideration. The ideal such criterion is *full recovery*, meaning the ability of the SN to function under faulty conditions in exactly the same way, i.e., give the same performance, as under normal conditions.

A less stringent criterion is *full-access retention*, meaning that under faulty conditions each outlet should still be accessible from each inlet. This criterion does not require, for example, that a set of paths realizable simultaneously under normal conditions be *simultaneously* realizable under faulty conditions.

The last element of the fault tolerance model is the fault tolerance size which refers to the maximum number of faults, K, that the system can tolerate, i.e. satisfy its fault tolerance criterion with. When $K = 1$, the SN is said to be single fault-tolerant. A network that can tolerate $K > 1$ specific (as opposed to arbitrary) faults is called *K-robust*. Recall that one fault corresponds to one disrupted path.

In the next four sections, four fault tolerance techniques will be presented.

4 Adding an Extra Stage

Adding an extra stage to the SN is one of the techniques to add fault tolerance to certain SNs. This technique has been introduced [1] through the binary banyan switching network (BBSN) [19]. This network is characterized by two salient features. First, it uses switches of only size 2×2, hence the "binary" in the name. Second, there is one and only one path between each inlet-outlet pair, hence the "banyan" in the name. A size N BBSN has $\nu = \log N$ stages, each with $N/2$ switches. Admittedly, the BBSN is the *simplest* SN due to the remarkable fact that it uses only the simplest switches as building blocks. Figure 1 shows a size 8 BBSN.

It turns out that the links of this network can be connected in any pattern provided the condition that there is one and only one path between each inlet-outlet pair is met. It is interesting to note in passing that this network has been rediscovered by researchers repeatedly and presented anew in the literature numerously, each time under a different name, just because a variation in the link pattern was made! Examples of these names are: the Cube [1], Shuffle Exchange [14], Delta [23], and Gamma [29]. The rediscovery phenomenon stopped when a proof was given [27] that the topology of the BBSN remains unchanged regardless of the changes made to the link pattern.

One of the primary advantages of the BBSN is its *distributed* routing [28]. This simply means that a packet sent from inlet S to outlet D; $S, D = 0, 1, \cdots, N$; can set up its own path dynamically as it moves from stage to

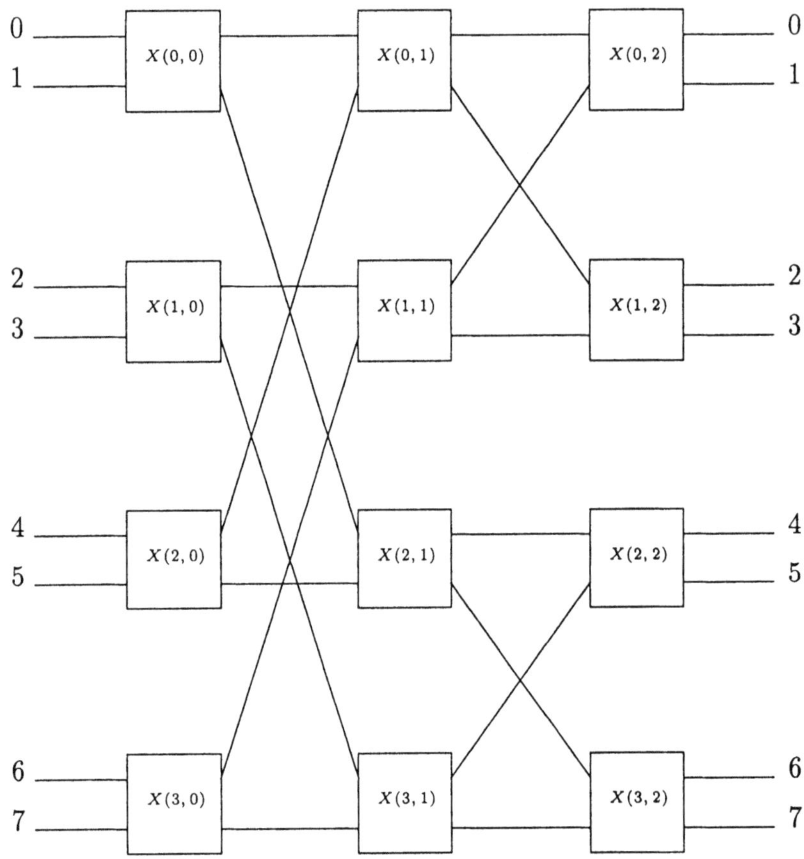

Figure 1: Binary banyan switching network (BBSN) of size 8

stage. Basically, S and D are used to generate a routing tag made up of ν bits each used to set one of the ν switches along the path.

The Extra Stage version of the BBSN of Figure 1 is shown in Figure 2. Note that each switch in the first and last stages is surrounded from both sides by multiplexers and demultiplexers in such a way that a switch can be bypassed if need be. In fact, with the help of the multiplexers and demultiplexers, the first and last stages can be made active (used) or inactive (bypassed). This facility is what makes this SN fault tolerant as explained next.

Operation and Fault-tolerance Model

In the Extra Stage BBSN of size N, stage 0 is normally made inactive and stage $\nu-1$ made active. This makes the network behave exactly as a normal BBSN, even in the way the routing tag is generated and used. If a fault occurs in stage $\nu - 1$ it is made inactive and stage 0 is made active instead. This configuration also makes the network behave as a normal BBSN.

Now, if a fault occurs in stage x, then both stages 0 and ν are made active. This leads to a network of $\nu + 1$ stages. Better yet, it leads to two (disjoint) paths, instead of one, for each inlet-outlet pair. The $(\nu + 1)$-bit routing tag is generated in such a way [1, 2] that the path affected by the fault is avoided and the other used.

The fault model in the Extra Stage BBSN is the typical model mentioned in Section 3. Its fault size is 1. Its fault tolerance criterion is full access retention. The criterion is not full recovery, since no N paths can be established simultaneously in the presence of a fault, as is the case in the normal BBSN. The network is 4-robust since two switches, such as $X(0,1)$ and $X(2,1)$ (counting as 4 faults since they route 4 data paths,) can fail and access retention can still be obtained.

The first advantage of the Extra Stage BBSN is that it is easy to operate. For one, the multiplexers and demultiplexers have to be adjusted only once after a fault occurs. For another, the routing tag is generated under faulty conditions in a way similar to that under normal conditions. The second advantage is that it works at the SN level. As such, it uses the same binary switches as the normal network. The third advantage is that it uses a minimum of extra links. Other than links associated with the multiplexers and demultiplexers, no additional links, interstage or intrastage, are needed.

The Extra Stage BBSN also has two disadvantages. First, it uses a large amount of extra switches and multiplexers/demultiplexers. A size N net-

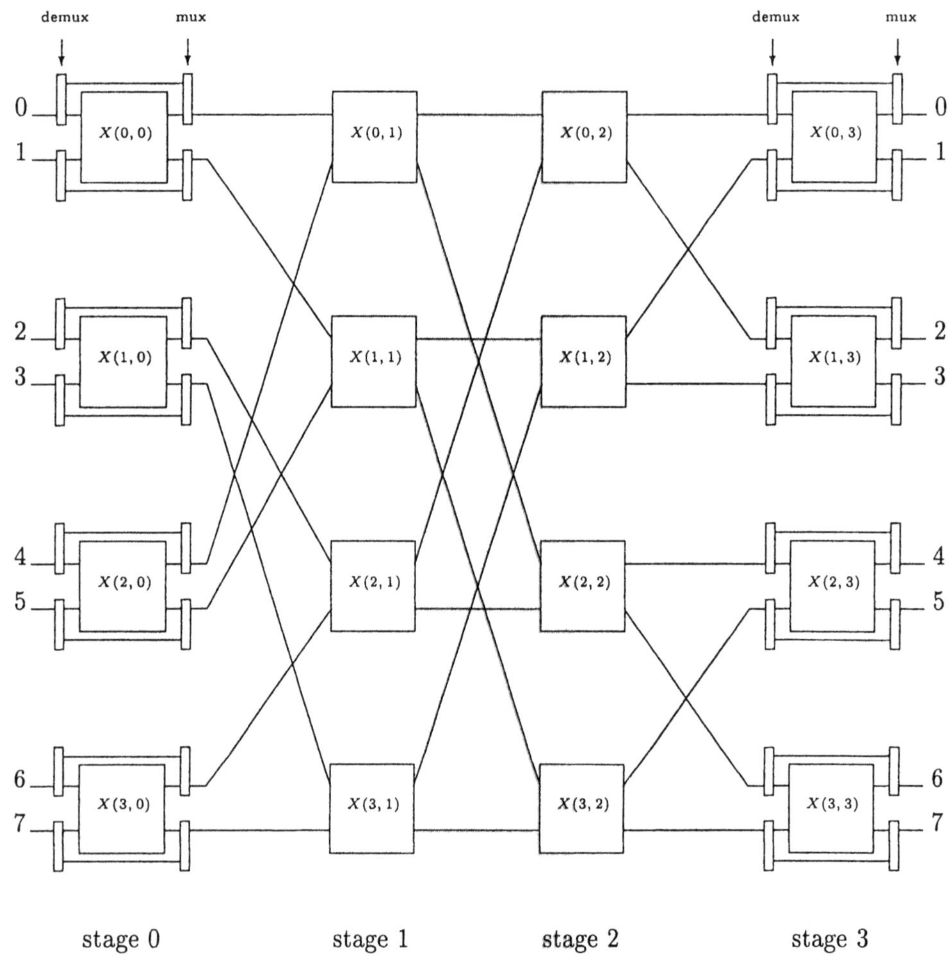

Figure 2: The Extra Stage Version of the SN of Figure 1

work needs $N/2$ extra switches in addition to $4N$ multiplexers/demultiplexers. Second, it does not provide full recovery under faulty conditions.

5 Modifying the Switch Design

Fault tolerance can be added to a BBSN at the switch level, by modifying the design of its switches. This technique is exemplified by the Modified Switch BBSN [14]. Before showing how this network modifies the switch, the concept of *conjugate subsets* should be introduced.

The switches of any stage of a BBSN can be grouped into disjoint subsets, each called a conjugate subset. A conjugate subset of switches is characterized by having access to a unique subset of the outlets no other switch in the stage has access to. For example, the switches of stage 1 of the BBSN of Figure 1 can be grouped into two conjugate subsets, the first containing $X(0,1)$ and $X(1,1)$, which have access to outlets $\{0,1,2,3\}$, and the second containing $X(2,1)$ and $X(3,1)$, which has access to outlets $\{4,5,6,7\}$. It can be easily noted that, in any BBSN, the first stage always has one conjugate subset, containing all the switches of that stage, and the second has two subsets, each containing half the switches of the stage, and so on. Furthermore, it can be noted that the switches of a subset can be always divided into two divisions, each having access to all the switches of exactly two conjugate subsets in the next stage. For example, the switches of stage 1, which form the only conjugate subset in that stage, can be divided into two divisions, one containing switches $X(0,0)$ and $X(1,0)$, with access to the two subsets of stage 1, and the other containing switches $X(2,0)$ and $X(3,0)$, also with access to the two subsets of stage 1.

The Modified Switch BBSN links together the switches of each division, with two or more switches, by *loop*. The switches involved in a loop should be of size 3×3, as shown in Figure 3, i.e. with one auxiliary input and one auxiliary output. A loop is used to route a path away from a faulty switch in the next stage, utilizing the fact that the looped switches have access to the same (two) conjugate subsets in the next stage. There are always two loops in the first stage, stage 0, four in the second stage, stage 1, and so on, as long as the loop is for a division of two or more switches. Note that each division in the stage before the last, stage $\nu - 2$, has only one switch, and therefore no loop is installed in that stage.

The input and output stages, stages 0 and $\nu - 1$, respectively, are made fault tolerant by using multiplexers and demultiplexers. Each inlet has ac-

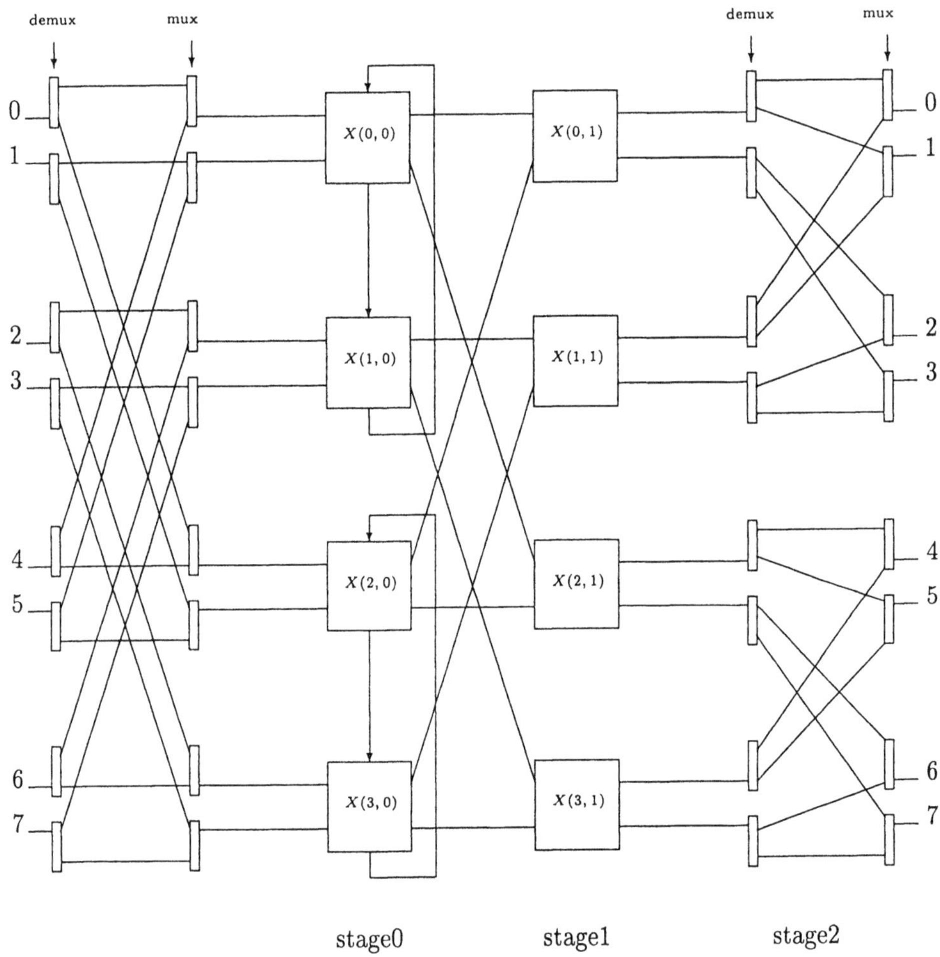

Figure 3: The Modified Switch version of the BBSN of Figure 1

cess to two loops so that if one loop is faulty the other is used. Similarly, each outlet is reachable from two distinct switches in stage $\nu - 2$ so that if a switch fails, the outlet can be reached from the other switch. Interestingly, this arrangement eliminates the switches of the last stage, stage $\nu - 1$.

Operation and Fault-tolerance Model

The Modified Switch works as follows. A source requests a path by putting the routing tag for a destination outlet. For each switch in stage j, $0 \leq j \leq \nu - 3$, the request may arrive on any of the three inputs of the switch. The switch uses its routing bit in the routing tag to extend the path to the next stage on one of its two normal outputs. If a switch cannot use a normal output, because of a fault in that output or in the switch connected to it in the next stage, it uses the auxiliary output.

To illustrate, consider the Modified Switch BBSN of Figure 3. Suppose now that switch $X(1,0)$ must establish a path to outlet 3, but finds switch $X(1,1)$ faulty. Then switch $X(1,0)$ can route the path through its auxiliary output to switch $X(0,0)$. As can be seen, switch $X(0,0)$ has access to outlet 3, and thus the path can be establish. Also, if switch $X(0,1)$ is faulty and a path is to be established from inlet 0 to outlet 0. Under normal conditions, this path was to go through switches $X(0,0)$, $X(0,1)$, and $X(0,2)$. Now, with the help of the loop, the path can go through switches $X(0,0)$, $X(1,1)$, and out to outlet 0 through the multiplexers and demultiplexers at the output stage.

The fault model in the Modified Switch BBSN is the typical model mentioned in Section 3. The fault size of the Modified Switch BBSN is 1, and the fault tolerance criterion is full access retention. The Modified Switch BBSN is 4-robust since the two upper or two lower switches in stages 0 or 1 can fail and the full access capability will still be retained.

Two advantages can be identified for the Modified Switch BBSN. First, the distributed routing capability is preserved. Second, the number of multiplexers and demultiplexers are half that used by the Extra Stage BBSN (excluding those used in eliminating the last stage.)

The disadvantages of the Modified Switch BBSN are as follows. First, specially designed (nonbinary) switches are needed. Recalling that switch complexity is square the size, then the complexity of the new switch is more than double that of the binary switch. Also noting that the new switch is to perform no ordinary routing, then intelligence is needed in it which adds to the complexity. Second, loops required in divisions of two or more switches add to the complexity of the link pattern. Recall that these links are not single lines, but complete buses incorporating data and signal lines.

In addition to these two disadvantages, it is not clear in the Modified Switch BBSN what a multiplexer at a given outlet would do if it receives two requests, now that the output stage switches have been eliminated. This is a very likely situation in environments where the sources are independent as far as their choice of the destinations is concerned. Also, the question is open as to how a multiplexer at a given input would resolve contention in case two requests were received simultaneously from the two sources connected to it.

6 Adding an Extra Row

In this technique, one switch is added to each stage of the SN, in effect adding an extra row of switches to the network. As a consequence, this technique is most amenable to SNs with a low number of stages, such as the Clos network [9]. This network is characterized by having only three stages regardless of its size. Incidentally, this advantage of having a few stages is mitigated by its problematic routing. A sophisticated routing algorithm [5] is needed to set the network switches to establish a set of paths. This algorithm is run by a central routing unit to generate the required settings for each switch in the network.

A Clos network of size N has no unique design. There can be several designs for such a network, each called a Clos (N, k), where k is a positive integer. A Clos (N, k) is built from

- k switches in stage 0, each of size $m \times m$,
- m switches in stage 1, each of size $k \times k$, and
- k switches in stage 2, each of size $m \times m$

where k and m are positive integers such that $km = N$. For nontrivial instantiations, k and N must satisfy $2 \leq k \leq N$. The network is then constructed by connecting each output of a stage 0 switch to a distinct switch in stage 1, and each output of a stage 1 switch to a distinct switch in stage 2. Figure 4 shows a Clos$(9, 3)$ SN.

Interestingly, the Clos network inherently has some degree of fault tolerance. Namely, it satisfies the full access retention criterion if a middle

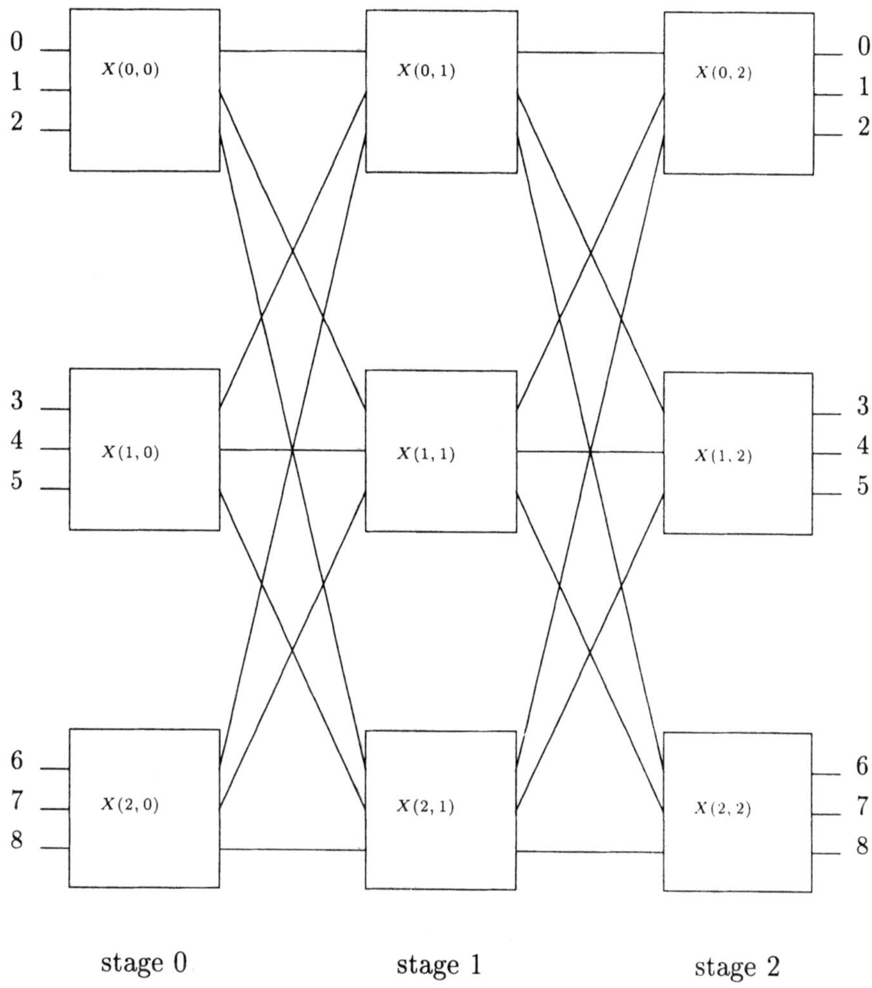

Figure 4: A Clos(9,3) Switching Network

stage switch is faulty, partially or totally. Precisely, a Clos (N, k) network is inherently fault tolerant with the following fault tolerance model. First, the fault model is

1. Any middle stage switch and any links connected to that stage can fail

2. Outer stage switches as well as inlets and outlets cannot fail

The fault size is $m - 1$, when the fault tolerance criterion is full access retention. That is because up to $m - 1$ inner links (i.e. outputs of any first stage switch or inputs of any last stage switch) can fail with the inlets still having access to the outlets. This network is also $k(m-1)$ robust under this fault tolerant model, since $m - 1$ inner links out of the m available for each outer switch can fail, and there can still be inlet-outlet accessibility.

But this inherent type of fault tolerance is not enough in Clos networks, since they are used mainly to realize permutations. That is, the fault tolerance criterion preferred here is full recovery, not just full access retention. Moreover, this inherent type of fault tolerance does not allow faults in the two outer stages, i.e. its fault model is very restrictive. These limitations are overcome by the *Fault Tolerant Clos (FTC)* network [20].

The FTC (N, k) network corresponding to a Clos (N, k) network is built from

- $k + 1$ switches in stage 0, each of size $m \times (m + 1)$
- $m + 1$ switches in stage 1, each of size $(k + 1) \times (k + 1)$, and
- $k + 1$ switches in stage 2, each of size $(m + 1) \times m$

where $km = N$ as in the Clos (N, k). Clearly, the FTC (N, k) has one more switch in each stage than its Clos (N, k) counterpart. Consider the last switch in each stage to be that additional switch, and refer to it as the *Extra Switch.*

With these switches, the FTC (N, k) is constructed as shown in the example in Figure 5, which is the fault tolerant version of the network of Figure 4. First, as in the Clos (N, k), each output of a stage 0 switch is connected to a distinct switch in stage 1, and each output of a stage 1 switch is connected to a distinct switch in stage 2, and so on. Second, the N inlets are connected to N demultiplexers. One of the two outputs of each demultiplexer is connected to an input of a stage 0 switch, the neighbor switch, and the other to the corresponding input of the Extra

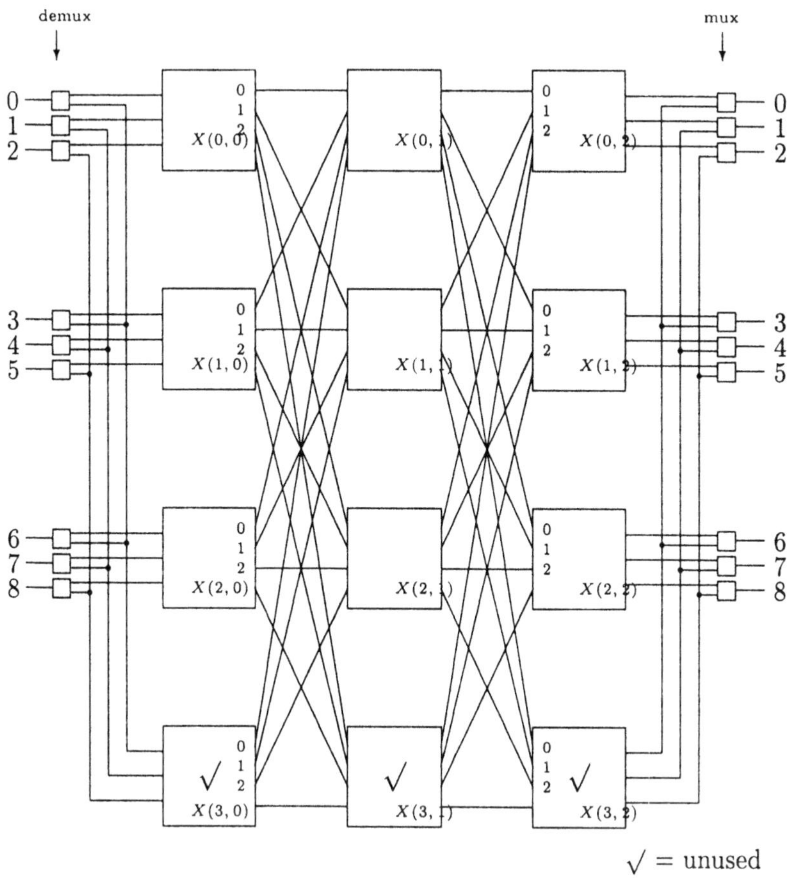

Figure 5: The FTC version of the Clos network of Figure 4

Switch. Thus, each demultiplexer will be connected to two switches: the neighbor switch and the Extra Switch. Finally, the N outlets are connected to N multiplexers, with the inputs of these multiplexers connected to the outputs of stage 2 switches. The Extra Switches in the three stages are marked in Figure 5.

Operation and Fault-tolerance Model

At any time, one switch in each stage of the FTC is *inactive*, while the rest are *active*. Under normal conditions, the Extra Switches are conveniently chosen to be inactive. The Extra Switch of stage 0 (stage 2) is made inactive by the demultiplexers (multiplexers), while the Extra Switch of the middle stage is made inactive by the routing algorithm.

Under faulty conditions, the FTC is reconfigured so that one or more of the three Extra Switches are activated. The *reconfiguration* is achieved using the multiplexers and demultiplexers in conjunction with the routing algorithm. Basically, it deactivates the switch(es) affected by the fault and activates the corresponding Extra Switch(es).

Similarly, at any time, one output (input) of each switch in stage 0 (stage 2) of the FTC is made inactive, while the rest are made active. These links are made active or inactive both by the FTC's routing unit. Under normal conditions, the last output (input) of the switch is made inactive, since it is the link connected to the middle stage Extra Switch.

Under faulty conditions, on the other hand, the FTC is reconfigured so that some inactive links are activated. Again, the reconfiguration deactivates the links affected by the fault and activates the corresponding redundant links.

Now, suppose the FTC is required to implement the mapping $\mathbf{P} = (P_0, P_1, \ldots, P_{N-1})$. That is, it is required to connect inlet 0 to outlet P_0, inlet 1 to outlet P_1, and so on. This $\mathbf{P}$ is not sent immediately to the routing unit, as would be the case in Clos network. Rather, $\mathbf{P}$ is translated first to another mapping $\mathbf{P}' = (P_0, P_1, \ldots, P_{N+m-1})$. Simply, this translation is done such that the original mapping $\mathbf{P}$ is satisfied and the inactive switches and links are avoided. The extra elements of $\mathbf{P}'$, which relate to the inactive components, are assigned arbitrarily. After obtaining $\mathbf{P}'$, the central routing unit uses it to generate the switch settings.

Once a fault occurs, the following two operations are performed to work around the fault.

1. *Setting the demultiplexers and multiplexers.* This operation is performed when an outer stage switch or link fails. The settings should be such that the faulty links or switches are rendered inactive.

2. *Translating the mapping.* This operation is performed when a fault occurs at any location in the FTC. The settings should be such that the faulty links or switches are rendered inactive.

The fault model of the FTC is the typical model mentioned in Section 3. The fault size of the FTC is $\min(m, k)$. Its fault tolerance criterion is full recovery. Under this criterion, the FTC is also $(2m + k)$-robust, since it is fully recoverable if one switch in each stage fails.

The main advantage of the FTC is the small amount of extra hardware—only three switches. Also, the extra switches are not specially designed. Moreover, the number of multiplexers and demultiplexers is less than $N/2$. Other than links associated with the multiplexers and demultiplexers, no additional links, interstage or intrastage, are needed. Finally, the extra hardware can be utilized [6, 15] under normal conditions to obtain considerable performance improvements.

The FTC has the same disadvantage as the Clos network, namely the problem in setting up the switches of the middle stage.

7 Adding an External Bus

Unfortunately, the fault tolerance techniques cited above have many drawbacks. First, they are SN-specific. Second, they invariably add a great amount of extra hardware, which surely increases cost. Third, they require changes in the original SN design, which typically negatively impacts the performance under *normal* conditions. Noting that faults tend to exist only temporarily, these techniques are too expensive a price for fault tolerance.

A technique free of the above drawbacks is adding an external bus parallel to the SN [21]. That is, each inlet is connected, through a demultiplexer, both to the network and to the bus. Similarly, each outlet is connected, through a multiplexer, both to the network and the bus. Under normal conditions, the SN is used as usual. Under faulty conditions, the bus is to used to establish a path which cannot be established through the SN. The mechanisms of sharing a bus among many sources and many destinations are well documented [17].

Although this technique can be implemented by any SN, it will be introduced here through a BBSN. The External Bus version of the BBSN of Figure 1 is shown in Figure 6.

Operation and Fault-tolerance Model

The operation of the External Bus BBSN is simple. Under normal conditions, the multiplexers and demultiplexers are set such that the network is used. Only when a fault occurs are the multiplexers and demultiplexers set such that the bus is used. Clearly, then, an External Bus SN is actually two networks in one: the shared bus "network" and the original SN.

For the External Bus BBSN [21], the fault model is the typical model mentioned in Section 3. If the fault tolerance criterion is *full recovery*, then

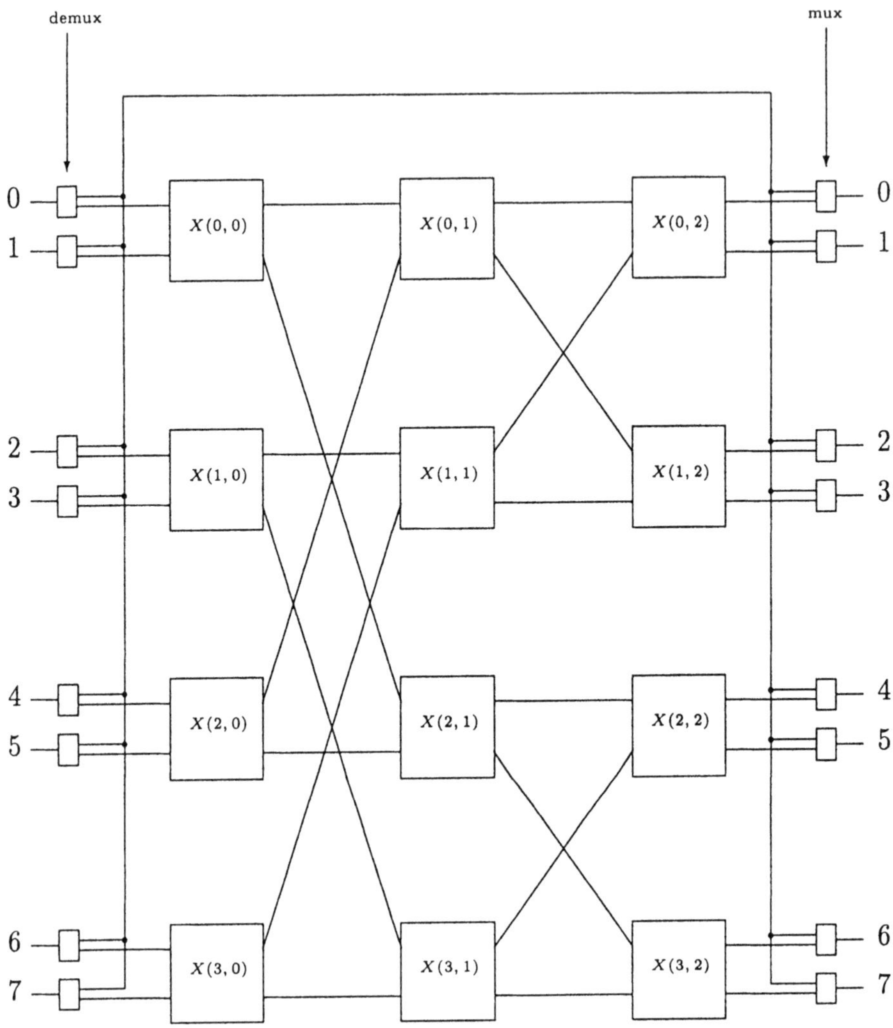

Figure 6: The External Bus version of the BBSN of Figure 1

the fault size is 1. Interestingly, if the criterion full access retention, then the fault size is N, since each input can still reach each output via the External Bus!

The technique of adding a bus can be generalized to multiple buses[18], yielding a higher fault tolerance size for the same criterion. For example, if a double bus is used, full recovery is attainable at a fault size of 2, rather than 1 when a single bus is used. But note that two buses entail the use of demultiplexers and multiplexers of size 1×4 and 4×1, respectively.

The external bus(es) can be used under normal conditions to obtain some benefits. First, it has been shown that using the external bus(es) under normal conditions enhance the performance of the BBSN greatly [21]. In addition, two useful functions can be obtained easily using the bus, *broadcasting* and *priority* paths. Broadcasting allows an inlet to access to all outlets simultaneously. Providing this capability on an ordinary BBSN requires switches capable of connecting one of their inputs to the two outputs simultaneously. It turns out that providing this capability adds considerably to the complexity of the switch. The other function that the bus can perform under normal conditions is establishing a path that is needed urgently but cannot be established over the BBSN due to contention from other paths.

The External Bus technique has advantages. First, there is no performance degradation under normal conditions. Second, it requires no modifications in the original network. Finally, it does not disturb the routing mechanism of the original network.

8 Fault Detection and Location

The work of any fault tolerance technique can start only after two necessary operations have been performed: fault detection and fault location. The methods proposed in the literature for these operations can be classified as either offline or online. The idea of offline methods [4, 10] is to periodically apply a test pattern at the inlets and analyze the pattern received at the outlets, to find if and where a fault exists. The online methods, on the other hand, can detect and locate faults while the system is running, through such mechanisms as parity checking [10] or data bits checking [16]. Admittedly, detecting and locating faults online is better than offline, as it does not disrupt the running of the system. However, adopting an online method requires special switch design with built-in hardware to carry out the dynamic checking.

9 Conclusions

A rundown of the most common techniques of fault tolerance has been given in this chapter. Namely, the following techniques have been discussed.

1. Adding extra stage
2. Modifying switch design
3. Adding extra row
4. Adding external bus

It should be noted, however, that these are by no means the only fault tolerance techniques available. There are other techniques such as duplication, dilation [13] and interstage link pattern modification [7]. All these techniques inevitably add extra hardware to the original network. A good design is one that minimizes the amount of this hardware.

All the techniques discussed in this chapter require the use of demultiplexers and multiplexers. However, if the sources and destinations using the network are dual ported, the number of these components can be either reduced, as in techniques (1) and (2), or totally eliminated, as in techniques (3) and (4).

References

[1] G. Adams and H. Siegel, The Extra Stage Cube: A Fault-Tolerant Interconnection Network for Supersystems, *IEEE Transactions on Computers*, vol. C-31, no. 5 (1982) pp. 443-454.

[2] ______________, Modifications to Improve the Fault Tolerance of the Extra Stage Cube Interconnection Network, *Proceedings of the 1984 International Conference on Parallel Processing*, 1984, pp. 169-173.

[3] G. Adams, D. Agrawal and H. Siegel, A Survey and Comparison of Fault-tolerant Multistage Interconnection Networks, *Computer*, June 1987, pp. 14-27.

[4] D. Agrawal, Testing and Fault Tolerance of Multistage Interconnection Networks, *Computer*, April 1982, pp. 41-53.

[5] J. D. Carpinelli and A. Y. Oruc, Parallel Set-Up Algorithms for Clos Networks Using a Tree-Connected Computer, *Proceedings of the 2nd International Conference on Supercomputing*, May, 1987 pp. 321-327.

[6] J. D. Carpinelli and C. B. Wang, Performance of a New Decomposition Algorithm for Rearrangeble Fault-tolerant Clos Interconnection Networks under Sub-maximal and No-fault Conditions, in D. -Z Du and F. K. Hwang (eds.) *Advances in Switching Networks* (American Mathematical Society, 1997) pp. 103-117.

[7] P. Chauang, CGIN: A Fault Tolerant Modified Gamma Interconnection Network, *IEEE Transactions on Parallel and Distributed Systems*, vol 7, no. 12 (1996) pp. 1301-1306.

[8] L. Ciminiera and A. Serra, A Connecting Network with Fault Tolerance Capabilities, *IEEE Transactions on Computers*, vol. C-35, no. 6 (1986) pp. 578-580.

[9] C. Clos, A Study of Nonblocking Switching Networks, *Bell Systems Technical Journal*, vol. 32, no. 2 (1953), pp. 406-424.

[10] T. Feng and C. Wu, Fault-Diagnosis for a Class of Multistage Interconnection Networks, *IEEE Transactions on Computers*, vol. C-30, no. 10 (1981), pp. 743-758.

[11] I. Gazit and M. Malek, Fault Tolerance Capabilities in Multistage Network-Based Multicomputer Systems, *IEEE Transactions on Computers*, vol. 37, no.7 (1988) pp. 788-798.

[12] M. Jeng and H. Siegel, A Fault-Tolerant Multistage Interconnection Network for the Multiprocessor Systems Using Dynamic Redundancy, *Proceedings of the 6th International Conference on Distributed Computing Systems*, 1986, pp. 70-77.

[13] C. Kruskal and M. Snir, The Performance of Multistage Interconnection Networks for Multiprocessors, *IEEE Transactions on Computers*, vol. C-32, no. 12, December 1983, pp. 1091-1098.

[14] V. Kumar and S. Reddy, Augmented Shuffle-exchange Multistage Interconnection Network, *Computer*, June 1987, pp. 30-40.

[15] H. Lee, F. K. Hwang and J. D. Carpinelli, A New Decomposition Algorithm for Rearrangeble Clos Interconnection Networks, *IEEE Transactions on Communications*, vol. 44, no. 11 (1996) pp. 1572-1578.

[16] W. Lin and C. Wu, Design of a 2×2 Fault-Tolerant Switching Element, *Proceedings of the 9th Annual Symposium on Computer Architecture*, 1982, pp. 181-189.

[17] M. Marsan, G. Blbo, G. Conte and F. Gregoretti, Modelling Bus Contention and Memory Interference in a Multiprocessor System, *IEEE Transactions on Computers*, vol. C-32, no. 1 (1983) pp. 60-72.

[18] H. Nassar, A Markov Model for Multibus Multiprocessor Systems under Asynchronous Operation, *Information Processing Letters*, vol. 54, no. 4 (1995) pp. 11–16.

[19] H. Nassar and J. Carpinelli, Generalized Model for Banyan Networks with Nonuniform Traffic, *IEE Proceedings on Computers and Digital Techniques*, vol. 145, no. 3 (1998) pp. 225–228.

[20] ---------------, Design and Performance of a Fault-Tolerant Clos Network, *26th Conference on Information Sciences & Systems*, Maryland, USA, March 22-24 (1995).

[21] ---------------, A Simple Fault Tolerant Technique for Interconnection Networks, *ISCA Intl Conference on Parallel and Distributed Computing & Systems*, Kentucky, USA, October 14-16 (1993) pp. 25–30.

[22] K. Padmanabhan and D. Lawrie, A Class of Redundant Paths Multistage Interconnection Networks, *IEEE Transactions on Computers*, vol. C-32, no. 12 (1983) pp. 1099-1108.

[23] J. Patel, Performance of Processor-Memory Interconnections for Multiprocessors, *IEEE Transactions on Computers*, vol. C-30, no. 10 (1981) pp. 771-780.

[24] C. Raghavendra and A. Varma, INDRA: A Class of Interconnection Networks with Redundant Paths, *Proceedings of the 1984 Real Time Systems Symposium*, pp. 153-164.

[25] S. Reddy and V. Kumar, On Fault Tolerant Multistage Interconnection Networks, *Proceedings of the 1984 International Conference on Parallel Processing*, pp. 155-164.

[26] J. P. Shen and J. P. Hayes, Fault-tolerant of Dynamic-full-access Interconnection Networks, *IEEE Transactions on Computers*, vol. C-33, no. 3 (1984) pp. 241-248.

[27] C. Wu and T. Feng, On a Class of Multistage Interconnection Networks, *IEEE Transactions on Computers*, vol. C-29, no. 8 (1980) pp. 694-702.

[28] ---------------, The Universality of the Shuffle-Exchange Network, *IEEE Transactions on Computers*, vol. C-30, no. 5, May 1981, pp. 324-332.

[29] K. Yoon and W. Hegazy, The Extra Stage Gamma Network, *Proceedings of the 13th Annual Symposium on Computer Architecture*, 1986, pp. 175-182.

SWITCHING NETWORKS: RECENT ADVANCES
D.-Z. Du and H.Q. Ngo (Eds.) pp. 25-61

Switching Networks for Generating Random Permutations

Artur Czumaj
Department of Computer and Information Science
New Jersey Institute of Technology, Newark, NJ 07102
E-mail: czumaj@cis.njit.edu

Przemka Kanarek and Krzysztof Loryś
Institute of Computer Science,
University of Wrocław, PL-51-151 Wrocław, Poland
E-mail: pka,lorys@ii.uni.wroc.pl

Mirosław Kutyłowski
Mathematical Institute, Wrocław University of Technology,
PL-50-370 Wrocław, Poland
and Department of Mathematics and Computer Science
University of Poznań, PL-60-769 Poznań, Poland
E-mail: mirekk@amu.edu.pl

Contents

1 Introduction

1.1 Preliminaries

Permuting at random is the following problem: *given n items on n input positions, choose uniformly at random a permutation $\pi \in S_n$ and deliver the items to n output positions so that the ith item is given at output position $\pi(i)$, for $i \leq n$.* For the sake of simplicity of notation, we shall assume throughout the paper that the collection of items to be permuted at random is $1, 2, \ldots, n$.

Efficient permuting at random is a problem that deserves a lot of attention due to its practical applications. An example of this kind is the technique of random changing transmission frequencies between cellular phones and their basis stations. Among other advantages, it makes extremely difficult to eavesdrop a certain connection. A vast application field is permuting at random in order to improve reliability of communication links: it helps to avoid with high probability bad configurations that lead to a poor performance. One of the classical examples of such an approach is the routing protocol due to Ranade (see [19]). At this point, let us remark that most communication services should be relatively fast, so we must relay on hardware rather than on software solutions. Even if some "software" algorithms are theoretically faster, many factors hidden from theoretical analysis make them practically less attractive.

Permuting at random is a component of many cryptographic protocols designed for achieving security in computer networks. Again, efficient and cheap circuits as a source of random permutations with probability distribution close to uniform (and contained within tamper-proof devices) might be very attractive for high-security applications. Again, any software solution implemented on an unprotected processor might be slow and, what is the key issue, vulnerable to an attack.

Algorithms for permuting at random that have to be implemented in hardware must meet certain design criteria. The operations executed must be simple enough to admit a fast and cheap hardware implementation. This provides a motivation for studying switching networks in this context, where the only operations executed (except of common clock pulses) are those of separate switches.

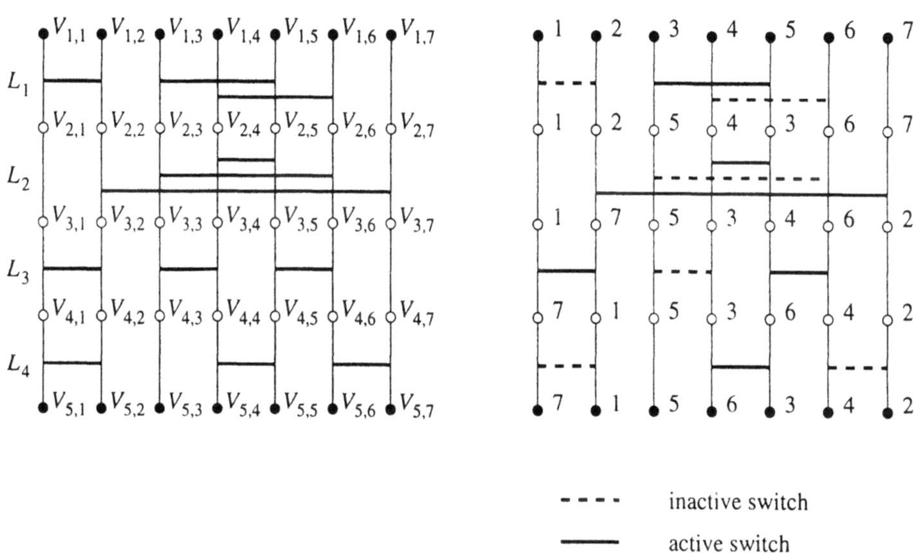

Figure 1: A switching network (left) and an example of its execution (right)

Switching networks

Definition 1.1 *A* switching network *(see Fig. 1) of depth d and with n inputs is described by a set of registers $V_{i,j}$ for $i = 1, \ldots, d+1$, $j =$*

$1, \ldots, n$ *and a set of* switches *organized in* layers $L_1, \ldots, L_d$. *Each register may hold a single item. The registers* $V_{0,1}, \ldots, V_{0,n}$ *are called* input registers, $V_{d+1,1}, \ldots, V_{d+1,n}$ *are* output registers. *For* $i \leq n$, *the registers* $V_{0,i}, V_{1,i}, \ldots, V_{d+1,i}$ *form the so called ith* wire. *Each layer* L_i *of a switching network is described by a matching (not necessarily a perfect one) over* $\{1, \ldots, n\}$. *That is, it is a set of pairs of numbers from* $\{1, \ldots, n\}$, *with each number occurring in at most one pair. If* $(m, k) \in L_i$, *then we say that there is a switch between the wires* m *and* k *at level* i.

Now, let us describe how a single switch works:

Definition 1.2 *A switch* $(m, k) \in L_i$ *may be set in two different states:* off *and* on. *If this switch is off, then it moves the item from* $V_{i,m}$ *into* $V_{i+1,m}$ *and from* $V_{i,k}$ *into* $V_{i+1,k}$. *If the switch is on, then it moves the item from* $V_{i,m}$ *into* $V_{i+1,k}$ *and from* $V_{i,m}$ *into* $V_{i+1,k}$.

Now, we may describe the work of the whole switching network:

Definition 1.3 *First, the states of all switches are set. Then, the input items are loaded into the input registers, the ith input item into* $V_{0,i}$. *At the next moment, the switches of the first layer move the input items from the input registers into* $V_{1,1}, \ldots, V_{1,n}$ *in some order given by the states of the switches. If some* $V_{0,i}$ *is not connected to any switch, then the item from* $V_{0,i}$ *is moved into* $V_{1,i}$. *Then, the switches of the second layer move the items into* $V_{2,1}, \ldots, V_{2,n}$. *This is continued, until the switches of layer* d *move the items into the output registers* $V_{d+1,1}, \ldots, V_{d+1,n}$.

Obviously, a setting of.switches determines a set of disjoint paths leading from the input registers to the output registers. So, after setting the switches, the network performs a fixed permutation of the input items.

One may easily see that we may pipeline the work of a switching network on different inputs: as soon as one set of input items is moved from registers $V_{i,1}, \ldots, V_{i,n}$ into registers $V_{i+1,1}, \ldots, V_{i+1,n}$, then a new set of input items may be moved from registers $V_{i-1,1}, \ldots, V_{i-1,n}$ into registers $V_{i,1}, \ldots, V_{i,n}$. Between consecutive usages of layer L_i, we may also change the setting of its switches, causing different moves to occur at this layer for different input sequences.

Permuting at random by switching networks Obviously, a switching network may be used for generating a random permutation via setting its switches at random. The main problems that arise are the following:

1. Is it possible for each permutation $\pi \in S_n$ to set the switches to obtain π?

2. By setting the switches uniformly at random, we obtain a certain probability distribution over S_n. How close is this distribution to the uniform distribution over S_n?

The first question is one of the classical ones in the theory of switching networks: a vast family of so called *permutation networks* has been designed. For the second goal, still there is no satisfactory solution in the literature.

Mixes It is interesting that the idea of composing random permutations out of single switches become one of basic techniques in computer security. In order to send messages to random locations, they are processed via *mixes* [6]: mix is a device with two input and two input links, just like a switch. After receiving two encrypted messages, a mix decodes them, permutes at random, encodes again, and sends to two output locations. The recoding ensures that an observer cannot find out in which order the incoming messages leave the mix. It has been believed that composing many mixes provides a certain degree of anonymity in the communication network: by analyzing the traffic it might be difficult to say who is communication with whom. For this reason, mixes became a standard tool in this area.

1.2 Permutation Networks

First, we recall the butterfly network, which is one of the most fundamental network architectures.

Definition 1.4 *Let $n = 2^k$, $k \in \mathbb{N}$. A* butterfly network *(see Fig. 2) with n inputs consists of k levels. At level i, for each $j \leq n$, a wire with index j is connected by a switch with a wire j' such that the binary representations of j and j' differ on the ith most significant bit.*

The butterfly network has many interesting properties. The first is that if we remove level 1, then we get two disjoint butterflies with $n/2$ inputs each. Similarly, below level m we get 2^m disjoint butterflies, each with $n/2^m$ inputs. The next important property is that for each $i, j \leq n$, there is exactly one path leading from the input register i into the output register j. This follows from the following observation: let $i_1 = i, i_2, \ldots, i_k$ be the indices of wires on a path starting at input node i and defined by the states

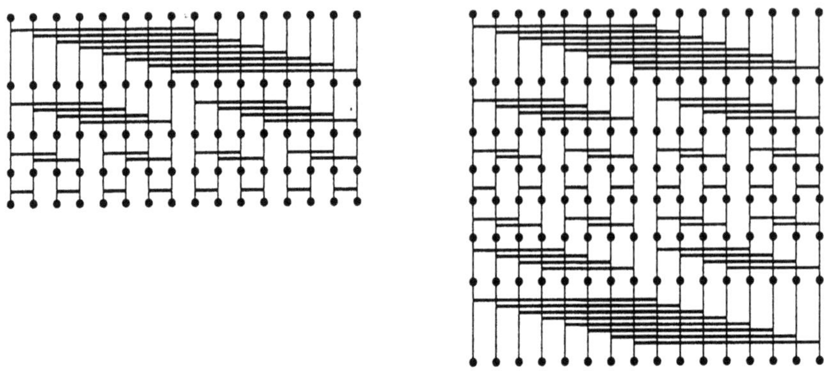

Figure 2: Butterfly network (left) and Beneš network (right)

of switches. By definition, for $m \leq k$, the mth most significant bit of $i_1, \ldots, i_{m-1}$ is the same. Similarly, the mth most significant bit of $i_m, \ldots, i_k$ is the same. It follows that setting of the switch on level m on a path from input node i to output node j depends only on the mth bits of the binary representations of i and j: it is off if the mentioned bits are equal, otherwise, it is on. It follows that there is at most one path from i to j. Moreover, we may construct such a path by starting from i and setting the switches as on consecutive levels as described above.

An immediate corollary of the above property is that if the switches of a butterfly are set uniformly at random and π denotes the permutation defined by this setting, then $\mathbf{Pr}[\pi(i) = j] = \frac{1}{n}$ for every $i, j \leq n$. This uniformness is however broken when we consider two input elements: if for instance $\pi(1) = 2^{k-1} + 1$, then the switch on level 1 connecting wire 1 with wire $2^{k-1} + 1$ must be on. Therefore, the most significant bit of $\pi(2^{k-1} + 1)$ is 0. So, we see that a butterfly may define permutations only of a very special form. Still, it is known that butterflies have some good "mixing" properties [20].

In order to be able to get an arbitrary permutation via a switching network, one may connect two butterflies, one of them in a reverse order. In that way we obtain the classical Beneš network (compare [19]):

Definition 1.5 *Let* $n = 2^k$, $k \in \mathbb{N}$. *A* Beneš network *(see Fig. 2) with* n *inputs consists of* $2k - 1$ *levels. The levels* 1 *through* k *are the same as in a butterfly. For* $i > k$, *the level* i *contains the same switches as the level*

$2k - i$.

It is easy to see that if we remove the first and the last level of the Beneš network, then we get two disjoint Beneš networks, each with $n/2$ inputs. Less obvious is the property that for each permutation $\pi \in S_n$, there is a setting of switches of the Beneš network with n inputs that defines π (for a proof see, for instance, [19]).

Variation distance If a switching network consists of k switches, then each setting of switches has probability $1/2^k$. It follows that any event has probability of the form $c/2^k$, where c is an integer $0 \leq c \leq 2^k$. It follows that no event may have probability $\frac{1}{n!}$. Therefore, all we can hope for is that the probability distribution over S_n defined by a switching network is *close* to the distribution with the constant value $\frac{1}{n!}$. The measure of closeness of two probability distributions that we shall use is *variation distance*: for two probability distributions ϑ and ν on a space Ω, it is defined as

$$\|\vartheta - \nu\| = \frac{1}{2} \sum_{\omega \in \Omega} |\vartheta(\omega) - \nu(\omega)| \ .$$

The variation distance is a standard measure of the global discrepancy between two probability distributions; observe that it assumes nothing where this discrepancy occurs.

Situation for permutation networks So far, not much is known how to build a permutation network such that each permutation would correspond to approximately the same number of settings of the switches. The classical networks, like Beneš network, are far from having this property: there are permutations that correspond to very few settings of switches, and there are privileged ones that may be obtained in relatively many ways. For instance, one can see that the identity permutation of n elements can be obtained for $(n/2)^{n/2}$ different settings of switches. This is much more than the average $2^{(2\log n-1)\cdot(n/2)}/n! \approx (e/\sqrt{2})^n/\sqrt{2\pi n}$.

Iteration technique Even if the known permutation networks yield probability distributions that are far from being uniform, there is a chance that they can be used efficiently for our purposes. Simply, one can pipeline several copies of a permutation network (or wrap the wires around in order to use one copy several times with different settings of switches). This corresponds to taking product of permutations generated by single networks, and the idea is that each permutation may be generated by composing "rare" and "frequent" permutations so eventually the differences between frequencies are significantly reduced. This hope can be somewhat backed by experimental results. However, one must be careful with interpreting experiments due to infeasible size of S_n even for moderate values of n. Hence, it is crucial (especially for security protocols) to analyze theoretically such networks. Unfortunately, so far no really efficient tool for such a theoretical analysis has been found. Designing a method, even for only one particular case, like pipelined butterfly networks, might receive a lot of attention.

1.3 Random Switching Networks

Even if no concrete switching network for generating random permutations has been designed, it is known that such networks do exist. These networks are based on *shuffling*:

Definition 1.6 *Consider an input consisting of two groups, each consisting of $n/2$ indistinguishable elements, say $n/2$ zeroes and $n/2$ ones.* Shuffling *is an operation that chooses a permutation $\pi \in S_n$ in some random way and, for $i \leq n$, puts the ith input item on output position $\pi(i)$. Shuffling is* perfect, *if the resulting probability distribution of the location of ones is the same as the probability obtained, when $n/2$ positions for $n/2$ ones are chosen uniformly at random.*

An important point is that perfect shuffling may loose its randomness, if we could distinguish the elements within a group. For instance, it is possible that perfect shuffling preserves the ordering of the elements of a group when scattering them on n output positions.

Let us consider the following simple strategy to permute n items at random:

Strategy 1.7 *Execute the following algorithm:*

1. *split the input elements arbitrarily into two equal groups,*
2. *permute each group at random using recursively the same algorithm,*
3. *shuffle perfectly both groups.*

It is clear that in this way we obtain each permutation with the same probability. In Section 3.2, we shall see that even if shuffling is not perfect, but the probability distribution it defines has small variation distance to the uniform probability, then Strategy 1.7 yields a probability distribution on S_n close to uniform. Therefore, it is crucial to have switching networks that perform shuffling that is close to the perfect shuffling.

In a paper [24] on the onion protocol designed for protecting anonymity in computer network communication, Rackoff and Simon show the following theorem:

Theorem 1.8 [Rackoff-Simon] *There is $c \in \mathbb{N}$ and $\tau = \mathcal{O}(\log^c n)$ for a small constant c such that almost every switching network of depth τ performs almost perfect shuffling. More precisely, for all but a fraction of $n^{-\Omega(1)}$ switching networks of depth τ the following property holds: for an input consisting of $n/2$ zeroes given on $n/2$ first positions and $n/2$ ones given on the remaining positions, the output probability distribution and the uniform distribution of $n/2$ ones over n position have variation distance $\mathcal{O}(\frac{1}{n})$.*

The proof of Theorem 1.8 is existential and gives no indication how to construct such switching networks. The proof in [24] is based on a quite complicated analysis of a corresponding Markov chain. The constant c that may be obtained in the proof of Theorem 1.8 is about 10 [25].

Using delayed path coupling method ([11], see also Section 2.3) one may substantially improve the claim of Theorem 1.8:

Theorem 1.9 [10] *Theorem 1.8 holds for $c = 1$.*

By Strategy 1.7, we obtain thereby the following corollary:

Corollary 1.10 *There are switching networks of depth $\mathcal{O}(\log^2 n)$ that generate permutations of n elements so that the variation distance between the probability distribution obtained and the uniform distribution on S_n is $\mathcal{O}(\frac{1}{n})$.*

1.4 New Construction

Our goal is to present an explicit architecture of a switching network of a polylogarithmic depth that permutes at random with probability distribution close to the uniform distribution over S_n. Namely, we show the following result:

Theorem 1.11 *One can construct a switching network $\mathcal{B}_n$ of depth $\log^{\mathcal{O}(1)} n$ with a simple architecture (based on pipelined butterflies) that generates permutations of n elements so that variation distance between the output probability distribution of $\mathcal{B}_n$ and the uniform distribution on S_n is $\mathcal{O}(\frac{1}{n})$.*

The key element in the approach presented in this paper is treating a computation on a switching network as a stochastic process. In this process, a step is transformation of the sequence of data items by a single layer or a group of layers of the switching network. Since the switches of each level are set independently at random, these transformations represent a random process with transition probabilities defined by the architecture of switches. In order to prove quality of our construction we have to prove certain properties of probability distribution of the output configurations. This is often a challenging problem, however we are able to use advanced techniques for showing rapid mixing of Markov chains.

The architecture we present is tightly connected with the proof technique we use - our feeling is that some of the features of its construction might be very specific to the proof method. On the other hand, constraints that we implement for the sake of the proof might be also crucial for achieving randomness of the output.

2 Preliminaries on Stochastic Processes

First, we recall some notions and results on stochastic processes. A detailed introduction to this subject can be found for instance in [23] or [21].

2.1 Markov Chains

Definition 2.1 *A* (homogeneous) Markov chain *is defined by a labeled directed graph $G = (V, E)$, where V is called the set of states, the labels are numbers from the range $[0, 1]$, and the sum of labels on the outcoming edges of a node is equal to 1, for every node of G.*

A (homogeneous) Markov chain *is a random walk through graph G, where transitions are made according to the probabilities given by the edges. More formally, a Markov chain $(V_t)_{t\in I\!N}$ is an infinite sequence $V_0, V_1, \ldots$ of random variables, where each of these variables have values in V. Given the values $V_0 = v_0, \ldots, V_m = v_m$, then V_{m+1} has the value in the set $\{w \in V | (v_m, w) \in E\}$, and $\mathbf{Pr}[V_{m+1} = u]$ is equal to the label of the edge (v_m, u).*

An important property of Markov chains are that they are "memoryless": the state visited at step $t+1$ is a random variable that depends only on the state visited at step t. For this transition, it is irrelevant how the state at step t has been reached. In other words, the Markov chain forgets all its history except its current state. The initial state V_0 of the Markov chain is unspecified, it might have an arbitrary probability distribution.

One of the most important properties of Markov chains is that in most cases the probability distribution of V_t converges with $t \to \infty$ to a certain probability distribution.

Definition 2.2 *Assume that μ is a probability distribution over the set of states of a Markov chain $\mathcal{Y}$. Assume that if the states of $\mathcal{Y}$ are initially distributed according to μ, then after executing one step of $\mathcal{Y}$ the probability distribution of the states is again μ. Then we call μ a* stationary distribution *of $\mathcal{Y}$.*

For Markov chains satisfying certain simple conditions, stationary distributions play a very important role:

Lemma 2.3 *Assume that a Markov chain $\mathcal{Y} = (\mathcal{Y}_t)_{t \in I\!N}$ has the following properties:*

- *Let s be a state of $\mathcal{Y}$, with a nonzero probability. If $\mathcal{Y}$ is in state s, then with a nonzero probability after executing one step the chain will be still in state s.*
- *For each pair of states $u \neq v$ of $\mathcal{Y}$, each with nonzero probability, there is a sequence of steps of the chain that originates in state u and terminates in state v.*

Then $\mathcal{Y}$ has a unique stationary distribution. Moreover, if μ_t denotes probability distribution of $\mathcal{Y}_t$, and μ is the stationary distribution of $\mathcal{Y}$, then

$$\mu(s) = \lim_{t \to \infty} \mu_t(s)$$

for each state s of $\mathcal{Y}$.

If the first of the above conditions is satisfied, we say that the chain is *aperiodic,*[1] if the second one holds, the chain is called *irreducible.*

[1] In fact, aperiodicity is a more general notion, but such a simple version will be sufficient for our purposes.

2.2 Application of Markov Chains for Designing Switching Networks

Let us assume that a switching network is periodic, that is, the same group of layers is applied repeatedly. Let this group be called a *period* of the network. We may consider the following Markov chain: its states are allocations of items to the wires of the switching network. A single step is an execution of the period with its switches set at random. In order to obtain an infinite Markov chain, we add infinitely many layers to the network by repeating its period.

Unless specified otherwise, we assume that the inputs of a switching network are the numbers 1 through n, number i given on wire i. Then, the state of the computation after applying layer j can be regarded as a permutation $\pi_j \in S_n$ such that immediately after applying layer j, number i is on wire $\pi_j(i)$ for $i \leq n$. So the question about quality of the switching network designed for permuting at random is the question about a stationary distribution of the Markov chain defined above. First, we have to show that there is a unique stationary distribution and that distribution is uniform over S_n. The next crucial property is that after a relatively small number of steps T, the probability distribution after step T should be close to the stationary distribution. Once we prove such a result, we may design the switching network by removing all layers corresponding to the steps of the Markov chain later than T.

The above scenario of construction and analysis of switching networks might be applied even if the network is not periodic. In this case, we may regard the original network as a single step of a Markov chain. In our analysis we have therefore to show that the Markov chain considered comes close to the stationary distribution already after one step! This looks strange and is rather untypical for the theory of stochastic processes, but with certain techniques (so called *delayed path coupling* [11]) such an approach is plausible. Another approach would be to regard every layer of a switching network as a step of a non-homogenous Markov chain.

First, we show that for each of the chains defined from switching networks the uniform distribution is stationary. Let $\pi \in S_n$ denote the state of the chain before applying the step and π^* be the state after one step of the chain. For each setting of the switches within a single step, there is a permutation μ, such that $\pi^* = \mu \circ \pi$. Let $\mu_1, \ldots, \mu_r$ denote permutations corresponding to the setting of switches. (We do not make any assumption about the probabilities of $\mu_1, \ldots, \mu_r$.) Let us assume that π has uniform probability distribution. Then for an arbitrary γ:

$$\mathbf{Pr}[\pi^* = \gamma] = \sum_{i=1}^{r} \mathbf{Pr}\left[\mu = \mu_i \wedge \pi = \mu_i^{-1} \circ \gamma)\right]$$

$$= \sum_{i=1}^{r} \mathbf{Pr}[\mu = \mu_i] \cdot \mathbf{Pr}\left[\pi = \mu_i^{-1} \circ \gamma\right] = \sum_{i=1}^{r} \mathbf{Pr}[\mu = \mu_i] \cdot \frac{1}{n!} = \frac{1}{n!}.$$

Therefore, after executing a single step, the probability distribution of the states remains uniform. A similar proof shows that if a switching network obtains the input consisting of h ones and $n - h$ zeroes, for a fixed $h < n$, then the uniform distribution over such inputs is stationary. Indeed, consider for example the case of $h = 1$. let q_j denote the probability that the switches corresponding to a step of a chain are set so that the item from wire j is directed to wire i. Obviously, $\sum_j q_j = 1$. Now, assume that the probability distribution before execution of a step is uniform and we consider the probability that after executing a step the one is on position i. This probability equals

$$\sum_j \left(q_j \cdot \frac{1}{n}\right) = \frac{1}{n} \cdot \sum_j q_j = \frac{1}{n},$$

as required.

Obviously, each switching network defines an aperiodic chain: we may set all switches off in order to preserve the state during a single step. Less obvious is that the chain is irreducible. For certain designs it might be easy to show. In such a case, one may add layers of a permutation network before the original layers of the period. For the coupling arguments that we use, insertion of such additional layers does not hurt. In this way, we guarantee that the Markov chain satisfies the assumptions of Lemma 2.3 and therefore has the properties stated in this lemma.

2.3 Rapid Mixing of Stochastic Processes

The key issue and the difficult part in the analysis of Markov chains of our interest is to show that their probability distributions converge fast to the stationary distribution. This property is known under the name of *rapid mixing*. Many techniques for showing rapid mixing have been developed over the last decades. In this paper, we use one of them, namely *coupling*.

Coupling. A *coupling* [1, 12, 21] is one of the standard methods for showing rate of convergence to the stationary distribution of a Markov chain. For a Markov chain $(\mathcal{Y}_t)_{t\in I\!N}$, coupling is a stochastic process $(\mathcal{X}_t, \mathcal{X}_t^*)_{t\in I\!N}$ such that each of $(\mathcal{X}_t)_{t\in I\!N}$, $(\mathcal{X}_t^*)_{t\in I\!N}$, considered independently, is a faithful copy of $(\mathcal{Y}_t)_{t\in I\!N}$. That is, the states of $(\mathcal{X}_t, \mathcal{X}_t^*)_{t\in I\!N}$ are pairs (π, π^*), where π (π^*) denotes a state of $\mathcal{X}_t$ (of $\mathcal{X}_t^*$). Furthermore, for each two states π, π' of Markov chain $(\mathcal{Y}_t)_{t\in I\!N}$, the probability of transition from a state of the form $(\pi, -)$ to the state of the form $(\pi', -)$ is the same as the probability of transition from π to π' in $(\mathcal{Y}_t)_{t\in I\!N}$. Similarly, the probability of transition from a state of the form $(-, \pi)$ to a state of the form $(-, \pi')$ is given by the probability of transition from π to π' in chain $(\mathcal{Y}_t)_{t\in I\!N}$. The coupling technique is based on the so-called *coupling inequality* (see for instance [1, Lemma 3.6]). It says that the variation distance between the stationary distribution and the distribution obtained after execution of T steps of the Markov chain $(\mathcal{Y}_t)_{t\in I\!N}$ does not exceed

$$\mathbf{Pr}\,[\,\mathcal{X}_T \neq \mathcal{X}_T^*\,]$$

for the worst choice of the initial states $\mathcal{X}_0$ and $\mathcal{X}_0^*$.

For a coupling proof, one has to define certain dependencies between $(\mathcal{X}_t)_{t\in I\!N}$, $(\mathcal{X}_t^*)_{t\in I\!N}$ so that with high probability $\mathcal{X}_T = \mathcal{X}_T^*$ for a sufficiently small T. We say that $\mathcal{X}$ and $\mathcal{X}^*$ *couple* at step T, if they reach the same state at step T. Usually, it is convenient to define the coupling process so that once $\mathcal{X}$ and $\mathcal{X}^*$ couple, they remain in the same states afterwards. The main technical problem of the coupling technique is to design dependencies between $\mathcal{X}$ and $\mathcal{X}^*$ so that they couple fast. This might be a challenging problem for some chains.

Path Coupling. One of the reasons why a coupling proof might be extremely tedious is that one has to analyze *every* possible pair of initial states. A technical breakthrough has been achieved by Bubley and Dyer through designing the *path coupling* method [3] (see also [4, 5, 8, 14, 15, 16]). This technique makes it possible to consider a coupling only for a very restricted set of initial states.

For the path coupling, one needs a metric Δ defined on the set of states of the Markov chain $(\mathcal{Y}_t)_{t\in I\!N}$. This metric should have values restricted to $\{0, \ldots, D\}$ for some (not too big) D. Let Γ denote the set of pairs of states $(\mathcal{X}, \mathcal{X}^*)$ of the chain $(\mathcal{Y}_t)_{t\in I\!N}$ such that $\Delta(\mathcal{X}, \mathcal{X}^*) = 1$. In order to apply path coupling argument, it is necessary that for each pair of states

$(\mathcal{X}, \mathcal{X}^*)$, if $\Delta(\mathcal{X}, \mathcal{X}^*) = r$, then there is a sequence (path) of states $\mathcal{X} = \Lambda_0, \Lambda_1, \ldots, \Lambda_r = \mathcal{X}^*$, with $(\Lambda_i, \Lambda_{i+1}) \in \Gamma$ for $0 \leq i < r$.

As always, let $\mathbf{E}[\, Z\,]$ stand for the expected value of a random variable Z. The following lemma is the core of the path coupling technique and shows that it is enough to consider a coupling only for $(\mathcal{X}, \mathcal{X}^*) \in \Gamma$:

Lemma 2.4 [Path Coupling Lemma of Bubley and Dyer [3], a simplified version]
Assume that there exists a coupling $(\mathcal{X}_t, \mathcal{X}_t^*)_{t \in I\!N}$ such that for some $\beta < 1$

$$\mathbf{E}[\, \Delta(\mathcal{X}_{t+1}, \mathcal{X}_{t+1}^*)\,] \leq \beta$$

for all $(\mathcal{X}_t, \mathcal{X}_t^*) \in \Gamma$ and $t \in I\!N$. Then for some T, if

$$T \leq \left\lceil \frac{\ln(D \cdot \varepsilon^{-1})}{\ln \beta^{-1}} \right\rceil,$$

then the variation distance between the stationary distribution and the probability distribution obtained after step T of the Markov chain is bounded by ε.

Delayed path coupling. In this paper, we use a simple extension of path coupling that leads to so called *delayed path coupling* [11]. In the path coupling, it is necessary to show a certain convergence in every single step (that is, to show inequality $\mathbf{E}[\, \Delta(\mathcal{X}_{t+1}, \mathcal{X}_{t+1}^*)\,] \leq \beta$ for a $\beta < 1$). However, for certain malicious pairs $(\mathcal{X}_t, \mathcal{X}_t^*) \in \Gamma$ this might be hardly possible. In this case, the path coupling lemma is useless.

A simple remedy to the problem mentioned above is considering a group of steps instead of a single step. That is, for a chosen $k \in I\!N$, we define a new Markov chain $(\mathcal{G}\mathcal{Y}_t)_{t \in I\!N}$ which has the same state space as $(\mathcal{Y}_t)_{t \in I\!N}$ with the exception that one step of $(\mathcal{G}\mathcal{Y}_t)_{t \in I\!N}$ consists of k steps defined for $(\mathcal{Y}_t)_{t \in I\!N}$. Obviously, $(\mathcal{G}\mathcal{Y}_t)_{t \in I\!N}$ has the same stationary distribution as $(\mathcal{Y}_t)_{t \in I\!N}$, since $\mathcal{G}\mathcal{Y}_t$ and $\mathcal{Y}_{kt}$ have the same probability distribution. Therefore, if we prove with path coupling that the probability distribution of $\mathcal{G}\mathcal{Y}_T$ is close to the stationary distribution, then the same applies to the probability distribution of $\mathcal{Y}_{kT}$. The key issue is that it might be easier to reach the same state of two stochastic processes considered in the coupling process, if we consider k steps of the original chain. In this way, we may escape some malicious configurations. The simplest version of delayed path coupling is based on the following lemma:

Lemma 2.5 (Simple Delayed Path Coupling Lemma) *Let $k \in I\!N$ and $\beta < 1$. Let $(\mathcal{X}_t, \mathcal{X}_t^*)_{t \in I\!N}$ be a coupling for a Markov chain $(\mathcal{Y}_t)_{t \in I\!N}$ such that for every $\tau \in I\!N$ and every $(\mathcal{X}_{\tau k}, \mathcal{X}_{\tau k}^*) \in \Gamma$ it holds that*

$$\mathbf{E}[\,\Delta(\mathcal{X}_{(\tau+1)k}, \mathcal{X}_{(\tau+1)k}^*)\,] \leq \beta.$$

Then for some T, if

$$T \leq k \cdot \left\lceil \frac{\ln(D \cdot \varepsilon^{-1})}{\ln \beta^{-1}} \right\rceil,$$

then the variation distance between the stationary distribution and the probability distribution of $(\mathcal{Y}_{kT})$ is bounded by ε.

Although delayed path coupling is a simple refinement of path coupling, it is a very powerful tool in proving convergence of Markov chains [10, 13].

3 Auxiliary Switching Networks

3.1 Permuting at Random in a Large Depth

Our aim is to construct a switching network of a small depth that permutes at random. Once we relax the assumption that the depth must be small it is well known how to solve this problem. One of possible ways is the following:

Lemma 3.1 *Let $r \in I\!N$. There is a switching network $\mathcal{S}_m$ with m inputs of depth $d = \mathcal{O}((r + \log m) \cdot m^2)$ such that variation distance between the probability distribution defined by the switching networks and the uniform distribution over S_m is $\mathcal{O}(1/2^r)$.*

Network $\mathcal{S}_m$, called Odd-Even Transposition Network, consists of two layers repeated $d/2$ times. The first of this layers has the form

$$(1,2), (3,4), (5,6), \ldots\ ,$$

the second one consists of the switches

$$(2,3), (4,5), \ldots\ .$$

The above result is a folklore, however we prove it in order to illustrate the path coupling technique for switching networks on a simple example. On the other hand, we use Odd-Even Transposition Network as a building block for the main construction in this paper, so it deserves some attention.

Proof of Lemma 3.1. For two permutations $\pi_1, \pi_2 \in S_m$, let $\Delta(\pi_1, \pi_2)$ denote the minimal number of transpositions necessary to convert π_1 into π_2. So, Δ defines a metric on S_m with $\Delta(\pi_1, \pi_2) \leq m-1$ for any $\pi_1, \pi_2 \in S_m$.

Now, we consider a stochastic process $\mathcal{Y}_t$ defined by the switching network $\mathcal{S}_m$. The state $\mathcal{Y}_t$ is defined as a permutation π such that $\pi(i)$ is the contents of the register at wire i after applying $2t$ layers of $\mathcal{S}_m$. Transitions of this chain are defined by the switches.

In order to prove Lemma 3.1 we have to show that the stationary distribution of $\mathcal{Y}_t$ is uniform and that after d steps of the Markov chain, its probability distribution is close to the stationary distribution. The first claim follows from considerations in Section 2.2: it is easy to see that the chain satisfies the assumptions of Lemma 2.3, and, as seen in Section 2.2, the uniform distribution is stationary for this chain.

For showing rapid mixing property of the chain $\mathcal{Y}_t$, we apply path coupling argument. So, we start with permutations π_1 and π_2 that differ by a single transposition and define a pair of processes $(\mathcal{X}_t, \mathcal{X}_t^*)_{t \in I\!N}$. Let $\lambda = \lambda(t)$ and $\delta = \delta(t)$ denote the positions where the states $\mathcal{X}_t, \mathcal{X}_t^*$ differ. We define coupling so that $\mathcal{X}_t$ and $\mathcal{X}_t^*$ never differ on more than two positions. For this purpose, $\mathcal{X}_t$ and $\mathcal{X}_t^*$ have the same settings of switches except when there is a switch between wires λ and δ. In this case, we allow the first process to set the switch between wires λ and δ at random. The second process sets this switch on if and only if the first process sets this switch off. So, each process makes the decision to activate the switch with probability 0.5. Simultaneously, after performing the step both processes achieve the same state (see Fig. 3).

Now we consider T steps of $(\mathcal{X}_t, \mathcal{X}_t^*)_{t \in I\!N}$ and estimate the probability that $\mathcal{X}_t$ and $\mathcal{X}^*$ couple. The states of $\mathcal{X}_t$ and $\mathcal{X}_t^*$ become identical when at some point of execution λ and δ are neighboring positions connected by a switch. Since we do not provide any means for steering the movements of λ and δ, this occurs randomly. We have to show that if T is big enough, there is a fair probability that such an event takes place. Let initially $\lambda < \delta$. This inequality remains fulfilled during the whole process until $\mathcal{X}_t$ and $\mathcal{X}^*$ couple. Indeed, getting $\delta \leq \lambda$ would require that at some step, wires λ and δ are connected by a switch. Consider the random walk of λ defined by $(\mathcal{X}_t)_{t \in I\!N}$. The corresponding Markov chain is described by random walk of a single item through n positions, where the transitions are defined as by the switches of $\mathcal{Y}_t$. It is well known [23] that for $T_0 = \mathcal{O}(m^2)$ the item visits the position m at least once within T_0 steps with a constant probability. It follows immediately that with a constant probability the processes $(\mathcal{X}_t)$ and

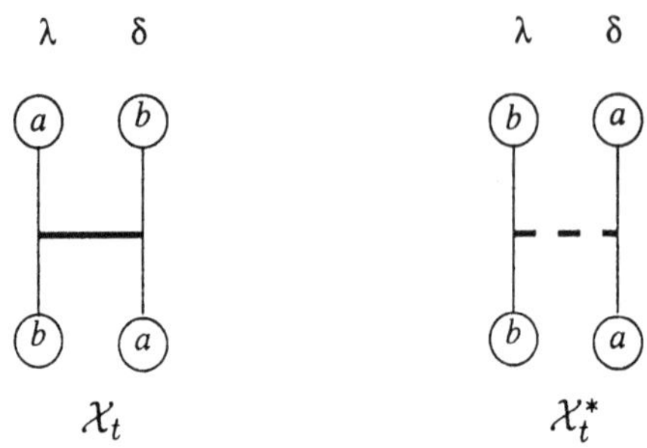

(a) the switch between positions λ and δ is active in $\mathcal{X}_t$

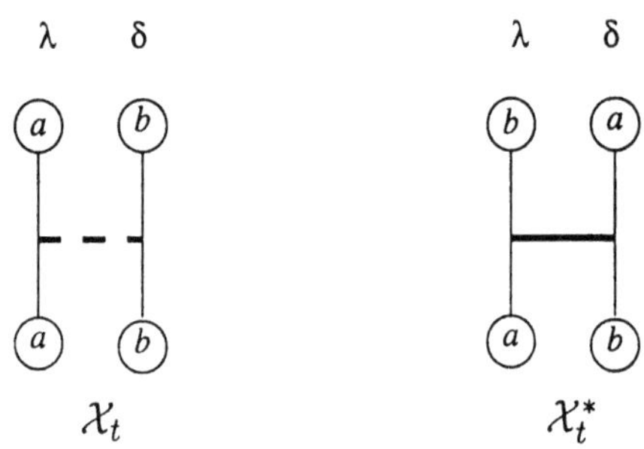

(b) the switch between positions λ and δ is inactive in $\mathcal{X}_t$

Figure 3: Coupling of processes $(\mathcal{X}_t)_{t\in\mathbb{N}}, (\mathcal{X}_t^*)_{t\in\mathbb{N}}$

$(\mathcal{X}_t^*)$ couple successfully within T_0 steps. Indeed, λ cannot reach m without meeting δ on its way, whatever moves are done by δ.

Now by applying path coupling inequality for $k = T_0$, we obtain that for some $T = \mathcal{O}(rm^2)$, the variation distance between the probability distribution of $\mathcal{X}_T$ and the uniform distribution is $\mathcal{O}(1/2^r)$. □

3.2 Approximating Product Algorithms

The process of generating random permutations consists often of some number of independent phases. Every of these phases is a random experiment, with the output having an appropriate probability distribution. The probability space corresponding to the whole computation is then the product space of the subspaces corresponding to independent phases. Strategy 1.7 is an example of such an approach, which we call a *product algorithm.*

When we implement a product algorithm on a switching network, it might be impossible to execute each phase of the algorithm with exactly the right probability distribution. A simple reason might be that the switching networks define only probabilities of the form $s/2^r$. So often all we can do is to approximate these probabilities. Now, even if for each phase this probability distribution is close to the ideal one, what is the guarantee that the whole process will yield the result with probability distribution close to the ideal one? An answer to this question is given by the following well-known lemma:

Lemma 3.2 *Let μ, μ' be probability distributions over probability space G, and σ, σ' be probability distributions over probability space H. Then variation distance between product distributions $\mu \times \sigma$ and $\mu' \times \sigma'$ over $G \times H$ is bounded by $\|\mu - \mu'\| + \|\sigma - \sigma'\|$.*

Proof. Obviously, $\|\mu \times \sigma - \mu' \times \sigma\| = \|\mu - \mu'\|$ and $\|\mu' \times \sigma - \mu' \times \sigma'\| = \|\sigma - \sigma'\|$. So by the triangle inequality, $\|\mu \times \sigma - \mu' \times \sigma'\| \leq \|\mu - \mu'\| + \|\sigma - \sigma'\|$. □

3.3 Pseudo-Butterfly Networks

As already seen, butterfly networks have the following nice property: if m is the number of wires in a butterfly, $i, j < m$ and the switches are set at random, then probability that the item from the ith input register goes to the jth output register equals $\frac{1}{m}$. For our construction, we need a network with such a property. However, in certain cases we are unable to apply butterflies. The problem is that the number of wires in a butterfly network must be a power of two. Note that we cannot hope to provide such a construction for an arbitrary number of wires n, since probabilities induced by a switching network are of the form $l/2^k$ and therefore we cannot get an event with probability *exactly* $1/n$, if n is not a power of 2. All we can do is an approximate solution.

Lemma 3.3 *Let $m \in I\!N$ be arbitrary. One can construct a network of depth $\mathcal{O}(\log^2 m)$ such that for each $i, j \leq m$, probability that the item from* *input register i goes to the output register j is a number in the interval* $[\frac{1}{m} - \frac{1}{m^2}, \frac{1}{m} + \frac{1}{m^2}]$.

For the rest of this paper, the network constructed in Lemma 3.3 for m wires will be called *pseudo-butterfly* and denoted by $\mathcal{PB}_m$.

Proof of Lemma 3.3. The network $\mathcal{PB}_m$ consists of a strip, which is repeated $\mathcal{O}(\log n)$ times. Let us assume that $k \in \mathbb{N}$ satisfies $2^k < m \leq 2^{k+1}$. The strip consists of the following parts (see Fig. 4): the first set of layers is composed of a butterfly on the first 2^k wires. The next part consists of a single layer in which each of the wires $i = 2^k + 1, \ldots, m$ is connected by a switch with wire $i - 2^k$.

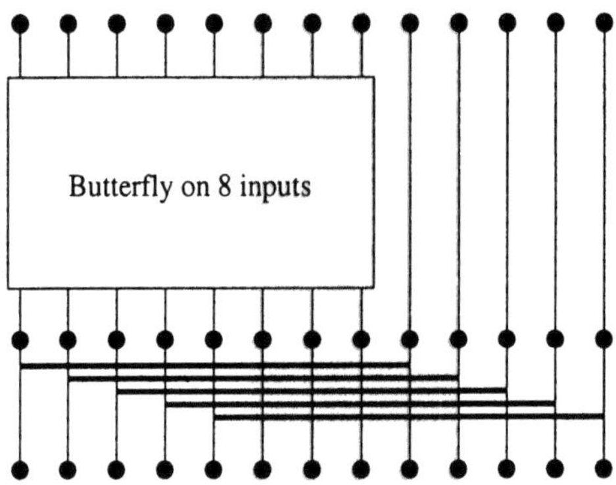

Figure 4: A strip of network $\mathcal{PB}_{13}$

For our analysis, two strips will be regarded as a step in a Markov chain $\mathcal{MPB}_m$. The inputs we consider here are the sequences containing a single one and $m-1$ zeroes. A step of $\mathcal{MPB}_m$ is an application of two consecutive strips with their switches set at random. It is easy to see that for any two configurations A, B of $\mathcal{MPB}_m$, it is possible to change the state from A to B in one step of $\mathcal{MPB}_m$. So, chain $\mathcal{MPB}_m$ is aperiodic and irreducible and the statement of Lemma 2.3 holds for $\mathcal{MPB}_m$. In Section 2.2, we have seen that the uniform distribution is stationary for $\mathcal{MPB}_m$. So all we need is to estimate the convergence rate to the uniform distribution. We prove the following claim:

Claim. *After $\mathcal{O}(\log m)$ steps of $\mathcal{MPB}_m$, the variation distance between the uniform distribution for sequences with a single one and $m-1$ zeroes and the probability distribution of $\mathcal{MPB}_m$ is not greater than $\frac{1}{m^2}$.*

Before we commence the proof of the claim, note that Lemma 3.3 follows from it. Indeed, if the variation distance is at most $\frac{1}{m^2}$, then probability of any state may not deviate by more than $\frac{1}{m^2}$ from the value for the uniform distribution, that is, $1/m$. Since convergence to the stationary distribution holds any initial distribution, it holds also if initially the one occupies always the ith input register. It follows that probability that the one reaches the jth output register is a number that deviates from $\frac{1}{m}$ by at most $\frac{1}{m^2}$.

Now, for the proof of rapid mixing of chain $\mathcal{MPB}_m$ we apply the coupling argument. It suffices to show that processes $\mathcal{MPB}_m$ and $\mathcal{MPB}^*_m$ may be coupled in a single step with probability greater than $\frac{1}{4}$. Assume that the first process has initially the one on position i and the second process has the one on position j. We have to consider some cases. The first one is that $i, j \leq 2^k$. Let us consider the first part of the strip containing the butterfly on the first 2^k wires. We set the switches in the butterfly in $\mathcal{MPB}^*_m$ so that the one from the wire j goes to the same wire as the one in $\mathcal{MPB}_m$. It is easy to see that this coupling is correct (it does not change probabilities of state transitions while regarding the second process alone). Obviously, in this case we are always able to couple $\mathcal{MPB}_m$ and $\mathcal{MPB}^*_m$. Now assume that the first case does not apply, for instance $j > 2^k$. Then we consider the layer immediately after the first butterfly. Let i' be the wire that contains the one of $\mathcal{MPB}_m$ at this moment. If $i' + 2^k = j$, then in $\mathcal{MPB}^*_m$ we set the switch (i', j) on if and only if, this switch is off in $\mathcal{MPB}_m$. In that way, we couple $\mathcal{MPB}_m$ and $\mathcal{MPB}^*_m$. If $j \neq i' + 2^k$, then we set the switches in both processes at random. So with probability $\frac{1}{2}$, in $\mathcal{MPB}^*_m$, the one gets into the first 2^k wires. The one of $\mathcal{MPB}_m$ either stays within the first 2^k wires, since there is no switch connected to its current wire, or with probability $\frac{1}{2}$ a switch leaves the one within the first 2^k wires. Concluding, we either couple both processes at this layer or with probability at least $\frac{1}{4}$ move both ones into the first 2^k wires. In the later case, we apply the technique used previously in the first butterfly in order to couple $\mathcal{MPB}_m$ and $\mathcal{MPB}^*_m$ with probability one. □

3.4 Properties of Networks of Small Depth

Let us consider a butterfly network with the switches set at random. We show in a relatively simple way that the output of this network achieves some degree of randomness. Assume that the input to the butterfly network with m inputs consists of $M = m/h$ ones and $m - M$ zeroes. Further, we consider a "bucket" B of output registers. The probability that a one from some fixed input position arrives in bucket B is $\frac{|B|}{n}$, since every output position is equally probable. Let J_j denote the indicator variable for the event that the jth one arrives in B. Then $\mathbf{E}[J_j] = \frac{|B|}{n}$. Further, by linearity of expectation,

$$\mathbf{E}[\sum_{j=1}^{M} J_j] = \sum_{j=1}^{M} \mathbf{E}[J_j] = \frac{M \cdot |B|}{n} .$$

Obviously, $\sum_{j=1}^{M} J_j$ is the number of ones that arrive in B. We see that the expected number of ones directed to B by the butterfly equals the expected number of ones in B in a random experiment, in which we throw M ones independently at random with probability of hitting B equal to $\frac{|B|}{n}$. Below, we show that not only the expected value of the number of ones in B is $\frac{|B|}{n}$, but also big deviations from the expected value have low probability. Surprisingly, the last phenomenon holds for all switching networks of a small depth.

Lemma 3.4 (Technical Lemma) *Let S be a switching network of depth d and m wires, $m \leq n$, B be a group of arbitrary output registers. Consider the inputs to S containing ones on some fixed positions and zeroes elsewhere. Let J be a random variable denoting the number of ones arriving in B when the switches of S are set at random.*
Let $r \in \mathbb{N}$ and assume that $16rd \cdot \log n \leq \mathbf{E}[J]$. Then

$$\mathbf{Pr}\left[|J - \mathbf{E}[J]| > \sqrt{\mathbf{E}[J] \cdot 16rd(\ln n + 1)} \right] \leq \frac{1}{n^r} .$$

In other words, with very high probability (greater than or equal to $1 - \frac{1}{n^r}$), the number of ones in B does not differ from the expected number by more than an additive factor depending on d, $\mathbf{E}[J]$, r and n. In the case of a butterfly on n wires and $\frac{n}{\log^\beta n}$ ones, we get

$$\mathbf{Pr}\left[J - \frac{|B|}{\log^\beta n} > \sqrt{\frac{|B|}{\log^\beta n} \cdot 16r \cdot \log n \cdot (\ln n + 1)} \right] < \frac{1}{n^r} .$$

So, for $|B| = \log^\gamma n$ and a constant r, the square root on the right side of inequality is $\mathcal{O}(\log^{1+0.5(\gamma-\beta)} n)$, whereas the expected number of ones in B equals $\log^{\gamma-\beta} n$.

Proof of Lemma 3.4. The proof is based on a variant of Talagrand's Inequality ([22], page 18: Talagrand's Inequality II). Let us recall the statement of this result.

Let U be a random variable determined by k independent trials $T_1, \ldots, T_k$, and satisfying the following conditions:

1. *changing the outcome of any trial can affect U by at most c, and*

2. *for any r, if $U \geq r$, then there are r trials, $T_{i_1}, \ldots, T_{i_r}$ whose outcomes certify that $U \geq r$.*

Then for any $0 < t \leq \mathbf{E}[U]$,

$$\mathbf{Pr}\left[|U - \mathbf{E}[U]| > t + 3c\sqrt{\mathbf{E}[U]}\right] \leq 2\exp\left(-\tfrac{t^2}{16\,c^2\,\mathbf{E}[U]}\right) .$$

Using this notation, we set $U \equiv d \cdot J$, $k = \frac{1}{2}\, d \cdot m$, and the independent trials to be the settings of the k switches of S. Then, it is easy to see that any change of a switch state can change J by at most 1, and therefore we can choose $c = d$. Further, if $J \geq \ell$, then there are always $d \cdot \ell$ switches which certify this event (we just pick any ℓ ones which arrive in B and fix the states of the switches defining their path to the output positions). Therefore, Talagrand's Inequality yields for any $0 < t \leq \mathbf{E}[U]$ and $T = \frac{t}{d}$:

$$\begin{aligned}
&\mathbf{Pr}\left[|dJ - \mathbf{E}[dJ]| > dT + 3d\sqrt{\mathbf{E}[dJ]}\right] \\
&= \mathbf{Pr}\left[|J - \mathbf{E}[J]| > T + 3\sqrt{d \cdot \mathbf{E}[J]}\right] \\
&\leq 2\exp\left(-\tfrac{(dT)^2}{16\,d^2\,\mathbf{E}[dJ]}\right) = 2\exp\left(-\tfrac{T^2}{16\,d\,\mathbf{E}[J]}\right) .
\end{aligned}$$

Now, if we set $T = 4\sqrt{\mathbf{E}[J] \cdot d \cdot r \cdot \log n}$, we get

$$T + 3\sqrt{d \cdot \mathbf{E}[J]} = \sqrt{d \cdot \mathbf{E}[J]} \cdot \left(3 + 4\sqrt{r \cdot \log n}\right) \leq \sqrt{d \cdot \mathbf{E}[J]} \cdot \sqrt{16r \cdot (\log n + 1)}.$$

On the other hand,

$$2\exp\left(-\tfrac{T^2}{16\,d\,\mathbf{E}[J]}\right) = 2\exp\left(-\tfrac{16\,\mathbf{E}[J] \cdot d \cdot r \cdot \log n}{16\,d\,\mathbf{E}[J]}\right) = 2\exp(-r\log n) = \frac{1}{n^r}.$$

$\square$

4 Construction of the Network

In this section, we present the network given in Theorem 1.11.

4.1 Outline of Permuting Strategy

The general strategy implemented by network $\mathcal{B}_n$ is as follows:

Strategy 4.1
Phase 1: *Randomly partition the input items into sets* $R_1, \ldots, R_{t+1}$*:*

- *Choose uniformly at random a number of items and put them into set* R_1*;*
- *For* $s = 1, \ldots, t-1$*, given sets* $R_1, \ldots, R_s$*, select set* R_{s+1} *by choosing uniformly at random a number of items not belonging to* $R_1 \cup \ldots \cup R_s$*;*
- *Set* R_{t+1} *contains all elements that do not belong to* $R_1 \cup \ldots \cup R_t$*.*

Phase 2: *Permute independently and uniformly at random items within each set* R_i*. For this purpose, apply recursively the same algorithm or any other algorithm for permuting uniformly at random.*

It is easy to see that Strategy 4.1 yields each permutation of the input items with the same probability.

The network $\mathcal{B}_n$ performs approximately Strategy 4.1. Once a set considered in (a recursive call to) Phase 2 is of a size $u < 2\log^7 n$, then the network $\mathcal{S}_u$ from Lemma 3.1 is used to permute this set. By Lemma 3.1, if we choose an appropriate $k = \Theta(\log n)$, then this algorithm yields a probability distribution that differs from the uniform distribution by less than $1/n^2$. The depth of $\mathcal{S}_u$ is then $\mathcal{O}(\log^{15} n \log\log n)$. The second important point is that for executing Phase 1 we use a network $\mathcal{C}_m$ constructed in the next subsection. It has depth $\mathcal{O}(\log^4 n)$. For m inputs, $\log^7 n \le m \le n$, network $\mathcal{C}_m$ selects m/h items almost uniformly at random and outputs them on positions 1 through m/h. The parameter h is chosen as $h = \log^2 n$. Again, the variation distance between the probability distribution obtained and the uniform one is bounded by $1/n^2$.

Now, we discuss some parameters in our algorithm. Phase 1 is based on the procedure of cutting off a fraction of $\frac{1}{h}$ items from a given set of m items. Therefore, each set R_i obtained by cutting off has size at most $\frac{m}{h}$. We continue cutting off until the number of the remaining items is bounded by $\frac{m}{h}$. One can easily verify that there are $\mathcal{O}(h \log h)$ sets R_i. Indeed, for $t = \frac{h \log h}{\log e}$,

$$\left(1 - \tfrac{1}{h}\right)^t = \left(\left(1 - \tfrac{1}{h}\right)^h\right)^{\log h / \log e} < \left(\tfrac{1}{e}\right)^{\log h / \log e} = \tfrac{1}{h}.$$

Phase 1 is called recursively $\mathcal{O}(\log_h n)$ times, since each time the size of the sets under consideration decreases at least h times. The number of sets R_i considered during the whole execution is $\mathcal{O}(\frac{n}{\log^7 n} \cdot \log_h n) = \mathcal{O}(n)$. Indeed, there are $\mathcal{O}(\frac{n}{\log^7 n})$ groups of wires, for which we finally apply Lemma 3.1. Elements of such a group belong to the same sets R_i that emerge during the recursive execution of the algorithm. Since there are $\mathcal{O}(\log_h n)$ levels of recursion, this makes $\mathcal{O}(\frac{n}{\log^7 n} \cdot \log_h n)$ sets R_i.

We see that the whole algorithm executed by switching network $\mathcal{B}_n$ consists of $\mathcal{O}(n)$ independent parts with probability space being the product of the probability spaces of these parts. Each part executed on the switching network approximates an appropriate uniform distribution so that the variation distance between them is bounded by $1/n^2$. Therefore, by Lemma 3.2, the variation distance between distribution defined by $\mathcal{B}_n$ and the uniform distribution is $\mathcal{O}(n \cdot 1/n^2) = \mathcal{O}(1/n)$.

Now, let us estimate the total depth of $\mathcal{B}_n$. During each execution of Phase 1 of Strategy 4.1, there are $\mathcal{O}(h \log h)$ subparts corresponding to the application of the network $\mathcal{C}_m$, each of depth $\mathcal{O}(\log^4 n)$. This sums up to $\mathcal{O}(\log^6 n \cdot \log \log n)$. There are $\mathcal{O}(\log_h n)$ recursive calls to Strategy 4.1, so Phases 1 of all recursive calls need depth $\mathcal{O}(\log^7 n$. Finally, application of the network $\mathcal{S}_u$ adds additional $\mathcal{O}(\log^{15} n \cdot \log \log n)$ layers. It follows that the total depth of $\mathcal{B}_n$ is $\mathcal{O}(\log^{15} n \cdot \log \log n)$.

¿From the above considerations we see that the core element of the construction is the switching network $\mathcal{C}_m$, which we describe in Section 4.2.

4.2 Network $\mathcal{SH}_m$

Lemma 4.2 *Let $m, n \in I\!N$, $\log^7 n \le m \le n$, $h = \log^2 n$. One can construct a switching network $\mathcal{SH}_m$ of depth $\log^{O(1)} n$ that has the following property: consider an input with $\frac{m}{h}$ ones in some fixed input registers, and zeroes in the remaining input registers. Then the variation distance between the probability distribution of the output positions of the ones for random setting of switches and the uniform distribution of $\frac{m}{h}$ ones over m positions is bounded by $\frac{1}{n^2}$.*

Obviously, the switching network $\mathcal{C}_m$ performing "cutting off" that we need for our construction may be obtained by reversing the direction of the flow of items in $\mathcal{SH}_m$. So, the rest of this section is devoted to the proof of Lemma 4.2.

Network $\mathcal{SH}_m$ is periodic and consists of $\mathcal{O}(\log n)$ identical strips, each made of a polylogarithmic number of layers. So $\mathcal{SH}_m$ is described by a Markov chain $\mathcal{MSH}_m$, where each strip of $\mathcal{SH}_m$ defines one step of $\mathcal{MSH}_m$. As seen in Section 2.2, the uniform distribution is stationary for $\mathcal{MSH}_m$. Obviously, the chain $\mathcal{MSH}_m$ is aperiodic. So, in order to apply Lemma 2.3 we have to show that chain $\mathcal{MSH}_m$ is irreducible. This is possible by a detailed analysis of the (still undefined) architecture of $\mathcal{SH}_m$. A simple trick already mentioned in Section 2.2, which allows to avoid this work, is to add an arbitrary permutation network of logarithmic depth before the first layer we define below.

Our aim is to design the strip so that we may apply path coupling arguments to chain $\mathcal{MSH}_m$. In this way, we get a bound on the number of steps of $\mathcal{MSH}_m$ necessary to reach a probability distribution close to the stationary distribution. This number of steps is finally chosen as the number of strips in $\mathcal{SH}_m$. We describe the construction of the strip simultaneously with the analysis of the corresponding Markov chain.

Before we go into details of the construction of $\mathcal{SH}_m$, let us take a closer look at the path coupling argument. We consider a coupling process $B = (B_t, B_t^*)_{t \in I\!N}$, where the input of each of the halves consists of $\frac{m}{h}$ ones and $m - \frac{m}{h}$ zeroes. We define the distance between two configurations of $\frac{m}{h}$ ones and $m - \frac{m}{h}$ zeroes as the number of transpositions necessary to transform one configuration into the other one. Since we are applying path coupling, we assume that the configurations B_0 and B_0^*, that is, the contents of the input registers, differ only at two positions. In order to avoid incrementing the distance between B_t and B_t^*, we set the switches of both processes to the same values, whenever the values on the nodes connected by a switch are the same for both processes. We only guide the choices made by B^* on the positions where it differs from B. Whenever $B_t \neq B_t^*$, we denote by $\lambda = \lambda_t$ and $\delta = \delta_t$ the positions on which B_t and B_t^* differ, and assume that B_t stores a 1 at position λ and a 0 at δ.

We split the set of wires into so called *buckets*, each bucket consisting of approximately $\log^7 n$ consecutive wires. The goal behind our definition of $\mathcal{SH}_m$ is to force λ and δ to go into the same bucket. At the moment, when they might be in the same bucket, we apply the switching network from Lemma 3.1 and use coupling construction from the proof of this lemma. So, if λ and δ are indeed in the same bucket, with high probability we may couple B_t and B_t^*.

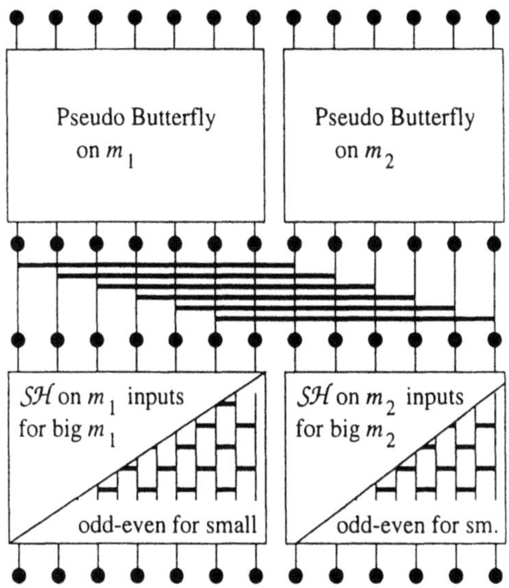

Figure 5: Architecture of network $\mathcal{SH}_m$

Network architecture We say that there are *butterfly switches* between sets of wires $\mathcal{S}$ and $\mathcal{S}'$, if for each $i \leq |\mathcal{S}|$ there is a switch between the ith wire of $\mathcal{S}$ and the ith wire of $\mathcal{S}'$. We assume that $|\mathcal{S}| = |\mathcal{S}'|$ or $|\mathcal{S}| + 1 = |\mathcal{S}'|$, so at most one wire is not connected by these switches.

The network $\mathcal{SH}_m$ has the following recursive structure:

Part 1. Let $m_1 = \lceil m/2 \rceil$, $m_2 = \lfloor m/2 \rfloor$. The first part of $\mathcal{SH}_m$ consists of pseudo-butterflies $\mathcal{PB}_{m_1}$, $\mathcal{PB}_{m_2}$ applied, respectively, to the first m_1 and the last m_2 wires.

Part 2. The next layer is composed of the butterfly switches between the first m_1 wires and the last m_2 wires.

Part 3. If $m_2 \geq 2\log^7 n$, the remaining layers of $\mathcal{SH}_m$ are composed from $\mathcal{SH}_{m_1}$ on the first m_1 wires and $\mathcal{SH}_{m_2}$ on the last m_2 wires. Otherwise, there are networks $\mathcal{S}_{m_1}$ and $\mathcal{S}_{m_2}$ on, respectively, the first m_1 and the last m_2 wires.

4.3 Properties of Network $\mathcal{SH}_m$

Each of the groups of wires to which a pseudo-butterfly is applied during the recursive definition of $\mathcal{SH}_m$ will be called a *cluster*. Totally, there are less than n clusters, since each stage of recursion doubles the number of clusters with respect to the number of clusters from the previous stage, and we stop once the clusters get size less than $\log^7 n$.

Our goal is to estimate the number of ones that arrive in a given cluster under assumption that there are totally $\frac{m}{h}$ ones given to $\mathcal{SH}_m$. The first step in this direction is to estimate the expected number of ones in the cluster. The problem we deal with is that the pseudo-butterflies distribute the ones quite evenly, but not exactly. So when a pseudo-butterfly is applied, some output positions get privileged. Since we apply many pseudo-butterflies, we must show there is no substantial cumulation of this effect.

Lemma 4.3 *Let C be a cluster, as defined above. Then the probability that an input item from a fixed position arrives in C belongs to the interval*

$$\left[\frac{|C|}{m}\cdot\left(1-\frac{1}{\log n+1}\right), \frac{|C|}{m}\cdot\left(1+\frac{1}{\log n}\right)\right]$$

(where $|C|$ denotes the number of wires in C).

Proof. Let $C_1, C_2, \ldots, C_k$ denote the clusters such that C_i emerges during phase i, $C \subseteq C_i$, for $i \leq k$, and $C = C_k$. Let x denote the input item under consideration. The probability that x arrives in C_1 is upper bounded by $\frac{|C_1|}{m}\cdot(1+\frac{1}{n^2})$. Indeed, by Lemma 3.3, probability that x arrives at any wire of C_1 does not exceed $\frac{1}{m}\cdot(1+\frac{1}{m^2})$. Assuming that x goes through C_1, probability that it goes into C_2 is at most $\frac{|C_2|}{|C_1|}\cdot(1+\frac{1}{|C_1|^2})$. Since the switches at different phases are set independently, the probability that x arrives in C_2 after stage is bounded by

$$\frac{|C_1|}{m}\cdot\left(1+\frac{1}{m^2}\right)\cdot\frac{|C_2|}{|C_1|}\cdot\left(1+\frac{1}{|C_1|^2}\right)=\frac{|C_2|}{m}\cdot\left(1+\frac{1}{m^2}\right)\cdot\left(1+\frac{1}{|C_1|^2}\right) .$$

Continuing in this way, we see that the probability that x arrives in C after stage k is non-greater than

$$\frac{|C|}{m}\cdot\prod_{i=0}^{k-1}\left(1+\frac{1}{|C_i|^2}\right)$$

where $|C_0|$ stands for m. Then,

$$\prod_{i=0}^{k-1}\left(1+\frac{1}{|C_i|^2}\right) \le \left(1+\frac{1}{|C|^2}\right)^k \le \left(1+\frac{1}{|C|^2}\right)^{|C|} \le e^{1/|C|}.$$

On the other hand,

$$\left(1+\frac{1}{\log n}\right)^{|C|} \ge \left(1+\frac{1}{\log n}\right)^{2\log n} \ge e\ .$$

So, $e^{1/|C|} \le 1+\frac{1}{\log n}$. Hence the probability that x arrives in C is not greater than $\frac{|C|}{m}\cdot(1+\frac{1}{\log n})$.

The second inequality stated in the lemma can be checked in an analogous way. □

The next step is to estimate deviations from the maximal number of ones in a cluster and show that large deviations are unlikely.

Lemma 4.4 *Consider a single strip of $\mathcal{SH}_m$. Let C be a cluster emerging during execution of this strip and let J be a random variable denoting the number of ones going through C when the switches are set at random. Then*

$$\mathbf{Pr}\,[\,J > 2\mathbf{E}[\,J\,]\,] < \frac{1}{n^2}\ .$$

Proof. Let d denote the depth on which cluster C emerges in $\mathcal{SH}_m$. Let J_i be the indicator variable for the event that the ith one goes through C after layer d. Then

$$\mathbf{E}[\,J\,] = \mathbf{E}[\sum_{i=1}^{m/h} J_i\,] = \sum_{i=1}^{m/h}\mathbf{E}[\,J_i\,] \ge \frac{m}{h}\cdot\frac{|C|}{m}\cdot\left(1-\frac{1}{\log n+1}\right) \ge 0.5\log^5 n\ .$$

Since $32d\log n = \mathcal{O}(\log^3 n)$, we get $32d\log n \le \mathbf{E}[\,J\,]$. Therefore, we may use Lemma 3.4 and obtain

$$\mathbf{Pr}\left[\,|J-\mathbf{E}[\,J\,]| > \sqrt{\mathbf{E}[\,J\,]\cdot 32d(\ln n+1)}\,\right] < \frac{1}{n^2}\ .$$

Let us take ζ that satisfies $|C| = \log^\zeta n$ (obviously, $\zeta \ge 7$ and, in general, ζ is not a constant). Since the depth of a strip of $\mathcal{SH}_m$ is bounded by $c_0\log^3 n$ and $\mathbf{E}[\,J\,] \le \frac{|C|}{h}\cdot(1+\frac{1}{\log n})$, we see that also

$$\mathbf{Pr}\left[\,J-\mathbf{E}[\,J\,] > \sqrt{\frac{\log^\zeta n}{\log^2 n}\cdot\left(1+\frac{1}{\log n}\right)\cdot 32\cdot c_0\log^3 n\cdot(\ln n+1)}\,\right] < \frac{1}{n^2}\ .$$

The expression, call it R, under square root is upper bounded by $c_1 \cdot \log^{\zeta+2} n$ for some constant c_1. So $\sqrt{R} \leq \sqrt{c_1} \cdot \log^{0.5\zeta+1} n$. Since $\zeta - 2 > 0.5(\zeta+1)$ for $\zeta \geq 7$ for large enough n, by Lemma 4.3:

$$\mathbf{E}[J] \geq \frac{\log^{\zeta} n}{\log^2 n} \cdot \left(1 - \frac{1}{\log n + 1}\right) \geq \sqrt{c_1} \cdot \log^{0.5\zeta+1} n.$$

Therefore,

$$\mathbf{Pr}[J > 2\mathbf{E}[J]] \leq \mathbf{Pr}\left[J > \mathbf{E}[J] + \sqrt{R}\right] < \frac{1}{n^2}.$$

□

4.4 Coupling for Network $\mathcal{SH}_m$

Now we define the way in which we try to couple the processes B_t and B_t^* within one strip. Recall that our strategy is to force λ and δ to go into the same bucket. This is achieved by forcing them to go to the same cluster during execution of Parts 1 and 2 at each stage of the recursive definition of the strip. If we succeed at each stage, then finally λ and δ arrive at the same bucket. Then coupling method from the proof of Lemma 3.1 is applied.In order to see how the coupling works at stage i of recursion assume that λ and δ are both inside the same cluster C after executing stage $i-1$. At stage i, we split cluster C into two clusters C' and C'' of approximately the same size. At Part 1, we apply pseudo-butterflies on, respectively, C' and C''. We define coupling so that all switches in these pseudo-butterflies are set in B_t^* just like in B_t. Consider the movements of λ during this part of computation. Due to the properties of pseudo-butterflies, λ goes to a random position within its cluster, with approximately the same probability to reach each position. The same applies to δ (possibly with some slight stochastic dependency between the positions of λ and δ, if they are in the same cluster). Hence, probability that there is a one on the same position as λ in the second cluster is $O(s/|C'|)$, where s denotes a bound on the number of ones in the cluster. The same holds for δ. Hence, we may bound the probability that either of δ and λ is connected with a one through butterfly switches at Part 2 of stage i or it is not connected to a butterfly switch (i.e., it is on the last wire of C' for $|C'| = |C''|+1$) by the sum of the probabilities of these events. The result is again $O(s/|C'|)$. If such an event occurs, coupling does not succeed, and we may set the switches in the same way in B_t^* as in B_t.

Now, consider the main case that λ and δ are connected with zeroes through butterfly switches. The first case is that λ and δ belong to different clusters C' and C''. Then the coupling is defined as follows:

> the switch connecting δ in the second process is set **on** if and only if the switch connecting λ in the first process is set **off**.

The switches adjacent to the wire λ in B_t^* and to the wire δ in B_t are set arbitrarily, since they connect the wires with zeroes. The remaining switches are set in the same way in both processes. It is easy to see that this definition guarantees that after application of the butterfly switches λ and δ are either both in C' or both in C''.

If λ and δ belong both to C' or both to C'', then we define coupling as follows:

> the switch connecting δ in the second process is set **on** if and only if the switch connecting λ in the first process is set **on**.

The remaining switches are set as in the previous case. It follows directly from the definition that λ and δ either both remain in their cluster or both move to the other cluster.

Estimation of success probability There are different events due to which coupling in a given strip may fail:

1. in some cluster, the number of ones exceeds the fraction of $2/\log^2 n$, and therefore the probability that λ or δ hits a one might be too high,
2. even if the first event does not occur, with some probability λ or δ may hit a one,
3. either λ or δ may fall into the wire of a cluster not connected by a butterfly switch,
4. coupling inside a bucket may fail.

Now let us estimate probabilities of these events. By Lemma 4.4, the probability that a fixed cluster contains a fraction of ones higher than $2/\log^2 n$ is bounded by $1/n^2$. Since there are less than n clusters, the probability that at least one cluster has this property is bounded by $1/n$. The probability that the second event occurs under condition that the first event does not occur is bounded by $4/\log n$, since at each of less than $\log n$ stages of recursion this probability is bounded by $2 \cdot \frac{2}{\log^2 n}$ (one of the factors 2 is due to the fact that we consider the probabilities for λ and δ separately).

For the third event, note that its probability is bounded by the number of stages multiplied with $2/s$, where s is the minimal size of a cluster. Hence, it is bounded by $2 \log n / \log^7 n \leq 2/\log^6 n$. The last event occurs with probability less than $1/\log n$ provided that we take $\mathcal{O}(\log^{14} n \log\log n)$ layers of odd-even network for permuting the buckets.

It follows that within a strip the processes B_t and B_t^* fail to couple with probability less than $1/\log n$. Hence

$$\mathbf{E}[\text{ variation distance between } B_t \text{ and } B_t^*] = \mathcal{O}\left(\frac{1}{\log n}\right)$$

and by path coupling inequality (Lemma 2.4), after $\mathcal{O}(\log n / \log\log n)$ steps of chain $\mathcal{MSH}_m$, the probability distribution obtained and the stationary distribution have variation distance bounded by $1/n$. □

5 Application: a fast EREW PRAM Algorithm

Construction of a switching network of polylogarithmic depth generating random permutations may be used to design a fast EREW PRAM algorithm. EREW PRAM (Exclusive Read Exclusive Write Parallel Random Access Machine) is one of the major theoretical models for a parallel computer with a shared memory [17]. It consists of a number of independent processors using a shared memory consisting of a number of memory cells. In one step of a PRAM, each processor may read an arbitrary memory cell, perform an arbitrary computation using its local memory, and, finally, write a value into a chosen cell of the shared memory. An algorithm for a PRAM is called EREW, if for no input data two different processors try to read or write the same memory cell during the same step. We describe this situation by saying that *no read and write conflict occurs.* In this model, it is assumed that the input data is given in the shared memory, scattered among many memory cells.

The EREW PRAM model has several advantages and disadvantages. It disregards the cost (and time) necessary to perform the read and write operations. On a real parallel machine, this is an important issue, since access time may vary substantially depending on which memory cells are accessed. Moreover, exclusive access to shared memory cells do not necessarily guarantee that there are no bottlenecks in the network that handles the read and write demands. On the other hand, by putting these issues aside, one may concentrate himself on communication demands of the problem independent on implementation details. This turns out that coordinating the processors via an appropriate choice of the algorithm is not an easy task. The problem is that any global control would demand a lot of time. Indeed, it turns out that [2] sending messages to k different processors from a processor that plays the role of a controller requires $\Theta(\log k)$ EREW PRAM steps. Therefore, in any fast EREW PRAM algorithm coordination chores must be performed in a distributed way with little communication between the processors.

It was shown that even such simple Boolean functions like logical AND or exclusive OR of n bits, with each input bit given in a separate shared memory cell, require $\Theta(n)$ steps on the EREW PRAM [7]. There are known extremely few algorithms on the EREW PRAM that achieve sublogarithmic runtimes; the bound $\Omega(\log n)$ is for most problems the barrier for runtime of EREW PRAM algorithms. Unexpectedly, we observe that this is not the case for such a complicated problem as permuting at random:

Theorem 5.1 *Given n different items stored in n shared memory cells, each item occupying a separate cell. There is an EREW PRAM algorithm that in $\mathcal{O}(\log\log n)$ steps chooses a permutation $\pi \in S_n$ at random and relocates the items so that the ith input item is stored in cell $\pi(i)$, for $i \leq n$. Moreover, the variation distance between the probability distribution of the permutations π generated by EREW PRAM and the uniform distribution over S_n is $\mathcal{O}(\frac{1}{n})$.*

The previous EREW PRAM algorithms for this problem have runtime $\Omega(\log n)$ [9, 18]. Our algorithm beats even the best previously known CREW PRAM algorithm [9] regarding the product of the number of the processors while preserving the runtime (on a CREW PRAM the processors are allowed to read the same memory cell at a time, only simultaneous writes are forbidden).

Proof. The fast EREW PRAM algorithm uses the switching network $\mathcal{B}_n$ from Theorem 1.11. The computation consists of three phases:

1. Setting of each switch of $\mathcal{B}_n$ is chosen by a separate processor by writing a random bit into a shared memory cell dedicated to this switch.

2. A random permutation corresponding to the chosen setting of switches of $\mathcal{B}_n$ is determined. For this purpose, EREW PRAM performs standard pointer doubling technique: we assign a separate processor $P(k, i)$

and a memory cell $C(k, i)$ to $r_{k,i}$, the kth position at level $i \leq \tau$ of $\mathcal{B}_n$. Processor $P(k, i)$ has a pointer $d(k, i)$ which after subphase j points to the position at level $i + 2^{j-1}$, to which $r_{k,i}$ is connected by the current setting of switches (provided that $i + 2^{j-1} \leq \tau$). The value of $d(k, i)$ is stored in $C(k, i)$. For $j = 1$ this is achieved by reading the setting of a single switch at level i. No read conflict occurs, since the processor corresponding to the left input line of the switch reads first the memory cell corresponding to the switch. At subphase $j + 1$, processor $P(k, i)$ reads $C(d(k, i), i + 2^{j-1})$. The value read is the position at level $i + 2^{j-1} + 2^{j-1} = i + 2^j$ connected to $r_{k,i}$ by the current setting of switches. Then $P(k, i)$ updates the pointer $d(k, i)$ to this value and stores it in $C(i, k)$. We may assume that τ is a power of two (if not we add less than τ dummy levels where all switches are set to value 1 in order to achieve a depth which is a power of two.) Then, after subphase $\log \tau$ of pointer doubling, each processor $P(k, 1)$ knows the position $\pi(k)$ at level τ to which the kth input line is connected.

3. For permutation π found during the second phase, for $i \leq n$ processor i reads the ith input item and writes it into the cell $\pi(i)$ in the shared memory.

Note that in the above algorithm read conflicts are avoided by the architecture of a switching network: each position k at level s is connected with exactly one position at level $s - 2^{j-1}$ ($s > 2^{j-1}$), so the cell $C_{k,s}$ is read by exactly one processor! □

Note that using Corollary 1.10 instead of Theorem 1.11 in the above construction we may prove existence of a non-uniform EREW PRAM algorithm for permuting at random (*non-uniform* means that for every fixed n there is a different algorithm performing this task, but we are not able to parameterize these algorithms so that for a given n the processors switch to an appropriate algorithm of this kind).

6 Conclusions

Theorem 1.11 provides a simple switching network that *provably* generates permutations almost uniformly at random. A challenging problem is to improve the construction in order to get a switching network of similar properties but of significantly smaller depth and simpler architecture so that the solution might be applied in practice for chips performing permuting at random.

Acknowledgment

This research has been partially supported by KBN grant 8 T11C 032 15, ALCOM EU ESPRIT Long Term Research Project 20244 (ALCOM-IT), DFG-Sonderforschungsbereich 376 "Massive Parallelität", and DFG Leibniz Grant Me872/6-1. Some of this work has been done when the first and the last authors were with Heinz Nixdorf Institute, University of Paderborn, and the last author was with Institute of Computer Science, University of Wrocław. A preliminary report on the construction described in this paper appeared in [10].

References

[1] D. Aldous, Random walks of finite groups and rapidly mixing Markov chains, in *Séminaire de Probabilités XVII 1981/82*, (Springer Verlag, Lecture Notes in Mathematics 986, Berlin, 1983), 243–297.

[2] P. Beame, M. Kik, and M. Kutyłowski, Broadcasting information by exclusive read PRAMs, *Parallel Processing Letters*, **4 (1&2)** (1994), 159–169.

[3] R. Bubley and M. Dyer, Path coupling: A technique for proving rapid mixing in Markov chains, in *Proceedings of the 38th Symposium on Foundations of Computer Science*, FOCS'97, (IEEE Computer Society Press, Los Alamitos, 1997), 223–231.

[4] R. Bubley and M. Dyer, Faster random generation of linear extensions, in *Proceedings of the 9th ACM-SIAM Symposium on Discrete Algorithms*, SODA'98, (SIAM, Philadelphia, 1998), 350–354.

[5] R. Bubley, M. Dyer, and C. Greenhill, Beating the 2Δ bound for approximately counting colourings: A computer-assisted proof of rapid mixing, in *Proceedings of the 9th ACM-SIAM Symposium on Discrete Algorithms*, SODA'98, (SIAM, Philadelphia, 1998), 355–363.

[6] D. Chaum, Untraceable electronic mail, return addresses, and digital pseudonyms, *Communications of the ACM* **24** (1985), 84–88.

[7] S. Cook, C. Dwork, R. Reischuk, Upper and lower time bounds for parallel random access machines without simultaneous writes, *SIAM Journal on Computing,* **15(1)** (1986), 87–97.

[8] A. Czumaj, Recovery time of dynamic allocation processes, in *Proceedings of the 10th Annual ACM Symposium on Parallel Algorithms and Architectures*, SPAA'98, (ACM Press, New York, 1998), 202–211.

[9] A. Czumaj, P. Kanarek, M. Kutyłowski, and K. Loryś, Fast generation of random permutations via network simulation, *Algorithmica* **21** (1998), 2–20. A preliminary version appeared in *Proceedings of the 4th Annual European Symposium on Algorithms*, ESA'96, (Springer Verlag, Lecture Notes in Computer Science 1136, Heidelberg, 1996), 246–260.

[10] A. Czumaj, P. Kanarek, M. Kutyłowski, and K. Loryś, Delayed path coupling and generating random permutations via distributed stochastic processes, *Proceedings of the 10th ACM-SIAM Symposium on Discrete Algorithms*, SODA'99, (SIAM 1999), 271–280.

[11] A. Czumaj and M. Kutyłowski, Generating random permutations and delayed path coupling method for mixing time of Markov chains, *Random Structures and Algorithms* **17** (2000), 238-259.

[12] P. Diaconis, *Group Representations in Probability and Statistics*, Lecture Notes Monograph Series, vol. 11, Institute of Mathematical Statistics, Hayward, CA, 1988.

[13] M. Dyer, L. A. Goldberg, C. Greenhill, M. Jerrum, and M. Mitzenmacher. An extension of path coupling and its application to the Glauber dynamics for graph colourings. *Proceedings of the 11th ACM-SIAM Symposium on Discrete Algorithms*, SODA'2000, (SIAM 2000), 616–624.

[14] M. Dyer and C. Greenhill, A genuinely polynomial-time algorithm for sampling two-rowed contingency tables, in *Proceedings of the 25th Annual International Colloquium on Automata, Languages and Programming*, ICALP'98, (Springer-Verlag, Lecture Notes in Computer Science, vol. 1443, Berlin, 1998), 339–350.

[15] M. Dyer and C. Greenhill, A more rapidly mixing Markov chain for graph colourings. *Random Structures and Algorithms*, **13(3–4)** (1998), 285–317.

[16] M. Dyer and C. Greenhill, Random walks on combinatorial objects, in J. D. Lamb and D. A. Preece, editors, *Surveys in Combinatorics*, *London Mathematical Society Lecture Note Series*, vol. 267, 101–136, Cambridge University Press, 1999.

[17] F. Fich, The Complexity of Computation on the Parallel Random Access Machine, in *Synthesis of Parallel Algorithms*, Morgan Kaufmann, San Mateo, CA, 1993, 843–900.

[18] T. Hagerup, Fast parallel generation of random permutations, in *Proceedings of the 18th Annual International Colloquium on Automata, Languages and Programming*, ICALP'91, (Springer Verlag, Lecture Notes in Computer Science 510, Heidelberg, 1991), 405–416.

[19] F. T. Leighton, *Introduction to Parallel Algorithms and Architectures*, Morgan Kaufmann Publishers, San Mateo, 1992.

[20] T. Leighton and C. G. Plaxton, Hypercubic sorting networks, *SIAM Journal on Computing* **27** (1998), 1–47. A preliminary version appeared in *Proceedings of the 31st Symposium on Foundations of Computer Science*, FOCS'90, (IEEE Computer Society Press, Los Alamitos, 1990), 264–274, under the title "A (fairly) simple circuit that (usually) sorts."

[21] T. Lindvall, *Lectures on the Coupling Method*, Wiley series in Probability and Mathematical Statistics, John Wiley & Sons, New York, 1992.

[22] M. Molloy, The probabilistic method, in M. Habib, C. McDiarmid, J. Ramirez-Alfonsin, and B. Reed (Eds.), *Probabilistic Methods for Algorithmic Discrete Mathematics*, Algorithms and Combinatorics, 1–35. Springer-Verlag, Berlin, 1998.

[23] R. Motwani and P. Raghavan, *Randomized Algorithms*, Cambridge University Press, New York, 1995.

[24] C. Rackoff and D. R. Simon, Cryptographic defense against traffic analysis, in *Proceedings of the 25th Annual ACM Symposium on Theory of Computing*, STOC'93, (ACM Press, New York, 1993), 672–681.

[25] D. R. Simon, private communication, October 1997.

SWITCHING NETWORKS: RECENT ADVANCES
D.-Z. Du and H.Q. Ngo (Eds.) pp. 63-88

Optical and Optoelectronic Interconnection Networks

Sartaj Sahni
Dept. of Computer and Information Science and Engineering
University of Florida, Gainesville, FL 32611
E-mail: sahni@cise.ufl.edu

Contents

1 Introduction

Several interconnection networks that are either wholly optical or combine optical and elctronic interconnect technology have been proposed recently for use in parallel computers. The more popular of these newly proposed interconnection networks may be classified as (1) networks that employ static pipelined optical buses, (2) networks that employ dynamically reconfigurable optical buses, (3) networks that employ the passive star network, and (4) optoelectronic interconnection networks that employ the optical transpose interconnect system. In this chapter, we provide the topology and properties of these networks. We also provide a few basic algorithms for parallel computers that employ these interconnection networks.

Single-mode waveguides are unidirectional and light pulses travel down the waveguide with a highly predictable delay [20]. As a result, waveguides support pipelining (i.e., at any instance, several messages encoded as light pulses can be traveling down a waveguide, one message behind the other). This means that several processors can simultaneously write different messages to the same optical bus; all messages can flow along the bus without interference. When electronic buses are used, only one message can be present on the bus at any given time. Therefore, when two or more processors attempt to write to a shared bus in the same cycle, we get a write conflict. Several pipelined bus models for parallel computers have been proposed and studied. Sections 2 and 3 review some of the more popular bus models.

The partitioned optical passive stars network (POPS) was proposed in [3, 6, 7, 13] as an optical interconnection network for a multiprocessor computer. The POPS network uses multiple optical passive star (OPS) couplers to construct a flexible interconnection topology. Each OPS (Figure 1) coupler can receive an optical signal from any one of its source nodes and broadcast the received signal to all of its destination nodes. The time needed to perform this receive and broadcast is referred to as a *slot*.

Although a single OPS can be used to interconnect n processors (in this case the n processors are both the source and destination nodes for the OPS), the resulting multiprocessor computer has very low bandwidth—only one processor may send a message in a slot. To alleviate this bandwidth problem a $POPS(d,g)$ network partitions the n processors into g groups of size d (d is also the degree of each coupler) each (so $n = dg$), and g^2 OPS couplers are used to interconnect pairs of processor groups. Section 4 describes the POPS network and provides a few basic algorithms for parallel

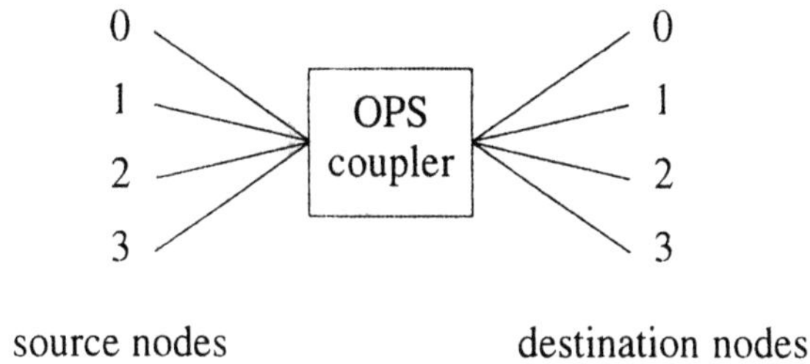

Figure 1: An optical passive star coupler with 4 source and 4 destination nodes

computers interconnected using this topology.

When designing a very large parallel computer system, processors are spread over several levels of the packaging hierarchy. For example, we might have several processors on a single chip, several chips on a wafer, several wafers on a printed circuit board, and so on. This means that, necessarily, the interprocessor distance cannot be kept small for all pairs of processors. When a connection must be run between processors on two different chips (for example), the connect distance is usually larger than when we connect two processors on the same chip. It is also known [5, 10] that optical connects provide power, speed, and crosstalk advantages over electronic interconnects when the connect distance is more than a few millimeters. Therefore, minimum overall delay is achieved when shorter interconnects are realized using electronics and longer interconnects are realized using optics. This realization leads to the concept of optoelectronic interconnection networks–interconnection networks that have a mix of optical and electronic interconnects. It is important to develop interconnection topologies that maximize the benefits of the two technologies and are manageable from the prespective of efficient algorithm design. The OTIS (optical transpose interconnection system) family of optoelectronic interconnection networks is a step in this direction.

2 Static Bus

2.1 A Unidirectional Bus

An n-processor unidirectional-bus computer is a synchronous computer in which n processors are connected to an optical bus (or waveguide). Figure 2 shows a 4-processor unidrectional-bus computer. The processors are labeled

0 through 3 and the optical bus is shown as a thick arrow. The processors are evenly spaced and the time required for an (fixed length) optical message (or signal) to travel the distance between two adjacent processors is denoted by τ. An optical bus is unidirectional–optical messages may travel in only one direction along the bus, that is, in the direction indicated by the arrow head. Each processor has a read/write connection to the optical bus, this is shown as a thin line in Figure 2.

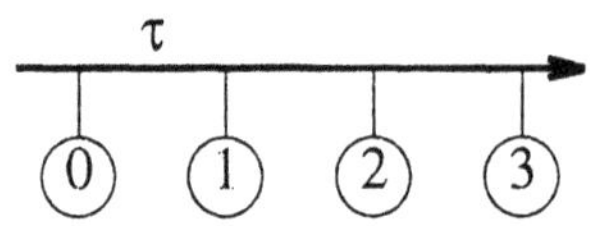

Figure 2: A unidirectional-bus computer.

If processor 0 writes a message to the bus, this message will arrive at processor 1 after τ time units and at processors 2 and 3 after 2τ and 3τ time units, respectively. If processors 0 through 3 write messages A, B, and C, respectively, to the bus at time 0, then message A arrives at processors 1, 2, and 3 at times τ, 2τ, and 3τ, respectively; message B arrives at processors 2 and 3 at time τ and 2τ; and message C arrives at processor 3 at time τ.

The *cycle time* of an n-processor unidirectional-bus computer is defined to be $n\tau$ [8]. The cycle time for the 4-processor configuration of Figure 2 is 4τ. A cycle may be regarded as composed of n slots, each of duration τ. It is generally assumed that a processor can write in only one slot in a cycle and can read from only one slot in a cycle. However, some models permit reading from several slots of a cycle. Several mechanisms have been proposed for how a processor knows which slot to write and which to read. Two of these mechanisms are:

1. *Use of a slot counter* All processors write in slot 0 of a cycle. If processor j, $j > i$, wants to read processor i's message, it starts a slot counter that is incremented by 1 every τ time units; when the counter reaches $j - i$, processor j reads from the bus. The drawback of this scheme [17] is that it requires an electronic counter that is at least as fast as the optical waveguide. This drawback is eliminated by replacing the arithmetic counter by the system clock which is advanced every τ units.

2. *Coincident Pulse Method* For this method, the optical bus is composed of three waveguides (see Figure 3). Call the original waveguide

the *message* or *data* waveguide and the new waveguides the *reference* and *select* waveguides. Delay units (essentially an optical fiber loop) are added between adjacent processors on the message and reference waveguides so that messages and reference signals (or pulses) experience a unit delay relative to select signals. Note that the time τ required for an optical signal to travel the distance between two adjacent processors is ≥ 1. Often, we assume that $\tau = 1$, and so the unit delay added by a delay unit equals one slot. If processor 0 writes a message to the message waveguide at time 0, this message gets to processor i at time $i(\tau + 1)$. To enable reading of this message by processor i, processor 0 also writes a reference signal to the reference waveguide at time 0 and a select signal to the select waveguide at time i. The reference signal arrives at processor i at the same time as does the message, that is, at time $i(\tau + 1)$. The select signal moves from one processor to the next in τ time and so takes $i\tau$ time (from initiation) to arrive at processor i. Therefore, the select signal also arrives at processor i at time $i(\tau + 1)$. When the reference and select signals arrive at a processor simultaneously (i.e., when the reference and select signals are coincident at a processor), the processor reads from the message waveguide. To enable reading of its data by all other processors, processor 0 must initiate a select signal at times 1, 2, 3, $\cdots$, $n-1$.

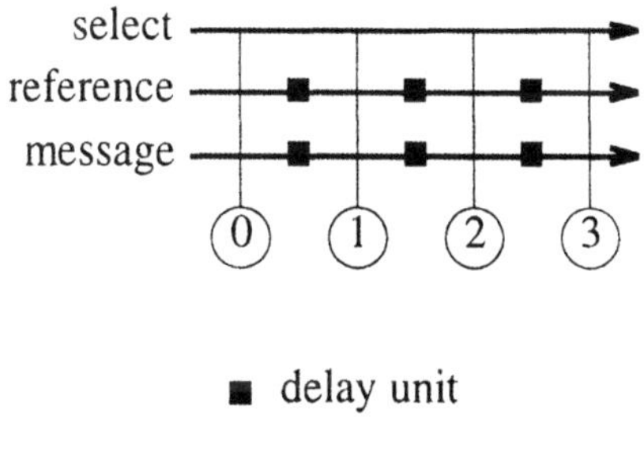

Figure 3: Reference and select buses.

As you can see, the coincident pulse method requires a clock that advances every time unit (a time unit equals the delay introduced between pairs of adjacent processors on the message and reference buses); this time unit is less than τ.

The complexity of algorithms for optical bus computers is usually measured in cycles. Although the cycle time for an n-processor bus varies with

n, for n up to a few thousand, the cycle time is no more than the time required to perform a CPU operation (such as an add or a compare) [18]. Typically, an algorithm will involve the CPU for $\Theta(1)$ steps between each bus cycle. So, the number of bus cycles times the CPU speed is a good measure of complexity for n up to a few thousand.

2.2 One-Dimensional Array

A unidirectional bus isn't of much use because there is no way for processor i to send a message to processor j when $i > j$, that is you cannot send a message to a processor on your left. This difficulty is overcome by adding an additional bus in which messages flow from right to left (Figure 4). The resulting parallel computer model is called *one-dimensional array with pipelined buses* (1D APPB) [8].

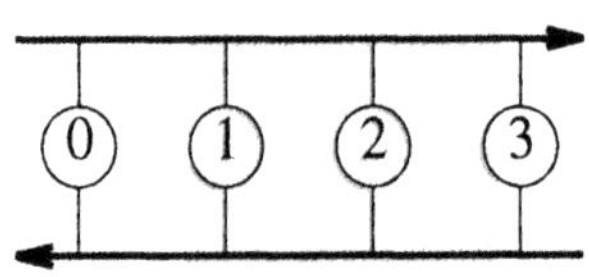

Figure 4: One-dimensional array with pipelined buses.

To see the power of a 1D APPB, suppose you want to permute the data in the n processors according to the permutation $p()$, that is processor j is to receive data from processor $p(j)$. Data from processor $p(j)$ will get to processor j in $|p(j) - j|$ slots (or $|p(j) - j|\tau$ time) provided it is written to the proper bus. So, processor j computes the number of slots $wait(j) = |p(j) - j|$ for which it must wait for its data. Following this computation, all processors write their data to the upper (i.e., the left to right) and lower buses at the start of a bus cycle. Processor j reads the desired data from the bus when its wait time is up. Amazingly, we can perform any data permutation in just one bus cycle. This is not possible using an electronic bus. For an electronic bus, define the bus cycle to be the time needed for an electronic signal written by a processor on a bus to become available at all processors on the bus. When an electronic bus is used, only one distinct data item can be transported on the bus in a bus cycle. Therefore, n cycles are required to perform a permutation.

An alternative one-dimensional model uses a folded message bus as in Figure 5 [20]. In this model, all writes are done to the upper bus segment and all reads are done from the lower bus segment. The cycle length for the folded bus is $2n\tau$.

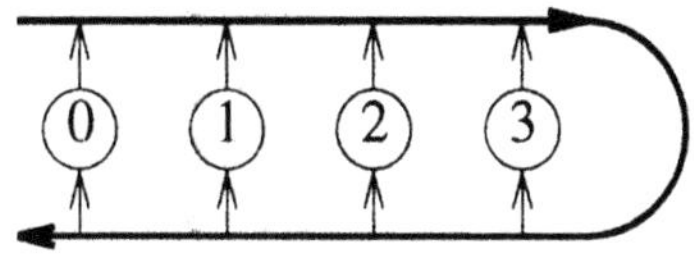

Figure 5: A folded one-dimensional bus computer.

When a folded bus is used, it takes $(2n - 2 - p(j) - j)\tau$ time for data to get from $p(j)$ to j. To perform the permutation $p()$, processor j computes $wait(j) = 2n - 2 - p(j) - j$ before the start of the bus cycle; all processors write to the upper bus segment at the start of a bus cycle; and processor j reads from the lower bus segment when its wait time is over. Again, the permutation is complete within one bus cycle.

2.3 Two-Dimensional Array

The 1D APPB model is easily generalized to higher dimensions. Figure 6 shows a two-dimensional version–the 2D APPB. The 2D APPB is quite similar to meshes with buses [26]–both have row and column buses. The essential difference is that the buses in a 2D APPB are optical whereas those in a mesh with buses are electronic; an electronic bus can carry only one distinct message at a time; whereas an optical bus can carry a distinct message in each slot.

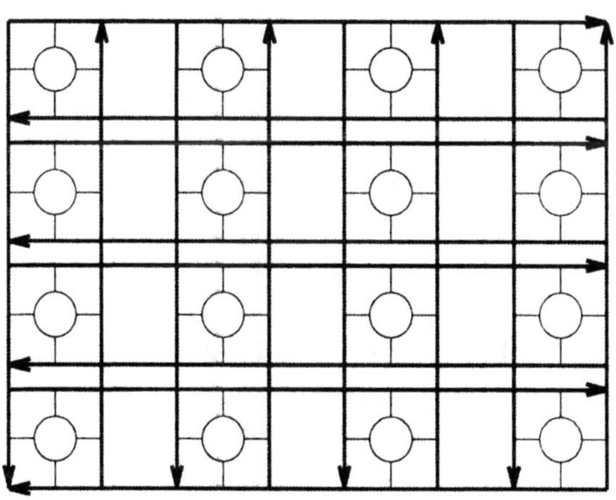

Figure 6: A 2D APPB.

A significant advantage afforded by two-dimensional arrays is the ability to build rather large computers while keeping the number of processors on an individual bus (and hence the bus cycle time) reasonable. If we limit the number of processors on a bus to a few thousand (see above), then the two-dimensional array allows us to build computers with up to a few million processors. The 2D APPB is harder to program than the 1D APPB. For example, to perform a data broadcast, we must first broadcast along a row bus (say) and then along all of the column buses. Even though an arbitrary permutation can be done in $O(1)$ bus cycles, the preprocessing needed by the permutation routing algorithm is excessive [8].

A two-dimensional array that uses folded row and column buses may also be developed.

3 Reconfigurable Bus

3.1 One-Dimensional Array

Given the success of reconfigurable architectures that employ electronic buses [14, 15], it is not surprising that reconfigurable optical bus architectures abound. In a one-dimensional reconfigurable bus, for example, processor i, $i > 0$ controls a *bus control* switch that enables it to break the optical bus at processor i. When the bus is broken at processors i_1, i_2, and i_3, $i_1 < i_2 < i_3$, for example, we get four independently operating one-dimensional bus computers. The first is comprised of processors 0 through $i_1 - 1$, the second of processors i_1 through $i_2 - 1$, the third has processors i_2 through $i_3 - 1$, and the fourth computer has processors i_3 through $n - 1$. This bus breaking into four segments is done by processors i_1, i_2, and i_3 by opening their bus control switch. Processors can open and close their switches dynamically while a program is executing. Hence, the computer may be reconfigured, as computation proceeeds, into a varying number of subcomputers.

The 1DAROB (one-dimensional array with reconfigurable optical bus) model of [18] and the LARPBS (linear array with a reconfigurable pipelined bus system) of [17] include a conditional delay unit between every pair of processors. This delay unit is on the upper segment of the select waveguide of the bus; processor i, $i > 0$, controls the delay unit to its left. When a delay unit is on, a one slot delay is introduced.

The conditional delay unit is useful in both the static and reconfigurable bus models. For example, the conditional delay unit may be used to find the binary prefix sum in $O(1)$ bus cycles [17, 19]. Suppose that b_i is a binary value that is stored in processor i and that processor i is to compute $\sum_{j=0}^{i} b_j$, $0 \leq i < n$. Processor i, $i > 0$ turns its delay unit on (i.e., sets the unit so that the select pulse will be delayed by one slot) iff $b_i = 1$. Then, the leader (i.e., processor 0 unless the bus has been broken into subbuses) writes a select signal in either slot 0 or 1 of a bus cycle; the writing is done in slot 0 iff $b_0 = 0$. The leader also writes a reference signal in each slot. Processor i receives a reference signal in each slot beginning with slot i; it receives the select signal in slot $i + \sum_{j=0}^{i} b_j$. So, the two signals are coincident at processor i in slot $i + \sum_{j=0}^{i} b_j$. By starting a slot counter at the beginning of the bus cycle and stopping the counter when the select and reference signals are coincident at the processor, processor i can determine the value of $\sum_{j=0}^{i} b_j$. At most 2 bus cycles are needed to perform the binary prefix sum operation.

3.2 Two-Dimensional Array

The 2DAROB of [18] is a reconfigurable mesh [14, 15] in which the elctronic buses have been replaced by optical ones. Figure 7 shows a 4×4 2DAROB. Each arrangement of circular arcs denotes a switch; each line segment denotes a bidirectional optical bus segment, and there is one processor (not shown) at the center of each arrangement of circular arcs. The permissible switch settings are shown in Figure 8 Each processor can set its switch dynamically and thereby determine the bus topology. The switch settings at any time result in a set of disjoint buses. Each of these disjoint buses is required to be a unidirectional chain. The first processor on a configured bus is called the *bus leader*.

An n^2-processor 2DAROB can simulate an $n \times n$ reconfigurable mesh with a constant factor slow down [19]. Since a 2DAROB can perform a column permutation in 1 cycle, whereas an $n \times n$ reconfigurable mesh requires $O(n)$ cycles to this, a 2DAROB is more powerful (in the asymptotic sense) than a reconfigurable mesh. Some fundamental 2DAROB algorithms are developed in [21].

An alternative two-dimensional reconfigurable model, the array with synchronous switches (ASOS) was proposed in [20] (Figure 9). This model uses folded row and column buses; each processor can write only to the upper segment of its row bus; and each processor can read (concurrently) from the lower segment of its row bus and the right segment of its column bus. The shown switches can be in one of two states. In the *straight* state, mes-

Figure 7: A 2DAROB.

sages move along row buses; and in the *cross* state, messages move from a row bus onto a column bus. Although [20] requires that all switches be set to the same state at any given time, we could permit each processor to independently control the state of its bottom left switch.

4 Partitioned Optical Passive Stars

4.1 The POPS(d,g) Network

The partitioned optical passive stars network (POPS) [3, 6, 7, 13] uses multiple optical passive star (OPS) couplers to construct a flexible interconnection topology. In a $POPS(d,g)$ network, the n processors are partitioned into g groups of size d (d is also the degree of each coupler) each (so $n = dg$), and g^2 OPS couplers are used to interconnect pairs of processor groups. Specifically the groups are numbered 0 through $g-1$ and the source nodes for coupler $c(i,j)$ are the processors in group j; the destination nodes for this coupler are the processors in group i, $0 \leq i < g$, $0 \leq j < g$. Figure 10 shows how a $POPS(4,2)$ network is used to connect 8 processors. Destination processor i is the same processor as source processor i, $0 \leq i < 8$. A $POPS(4,2)$

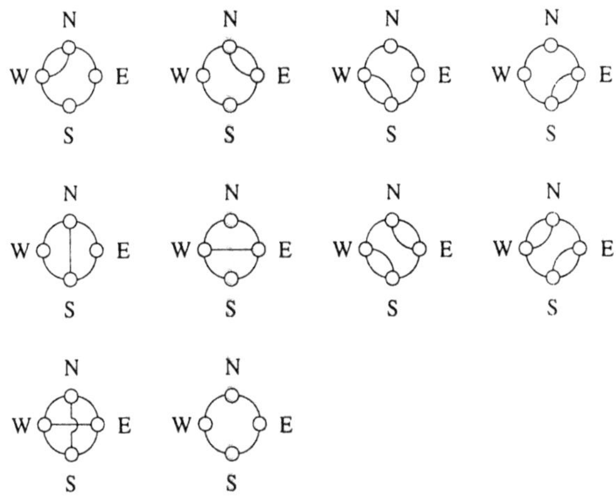

Figure 8: Permissible switch settings.

network comprises $2^2 = 4$ degree $d = 4$ OPS couplers.

When 8 processors are connected using a $POPS(8,1)$ network only one degree 8 OPS coupler is used. A 32 processor computer may be built using any one of the following networks: $POPS(32,1)$, $POPS(16,2)$, $POPS(8,4)$, $POPS(4,8)$, $POPS(2,16)$, and $POPS(1,32)$. A 25 processor computer may be built using a $POPS(25,1)$, $POPS(5,5)$, or $POPS(1,25)$ network. A multiprocessor computer that employs the POPS interconnection network is called a *POPS computer*.

The choice of the POPS network that is used to interconnect the processors affects both the interconnection cost as well as the bandwidth. When a $POPS(d,g)$ network is used to connect $n = dg$ processors, each processor must have g optical transmitters (one to transmit to each of the g OPSs for which it is a source node) and g receivers. The total number of transmitters and receivers is $2ng = 2n^2/d$, the number of OPSs is g^2, and each OPS has degree d. In one slot each OPS can receive a message from any one of its source nodes and broadcast this message to all of its destination nodes. In particular, in a single slot, a processor can send the same message to all of the OPSs for which it is a source node. In a single slot, a processor can receive a message from only one of the OPSs for which it is a destination node. (Melhem et al. [13] note that allowing a processor to receive different messages from different OPSs in the same slot permits faster all-to-all broadcast.)

A major advantage of the POPS network is that its diameter is 1. A

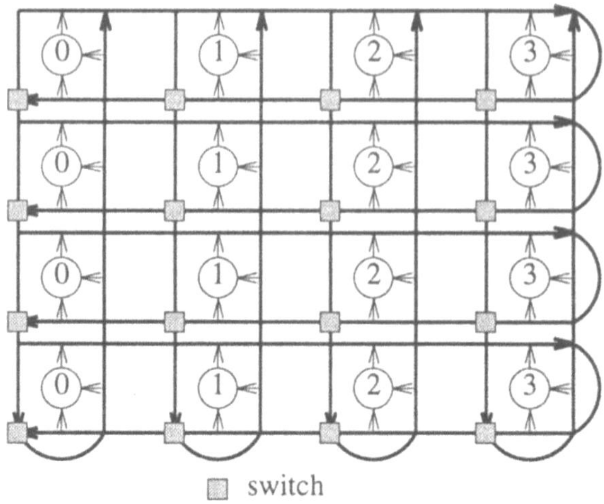

Figure 9: A 4×4 ASOS.

message can be sent from processor i to processor j, $i \neq j$, in a single slot. Let $group(i)$ be the group that processor i is in. To send a meesage to processor j, processor i first sends the message to coupler $c(group(j), i)$. This coupler broadcasts the received message to all its destination processors; that is, to all processors in $group(j)$ [7, 13]. A one-to-all broadcast can also be done in one slot [7, 13]. Suppose that processor i wishes to broadcast a message to all other processors in the system. Processor i first sends the message to all couplers $c(*, group(i))$ for which it is a source node. Next all couplers of the form $c(*, group(i))$ broadcast the received message to their destination nodes. Since processor j is a destination node of coupler $c(group(j), group(i))$, processor j, $0 \leq j < n$ receives the message broadcast by processor i. This algorithm is easily extended to perform an all-to-all broadcast in n slots (or in 1 slot when a processor can receive, in a single slot, messages from all couplers that it is a destination node of). [7, 13] also give an algorithm for all-to-all personalized communication.

[7] shows how to embed rings and tori into POPS networks. [1] shows that POPS networks may be modeled by directed stack-complete graphs with loops. This modeling is used to obtain optimal embeddings of rings and de Bruijn graphs into POPS networks. An alternative multiprocessor interconnection network using multiple OPSs is proposed in [4].

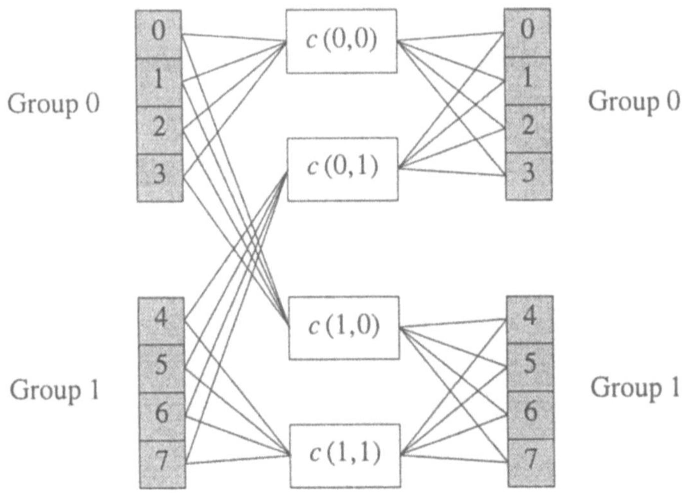

Figure 10: An 8-processor computer connected via a $POPS(4,2)$ network

4.2 SIMD Hypercube Simulation

Let $i^{(b)}$ be the number whose binary representation differs from that of i only in bit b. The primitive communication in an $n = 2^D$ processor SIMD hypercube is for every processor i, $i \leq 0 < n$, to send data to processor $i^{(b)}$, for some fixed b, $0 \leq b < D$. This communication can be simulated by a $POPS(d, g)$ $(dg = n)$ computer using $2\lceil d/g \rceil$ moves. So every n processor SIMD hypercube algorithm can be run on an n processor $POPS(d, g)$ with a slowdown of a factor of at most $2\lceil d/g \rceil$.

Theorem 4.1 *An n processor $POPS(d, g)$ can simulate every move of an n processor SIMD hypercube using 1 slot when $d = 1$ and $2\lceil d/g \rceil$ slots when $d > 1$.*

Proof. First consider the case $d = 1$. A $POPS(1, g)$ can perform any permutation π in 1 slot using the routing

$$p(i, 1) \to c(\pi(i), i) \to p(\pi(i), 1)$$

In particular, the data routing done in 1 hypercube move can be done in 1 slot.

Next consider the case $d > 1$. Processor i of the SIMD hypercube is mapped onto processor i of the $POPS(d, g)$. Let b be the bit along which the hypercube communication is to be done. That is, processor i of the hypercube is to send data to processor $i^{(b)}$ of the hypercube for all i. Let e and f be such that $ed + f = i$. That is $p(i)$ and $p(e, f)$ denote the same $POPS(d, g)$ processor.

First consider the case when $d = g = \sqrt{n} = 2^{D/2}$. When $b < D/2$, b is a bit of f and the communication can be done in two slots as below:

$$p(i) = p(e, f) \rightarrow c(f, e) \rightarrow p(f, e) \rightarrow c(e, f) \rightarrow p(e, f^{(b)}) = p(i^{(b)})$$

When $b \geq D/2$, b is a bit of e and the the communication is done in two slots as below:

$$p(i) = p(e, f) \rightarrow c(f, e) \rightarrow p(f, e) \rightarrow c(e^{(q)}, f) \rightarrow p(e^{(q)}, f) = p(i^{(b)})$$

The next case to consider is $d < \sqrt{n} < g$. When b is a bit of f we use the following 2 slot routing:

$$\begin{aligned} p(i) = p(e, f) &\rightarrow c(i \bmod g, e) && (1)\\ &\rightarrow p(i \bmod g, \lfloor i/g \rfloor) && (2)\\ &\rightarrow c(e, i \bmod g) && (3)\\ &\rightarrow p(e, f^{(b)}) = p(i^{(b)}) && (4) \end{aligned}$$

To establish the correctness of this routing we note that data that start in two different processors $i_1 = e_1 d + f_1$ and $i_2 = e_2 d + f_2$ use different couplers in slot 1 (i.e., for the source node to coupler routing of line 1). To see this observe that when $e_1 \neq e_2$ the slot 1 couplers differ in their second index. When $e_1 = e_2$, i_1 and i_2 differ by at most $d - 1 < g$ (and at least 1). So $i_1 \bmod g \neq i_2 \bmod g$, and the slot 1 couplers differ in their first index. Since $i = \lfloor i/g \rfloor g + i \bmod g$, data that originate in different processors are routed to different destination processors of line (2). For the slot 2 couplers (line (3)) we see that if $e_1 \neq e_2$, these couplers differ in their first index. When $e_1 = e_2$, i_1 and i_2 differ by at most $d - 1 < g$ (and at least 1). So the couplers differ in their second index.

When b is a bit of e the following 2 slot routing may be used:

$$\begin{aligned} p(i) = p(e, f) &\rightarrow c(i \bmod g, e)\\ &\rightarrow p(i \bmod g, \lfloor i/g \rfloor)\\ &\rightarrow c(e^{(b)}, i \bmod g)\\ &\rightarrow p(e^{(b)}, f) = p(i^{(b)}) \end{aligned}$$

The correctness proof for this routing is similar to that for the case when b is a bit of f. Notice that the 2 slot routing for the case $d < \sqrt{n} < g$ works when $d = g = \sqrt{n}$ also.

The final case to consider is $d > \sqrt{n} > g$. This case is handled using $\lceil d/g \rceil$ passes of the simulation strategy for the case $d < \sqrt{n} < g$. Notice that when the 2 slot routing of lines 1–4 is used up to $\lceil d/g \rceil$ source processors from the same group e attempt to route data to the same coupler in lines 1 and 3. This conflict is resolved by dividing the processors in each group into $\lceil d/g \rceil$ subgroups and doing the routing for each subgroup in a different pass. □

A SIMD hypercube algorithm may restrict a particular move along bit b to a subset of the processors. When the restriction is such that only one processor of each $POPS(d, g)$, $d > 1$, group is to send data, the simulation of Theorem 4.1 is not optimal. Suppose that a hypercube routing step is restricted in this way. When b is a bit of f, the data transfer along bit b may be accomplished using a single slot using the following routing:

$$p(i) = p(e, f) \rightarrow c(e, e) \rightarrow p(e, f^{(b)}) = p(i^{(b)})$$

Since for each $POPS(d, g)$ group e only one processor $p(e, f)$ is active, there is no coupler conflict in the above route. When b is a bit of e, the data transfer along bit b may be accomplished using a single slot as below:

$$p(i) = p(e, f) \rightarrow c(e^{(b)}, e) \rightarrow p(e^{(b)}, f) = p(i^{(b)})$$

Most SIMD hypercube data transfer steps along bit b will, however, require two or more processors of the same $POPS(d, g)$ group to transmit data. In this case the data transfer cannot be accomplished in a single slot and the 2 slot simulation of Theorem 4.1 is optimal. To see this observe that when b is a bit of f the only single slot routes for data in $p(e_1, f_1)$ and $p(e_1, f_2)$, $f_1 \neq f_2$, are:

$$p(e_1, f_1) \rightarrow c(e_1, e_1) \rightarrow p(e_1, f_1^{(b)})$$

and

$$p(e_1, f_2) \rightarrow c(e_1, e_1) \rightarrow p(e_1, f_2^{(b)})$$

These routes have a conflict because they use the same coupler. So no single slot routing is possible. When b is a bit of e, the only single slot routes for the two data are:

$$p(e_1, f_1) \rightarrow c(e_1^{(b)}, e_1) \rightarrow p(e_1^{(b)}, f_1)$$

and

$$p(e_1, f_2) \to c(e_1^{(b)}, e_1) \to p(e_1^{(b)}, f_2)$$

These routes use the same coupler and so are in conflict.

4.3 SIMD Mesh Simulation

In an $N \times N$ SIMD mesh data may be moved one processor right/left/up/down along the rows/columns of the mesh in one step. The direction of data movement in a single step is the same for all data. In a mesh with wraparound, a rightward move by 1 causes data in the rightmost processor of each row to reach the leftmost processor of the row (left, up, and down moves handle end elements similarly). A communication step of a mesh with wraparound (and hence of a mesh with no wraparound) can be simulated by a $POPS(d, g)$ ($dg = N^2$ and either d divides N or g divides N) computer using $2\lceil d/g \rceil$ slots. So every n processor SIMD mesh algorithm can be run on an n processor $POPS(d, g)$ with a slowdown of a factor of at most $2\lceil d/g \rceil$.

Theorem 4.2 *An $n = N^2$ processor $POPS(d, g)$ (d or g divides N) can simulate every move of an $N \times N$ processor SIMD mesh with wraparound using 1 slot when $d = 1$ and $2\lceil d/g \rceil$ slots when $d > 1$.*

Proof. The proof for the case $d = 1$ is the same as that for the simulation of a hypercube move. So consider the case $d > 1$. Processor (i, j) of the mesh is mapped onto processor $p(e, f)$ of the $POPS(d, g)$, where $ed + f = iN + j$. As in the case of the proof of Theorem 4.1, we consider three cases. The first case has $d = g = N$. When $d = g = N$, a rightward row move by 1 is accomplished using the 2 slot routing:

$$p(e, f) \to c(f, e) \to p(f, e) \to c(e, f) \to p(e, (f + 1) \bmod d)$$

A downward column move by 1 is done by the following 2 slot routing:

$$p(e, f) \to c(f, e) \to p(f, e) \to c((e + 1) \bmod g, f) \to p((e + 1) \bmod g, f)$$

Leftward row moves and upward column moves are done similarly.

When $d < N < g$ each row of the mesh occupies more than one group of the $POPS(d, g)$. Since d divides N, each row is housed in an integral number N/d of groups. A rightward shift by 1 along rows of the mesh requires data in processor $p(e, f)$ of the $POPS(d, g)$ to be routed to $p(e', f')$, where

$$e' = \begin{cases} e & \text{if } f \neq d - 1 \\ e + 1 & \text{if } f = d - 1 \text{ and } (e + 1) \bmod N/d \neq 0 \\ e + 1 - N/d & \text{if } f = d - 1 \text{ and } (e + 1) \bmod N/d = 0 \end{cases}$$

and

$$f' = (f+1) \bmod d$$

Let $r = ed + f$. The following 2 slot routing moves data from $p(e, f)$ to $p(e', f')$ for all e and f.

$$\begin{aligned} p(e,f) &\rightarrow c(r \bmod g, e) & (5)\\ &\rightarrow p(r \bmod g, \lfloor r/g \rfloor) & (6)\\ &\rightarrow c(e', r \bmod g) & (7)\\ &\rightarrow p(e', f') & (8) \end{aligned}$$

The proof that different source processors use different slot 1 couplers (i.e., in line (5)) and different slot 1 destination processors (line (6)) is the same as for the corresponding lines of the proof of Theorem 4.1. Suppose that data originates in two different processors $p(e_1, f_1)$ and $p(e_2, f_2)$ in line (5). If $e'_1 \neq e'_2$ there is no coupler conflict in line (7). If $e'_1 = e'_2$, then $p(e_1, f_1)$ and $p(e_2, f_2)$ represent the same row of the mesh. Hence, $0 < |r_1 - r_2| < N < g$. So $r_1 \bmod g \not\equiv r_2 \bmod g$ and there is no coupler conflict in line (7).

For a downward column move $e' = (e + N/d) \bmod g$ and $f' = f$. A downward move may also be done using the routing of lines 5–8. Leftward row moves and upward column moves are similarly dome in 2 slots per move.

As was the case in the proof of Theorem 4.1, the case $d > N > g$ is handled using $\lceil d/g \rceil$ passes of the routing of lines 5–8. □

The comments made following the proof of Theorem 4.1 regarding the optimality of the simulation apply to our simulation of a mesh also. When $d \geq 2\sqrt{n}$, a similarly optimal simulation of a 2D-mesh may be obtained using the embedding of a tori described in [2, 7].

4.4 Intragroup Data Permutations

Although [7, 13] give a data permutation algorithm for the $POPS(d, g)$ network, this algorithm is optimal only when we require that each data item be routed from its source processor to its destination processor using a single slot (or hop). Under this assumption datum initially in $p(i)$ and destined for $p(\sigma(i))$, where $\sigma(i)$ is the permutation that is to be performed and $i \neq \sigma(i)$, must be routed using the path $p(i) \rightarrow c(group(\sigma(i)), group(i)) \rightarrow p(\sigma(i))$.

When an intragroup permutation with $i \neq \sigma(i)$ for all i in the group (an example of such a permutation would be a nonzero intragroup shift), for example, is performed we get a coupler conflict—all d processors of a group need the same coupler to get their source data to the destination processors. This coupler conflict is eliminated by scheduling the use of the couplers so that no coupler has more than one source node transmitting data to the coupler. For an intragroup pemutation with $i \neq \sigma(i)$ for all i d time slots are taken to complete the permutation with each data making a single hop. We refer to the permutation algorithm of [7, 13] as the *single-hop* algorithm.

First we consider the case when a single intragroup permutation is to be performed. As noted above, when $i \neq \sigma(i)$ for all i in the group being permuted, the single hop algorithm takes d slots to perform an intragroup permutation. Without loss of generality, assume that the data in group 0 are to be permuted as per the permutation σ_0. That is, data from $p(0, i)$ is to be moved to $p(0, \sigma_0(i))$, $0 \leq i < d$. The single intragroup permutation σ_0 may be performed in $2\lceil d/g \rceil$ slots by routing g elements to their destination processors in every pair of slots. The number of slot pairs is d/g and in the qth pair, $0 \leq q < d/g$, processors $qg + i$, $0 \leq i < g$ send their data to the destination processors using two slots as follows:

1. In the first slot of a slot pair $p(0, qg + i)$ sends its data to $p(i, 0)$ using the path $p(0, qg + i) \rightarrow c(i, 0) \rightarrow p(i, 0)$, $0 \leq i < g$.

2. In the second slot of the slot pair $p(i, 0)$ sends the data it received in the first slot using the path $p(i, 0) \rightarrow c(0, i) \rightarrow p(0, \sigma_0(qg + i))$, $0 \leq i < g$.

The correctness of the above algorithm is easily established. There is no advantage to having $d < \sqrt{n}$ because when $d < \sqrt{n}$, $g > \sqrt{n}$ and the intragroup permutation takes 2 slots; the same as when $d = g = \sqrt{n}$.

Although the above intragroup algorithm is simple, it is not optimal (except when $d \leq \sqrt{n}$). Figure 11 gives an optimal algorithm.

Once again, correctness is easily established. The number of slots is seen to be $\lceil (d-1)/g \rceil + 1$. Notice that each data item is routed to its destination using either one or two hops. Under the assumption that $i \neq \sigma_0(i)$ for any i in group 0, the single hop algorithm takes d slots. So the algorithm of Figure 11 provides a speedup of $d/(\lceil (d-1)/g \rceil + 1)$. When $d = g = \sqrt{n} > 1$ the speedup is $d/2$. When $g = 1$ the single hop algorithm uses the same number of slots (i.e., d) as used by the algorithm of Figure 11.

1. In slot 0 $p(0,0)$ uses the path $p(0,0) \to c(0,0) \to p(0,\sigma_0(0))$ to get its data to its destination processor. Processors $p(0,i)$, $0 < i < g$ use the paths $p(0,i) \to c(i,0) \to p(i,0)$ to get their data to intermediate processors in groups other than group 0.

2. In slot q, $q > 0$ we do the following:

   (a) $p(0,qg)$ uses the path $p(0,qg) \to c(0,0) \to p(0,\sigma_0(qg))$ to get its data to its destination processor.

   (b) Processors $p(i,0)$, $i \neq 0$, use the following paths to get the data they received in slot $q-1$ to destination processors: $p(i,0) \to c(0,i) \to p(0,\sigma_0((q-1)g+i))$.

   (c) Processors $p(0,qg+i)$, $1 \leq i < g$ and $qg+i < d$, use the following paths to get their data to intermediate processors: $p(0,qg+i) \to c(i,0) \to p(i,0)$.

Figure 11: An optimal algorithm for a single group intragroup permutation

A circular shift of the elements of group 0 by one unit takes d slots when the single hop algorithm is used and $\lceil (d-1)/g \rceil + 1$ slots when we use the algorithm of Figure 11.

Theorem 4.3 *The algorithm of Figure 11 is optimal when $i \neq \sigma_0(i)$, $0 \leq i < d$.*

Proof. Since all the data that is to be routed begins and ends in group 0, at most 1 datum can be routed to its final destination in the first slot of any algorithm. Since group 0 has g couplers, at most g different data can be routed to the group 0 processors in any slot. So every algorithm routes at most 1 datum to the final destination in the first slot and at most g in each of the remaining slots. Since d data are to be routed to their destination processors, at least $\lceil (d-1)/g \rceil + 1$ slots must be used. □

Now we consider the case when multiple intragroup permutations are to be performed. Suppose that data from $p(i,j)$ is to be routed to $p(i,\sigma_i(j))$, $0 \leq i < g$, $0 \leq j < d$. The $2\lceil d/g \rceil$-slot algorithm described above for a single permutation may be run, in parallel, on all groups. Therefore, data in all groups may be permuted in $2\lceil d/g \rceil = 2\lceil n/g^2 \rceil$ slots. We can reduce

1. Do the following for $0 \le q < \lceil n/(g+g^2) \rceil$

2. // slot 1 of slot pair q
   Let $h = g + 1$.
   $p(i, qh)$ uses the path $p(i, qh) \rightarrow c(i, i) \rightarrow p(i, \sigma_i(qh))$ to route its data to the destination processor.
   Processors $p(i, qh+s)$, $1 \le s < g$ use the following paths to route their data to intermediate processors (let $t = (i+s) \bmod g$): $p(i, qh+s) \rightarrow c(t, i) \rightarrow p(t, i)$ when $\sigma_t(qh) \ne i$ and $p(i, qh+s) \rightarrow c(t, i) \rightarrow p(t, t)$ when $\sigma_t(qh) = i$.

3. // slot 2 of slot pair q
   $p(i, qh+g)$ uses the path $p(i, qh+g) \rightarrow c(i, i) \rightarrow p(i, \sigma_i(qh+g))$ to route its data to the destination processors.
   Processors that received data from a different group in slot 1 of this slot pair route the received data via the coupler in the destination group to the destination processor.

Figure 12: Optimal algorithm to permute in all groups

the number of slots to $2\lceil n/(g+g^2) \rceil$ by moving batches of $g + g^2$ elements to their destination processors using 2 slots per batch. In the first slot g elements are moved to their destination processors and $g^2 - g$ elements are moved to intermediate processors. In the second slot g^2 elements are moved to their destination processors. Figure 12 gives the algorithm.

Using the algorithm of Figure 12 we can, for example, perform a circular shift by 1 of data in each group using $2\lceil n/(g+g^2) \rceil$ slots. Shifting all groups by 1 takes $d = n/g$ slots when the single hop algorithm is used.

Theorem 4.4 *The algorithm of Figure 12 is within 1 slot of being optimal when $j \ne \sigma_i(j)$, $0 \le i < g$, $0 \le j < d$.*

Proof. Suppose that an algorithm completes the group permutations in k slots. A total of n elements are to be moved from their initial processors to their destination processors. At most kg of these elements can be so moved using a single hop. Therefore, at least $n - kg$ elements are moved using 2 or more hops each. So the total number of data moves (in one data move a data item makes 1 hop) is $\ge 2(n - kg) + kg = 2n - kg$. In a $POPS(d, g)$ at most g^2 data moves can be made in one slot. Therefore, $kg^2 \geq 2n - kg$. So $k \geq 2n/(g + g^2)$. Since k is an integer, $k \geq \lceil 2n/(g + g^2) \rceil$.

The number of slots used by the algorithm of Figure 12 is $2\lceil n/(g + g^2) \rceil \leq \lceil 2n/(g + g^2) \rceil + 1$. □

See [24, 25] for more algorithms that use the POPS network.

5 OTIS

5.1 OTIS Topology

The OTIS (optical transpose interconnect system) family of parallel computer models was proposed in [12, 9, 29]. In an OTIS parallel computer, the processors are divided into groups and each group of processors is realized using an electronic package (such as a high density chip or wafer). Intragroup connections are electronic and intergroup connections are realized using free space optics. Thus an OTIS system is an optoelectronic system. By contrast, all interprocessor connections in a pipelined bus system are optical.

Since optical connects provide power, speed, and crosstalk advantages over electronic interconnects when the connect distance is more than a few millimeters [5, 10], an OTIS system attempts to get the best of both worlds–electronic interconnect is used for the short-distance intragroup (or intra package) connections and optical interconnect is used for the longer-distance interpackage connections.

The bandwidth of an OTIS system is maximized and power consumption is minimized when the number of groups equals the number of processors in a group [11]. This means that an optimal N^2 processor OTIS system has N groups of N processors each. Let (G, P) denote processor P of group G (the processors in each group and the groups are numbered from 0 to $N - 1$). In an OTIS system, processor (G, P) is connected via an optical link to processor (P, G), for all G and P. If you regard (G, P) as a matrix index, then the matrix transpose operation moves element(G, P) to position (P, G), hence the name optical transpose interconnect system.

Figure 13 shows the topology of a 16-processor OTIS computer; processors are shown as small shaded boxes; the (G, P) index of each processor is given; each group of 4 processors is enclosed by a large box; and the OTIS (i.e., the optical transpose connections) are shown as bidirectional arrows.

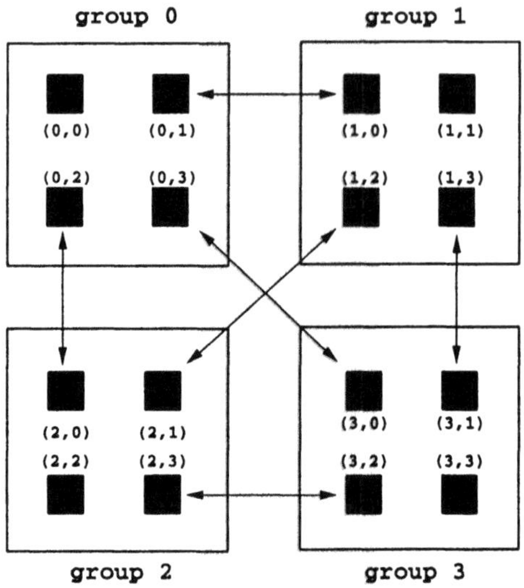

Figure 13: A 16-processor OTIS computer

For the intragroup interconnect topology, we may choose any of the electronic topologies proposed for parallel computers–mesh, hypercube, mesh of trees, etc. The selection of the intragroup topology identifies the specific OTIS model within the family of OTIS models. For example, an OTIS-mesh is an OTIS computer in which the intragroup interconnections correspond to a square mesh and in an OTIS-hypercube, the processors within each group are connected using the hypercube topology.

When analyzing the complexity of an OTIS computer we count the number of OTIS (i.e., intergroup or optical) data move steps and the number of electronic (or intragroup) data move steps.

5.2 OTIS-Mesh

Figure 14 shows a 16-processor OTIS-Mesh computer.

An OTIS-mesh can simulate each move of a four-dimensional $\sqrt{N}\times\sqrt{N}\times\sqrt{N}\times\sqrt{N}$ mesh using either one intragroup (i.e., electronic) move or one intragroup and two intergroup (i.e., OTIS) moves [29]. For the simulation, processor (i,j,k,l) of the 4D mesh is mapped on to processor (G,P) of the OTIS-Mesh, $G = i\sqrt{N}+j$ and $P = k\sqrt{N}+l$. It is easy to see that the 4D mesh moves $(i,j,k\pm 1,l)$ and $(i,j,k,l\pm 1)$ can be done with one intragroup move of the OTIS mesh. Moves of the form $(i\pm 1,j,k,l)$ and $(i,j\pm 1,k,l)$ can be done with two OTIS and one electronic move as follows. First an

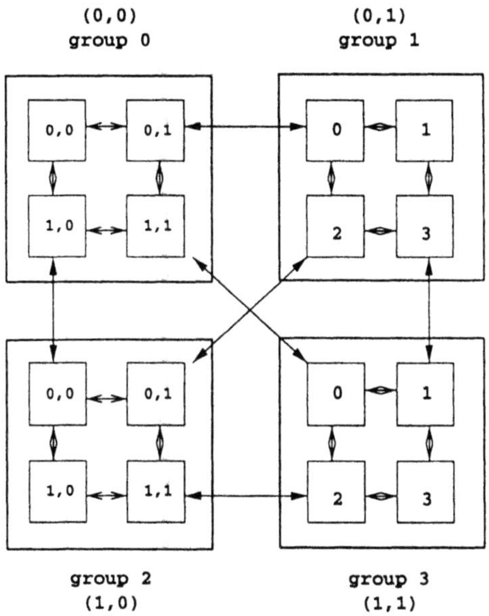

Figure 14: A 16-processor OTIS-Mesh

OTIS move is made to get data from (i, j, k, l) to (k, l, i, j); then an electronic move gets the data to $(k, l, i + 1, j)$ (say); and a final OTIS move gets it to $(i + 1, j, k, l)$.

Many OTIS-mesh and OTIS-hypercube algorithms appear in [22, 23, 16, 27, 28].

6 Acknowledgement

This work was supported, in part, by the National Science Foundation under grant CCR-9912395.

References

[1] P. Berthomé and A. Ferreira. Improved embeddings in POPS networks through stack-graph models. *Third International Workshop on Massively Parallel Processing Using Optical Interconnections*, IEEE, 130-135, 1996.

[2] P. Berthomé, J. Cohen, and A. Ferreira. Embedding tori in Partitioned Optical Passive Star networks. *Fourth International colloquium*

on Structiral Information and Communication Complexity-Sirocco'97, volume 1 of *Proceedings in Informatics*, Carleton Scientific, 40-52, 1997.

[3] D. Chiarulli, S. Levitan, R. Melhem, J. Teza, and G. Gravenstreter. Multiprocessor interconnection networks using partitioned optical passive star (POPS) topologies and distributed control. *First International Workshop on Massively Parallel Processing Using Optical Interconnections*, IEEE, 70-80, 1994.

[4] D. Coudert, A. Ferreira, and X. Muñoz. Multiprocesor architectures using multi-hop multi-ops lightwave networks and distributed control *12th International Parallel processing Symposium and 9th Symposium on Parallel and Distributed Processing*, IEEE, 151-155, 1998.

[5] M. Feldman, S. Esener, C. Guest, and S. Lee, Comparison between electrical and free-space optical interconnects based on power and speed considerations, *Applied Optics*, 27, 9, 1988.

[6] G. Graventreter, R. Melhem, D. Chiarulli, S. Levitan, and J. Teza. The partitioned optical passive stars (POPS) topology. *9th International Parallel processing Symposium*, IEEE, 4-10, 1995.

[7] G. Graventreter and R. Melhem. Realizing common communication patterns in partitioned optical passive star (POPS) networks. *IEEE Transactions on Computers*, 998-1013, 1998.

[8] Z. Guo, R. Melhem, R. Hall, D. Chiarulli, and S. Levitan, Pipelined communications in optically interconnected arrays, *JPDC*, 12, 3, 1991, 269-282.

[9] W. Hendrick, O. Kibar, P. Marchand, C. Fan, D. Blerkom, F. McCormick, I. Cokgor, M. Hansen, and S. Esener, Modeling and optimization of the optical transpose interconnection system in Optoelectronic Technology Center, Program Review, Cornell University, Sept. 1995.

[10] F. Kiamilev, P. Marchand, A. Krishnamoorthy, S. Esener, and S. Lee, Performance comparison between optoelectronic and vlsi multistage interconnection networks, *Journal of Lightwave Technology*, 9, 12, 1991.

[11] A. Krishnamoorthy, P. Marchand, F. Kiamilev, and S. Esener, Grainsize considerations for optoelectronic multistage interconnection networks, *Applied Optics*, 31, 26, 1992.

[12] G. Marsden, P. Marchand, P. Harvey, and S. Esener, Optical transpose interconnection system architectures, *Optics Letters*, 18, 13, 1993, 1083–1085.

[13] R. Melhem, G. Graventreter, D. Chiarulli, and S. Levitan. The communication capabilities of partitioned optical passive star networks. In *Parallel computing using optical interconnections*, K. Li, Y. Pan, and S. Zheng, Editors, Kluwer Academic Publishers, 77-98, 1998.

[14] R. Miller, V. K. Prasanna Kumar, D. Reisis, and Q. Stout, Efficient parallel algorithms for intermediate level vision analysis on the reconfigurable mesh, in *Parallel Architectures and Algorithms for Image Understanding*, V. K. Prasanna Kumar, ed., Academic Press, 1991, 185-207.

[15] R. Miller, V. K. Prasanna Kumar, D. Reisis, and Q. Stout, Meshes with reconfigurable buses, *IEEE Transactions on Computers*, 42, 1993, 678-692.

[16] A. Osterloh, Sorting on the OTIS-Mesh, *Proceedings 14th Annual International Parallel & Distributed processing Symposium*, IEEE Computer Society, 2000, 269-274.

[17] Y. Pan, Basic data movement operations on the LARPBS model, in *Parallel Computing Using Optical Interconnections*, K. Li, Y. Pan, and S. Zheng, editors, Kluwer Academic Publishers, Boston, 1998, 227-247.

[18] S. Pavel and S. Akl, Matrix operations using arrays with reconfigurable optical buses, *Journal of Parallel Algorithms and Applications*, 8, 1996, 223-242.

[19] S. Pavel and S. Akl, On the power of arrays with optical pipelined buses, *Proceedings 1996 International Conference on Parallel and Distributed Processing Techniques and Applications*, 1996, 1443-1454.

[20] C. Qiao and R. Melhem, Time-division optical communications in multiprocessor arrays, *IEEE Transactions on Computers*, 42, 5, 1993, 577-590.

[21] S. Rajasekeran and S. Sahni, Fundamental algorithms for the array with reconfigurable optical buses, in *Parallel Computing Using Optical Interconnections*, Kluwer, 1998, 185-204. Ed. Li and Zhenq.

[22] S. Sahni and C. Wang, BPC permutations on the OTIS-hypercube optoelectronic computer, *Informatica*, 22, 1998, 263-269.

[23] S. Sahni and Chih-Fang Wang, BPC permutations on the OTIS-mesh optoelectronic computer, *IEEE Conference on Massively Parallel Programming with Optical Interconnect*, 1997, 130-135.

[24] S. Sahni, Matrix Multiplication And Data Routing Using A Partitioned Optical Passive Stars Network, *IEEE Transactions on Parallel and Distributed Systems*, 11, 7, 2000, 720-728.

[25] S. Sahni, The Partitioned Optical Passive Stars Network: Simulations And Fundamental Operations, *IEEE Transactions on Parallel and Distributed Systems*, 11, 7, 2000, 739-748.

[26] Q. Stout, Mesh connected computers with broadcasting, *IEEE Transactions on Computers*, 1983, 826-830.

[27] Chih-fang Wang and S. Sahni, OTIS optoelectronic computers, in *Parallel Computing Using Optical Interconnections*, Kluwer, 1998, 99-116. Ed. Li and Zhenq.

[28] C. Wang and S. Sahni, Basic operations on the OTIS-Mesh optoelectronic computer, *IEEE Trans. on Parallel and Distributed Systems*, 9, 12, 1998, 1226-1236.

[29] F. Zane, P. Marchand, R. Paturi, and S. Esener, Scalable network architectures using the optical transpose interconnection system (OTIS), *Proceedings of the second International Conference on Massively Parallel Processing Using Optical Interconnections (MPPOI'96)*, 1996, 114–121.

SWITCHING NETWORKS: RECENT ADVANCES
D.-Z. Du and H.Q. Ngo (Eds.) pp. 89-100

Wide-sense Nonblocking for 3-stage Clos Networks

Ding-Zhu Du
Department of Computer Science and Engineering
University of Minnesota, Minneapolis, MN 55455
E-mail: dzd@cs.umn.edu

P. C. Fishburn
AT&T Shannon Laboratory, Florham Park, NJ 07932
E-mail: fish@research.att.com

Biao Gao
Lattice Semiconductor Corp., Milpitas CA 95035
E-mail: liao@lscsjc.latticesemi.com

F. K. Hwang
Department of Mathematics
Chiao Tung University, HsinChu, Taiwan, ROC
E-mail: fhwang@math.nctu.edu.tw

Contents

Abstract

Let $C(n, m, r)$ denote the symmetric 3-stage Clos network with r $n \times m$ crossbars in the first and third stage, and m $r \times r$ crossbars in the second stage. A network is wide-sense nonblocking if it is nonblocking under a given routing algorithm. Packing is a routing algorithm much heralded in the folklore, and was shown to save $n/2$ center switches for $r = 2$. We prove the surprising result that packing does not save anything for $r \geq 3$. We also show that $C(n, m, r)$ is not wide-sense nonblocking (under any algorithm) for $r \geq 3$ if $m < \lfloor 7n/4 \rfloor$. We also extend some results to asymmetric 3-stage Clos networks

1 Introduction

In classical circuit switching, three types of nonblocking properties have been extensively studied [1]. A request for the connection between an idle pair (input, output) is routable if there exists a path connecting them such that no link on the path is used by any other connection path. A network is strictly nonblocking if regardless of the routing of existing connections in the network, a new request is always routable. A network is wide-sense nonblocking (WSNB) if a new request is always routable as long as all previous requests were routed according to a given routing algorithm. A network is rearrangeably nonblocking, or simply rearrangeable, if a new request is always routable given that we can reroute existing connections. Clearly, strictly nonblocking implies WSNB implies rearrangeable.

The 3-stage Clos network $C(n, m, r)$ is generally considered the most basic multistage interconnection network. A result obtained for $C(n, m, r)$ is often extendible to a multistage interconnection network with more than three stages. $C(n, m, r)$ is symmetric with respect to the center stage. The first stage, or the input stage (hence the third stage or the output stage), has r $n \times m$ crossbars; the center stage has m $r \times r$ crossbars. The n inlets (outlets) on each input (output) crossbar are the inputs (outputs) of the network. There exists exactly one link between every center crossbar and every input (output) crossbar.

Important and powerful results have been obtained for $C(n, m, r)$ with respect to strictly nonblocking and rearrangeably nonblocking. For example, Clos [4] proved that $C(n, m, r)$ is strictly nonblocking if and only if $m \geq 2n - 1$. Slepian [6] and Duguid [5] proved that $C(n, m, r)$ is rearrangeably nonblocking if and only if $m \geq n$. However, very little is known about WSNB. This is probably due to two factors which complicate the problem.

The first is that results on WSNB are algorithm-dependent. The second is that, unlike the other two cases, the results also depend on r.

We now define some classes of routing algorithms which have been discussed in the literature:

STU (save-the-unused). A request will be routed through a used center switch whenever possible.

P (packing). A request will be routed through an available center switch which already carries the most calls.

MI (minimum-index). The center switches are indexed by the natural numbers. A request is routed through the center switch with the minimum index which can accommodate the request.

Note that for $r = 2$, P and STU are identical. For $r \geq 3$, P clearly implies STU. Folklore has heralded P as a very good routing algorithm.

Let A denote a class of algorithms. We say that a network is SNB under A if it is WSNB under every algorithm in A, and is WSNB under A if it is WSNB for an algorithm in A. Benes [1], [2] proved that $C(n, m, 2)$ is WSNB under STU if and only if $m < \lfloor 3n/2 \rfloor$. Smith [7] proved that $C(n, m, r)$ is not WSNB under P or MI if $m < \lfloor 2n - n/r \rfloor$. In this paper we first improve Smith's bound to $m < \lfloor 2n - n/2^{r-1} \rfloor$. We then show that for $r = 3$, this bound holds for all routing algorithms, not just P and MI. In other words, $C(n, m, 3)$ is not WSNB if $m < \lfloor 7n/4 \rfloor$. We also show that $C(n, m, r)$ is not SNB under P for $n \geq 3$ if $m < 2n - 1$. Recall that $C(n, 2n-1, r)$ is strictly nonblocking. Thus we reach the surprising conclusion that P is useless for $r \geq 3$. This improves a recent result of Yang and Wang [8] which gave the weaker condition $m \leq 2n - \lceil n/F_{2n-1} \rceil$, where F_k is the kth Fibonacci number. Finally, we extend our results to asymmetric 3-stage Clos networks.

2 Lower bounds

A switch is itself nonblocking, i.e.. any idle inlet can reach any idle outlet regardless of other connections through the switch. We can therefore represent a pair (x, y) by the pair (X, Y) where $X(Y)$ is the input (output) switch the input x (output y) belongs to. In particular we denote by (i, j) the pair (I_i, O_j), the ith input switch and the jth output switch. Let C_i denote the ith center switch.

Theorem 1 *$C(n, m, r)$ is not WSNB under P or MI if $m < \lfloor 2n - n/2^{r-1} \rfloor$.*

Proof. For easier presentation we assume 2^{r-1} divides n. The nondivisibility case can be similarly argued.

At step 1, consider n requests (i,i) for each $i = 1,\ldots,r$. By P or MI, we may assume that each set of these n requests are routed by $C_1,\ldots,C_n$.

At step 2, suppose the $n/2$ $(1,1)$ calls carried by $C_{n/2}+1,\ldots,C_n$ and the $n/2$ $(2,2)$ calls carried by $C_1,\ldots,C_{n/2}$ hang up. Furthermore, $n/2$ new $(1,2)$ calls occur. Since each of $C_1,\ldots,C_n$ already carries either a $(1,1)$ or a $(2,2)$ call, the new $(1,2)$ calls have to be routed through $C_{n+1},\ldots,C_{3n/2}$.

At step 3, suppose the $n/4$ $(1,1)$ calls carried by $C_{n/4+1},\ldots,C_{n/2}$, and the $n/4$ $(3,3)$ calls carried by $C_1,\ldots,C_{n/4}$ hang up. Furthermore, $n/4$ new $(1,3)$ calls occur. Since each of $C_1,\ldots,C_{3n/2}$ carries either a call from I_1 or a call to O_3, the new $(1,3)$ calls have to be routed through $C_{3n/2},\ldots,C_{7n/4}$.

Continuation of the argument shows that, at step r, $2n - n/2^{r-1}$ center switches are needed.

Corollary 1 *$C(n,m,r)$ is not WSNB under P or MI if $m < \lfloor 2n - n/2^{k-1} \rfloor$ and $r \geq (k-1)\binom{m}{n} + 1$.*

Proof. Let S_i denote the set of n center switches each carrying an (i,i) call, $i = 1,\ldots,r$. For r large enough, there exist k identical S_i and the argument in Theorem 1 applies.

We show next that for $r = 3$ the bound in Theorem 1 holds for every routing algorithm. We first prove a lemma.

Lemma 1 *Let I denote a set of s center switches each carrying a call from I_i, let J denote a set of s center switches each carrying a call to O_j, and suppose there is no call from I_i to O_j. Then $|I \cap J| \leq x$ implies that at least $n + s - \lceil x/2 \rceil$ center switches are needed.*

Proof. Split $I \cap J$ into two halves with sizes $\lceil x/2 \rceil$ and $\lfloor x/2 \rfloor$ respectively. Hang up $\lceil x/2 \rceil$ call from I_i carried by one half of $I \cap J$, and hang up $\lfloor x/2 \rfloor$ call to O_j carried by the other half. Then I' and J', the updated I and J, are disjoint and $|I'| + |J'| = 2s - x$. Now add $n - s + \lfloor x/2 \rfloor$ new (i,j) calls, which must use center switches disjoint from I' and J'. Lemma 1 follows immediately.

Theorem 2 *$C(n,m,3)$ is not WSNB if $m < \lfloor 7n/4 \rfloor$.*

Proof. At step 1 assume that S_i carries n (i,i) calls, $i = 1,2,3$, as in the proof of Corollary 1. By Lemma 1 we may assume that $S_1 = \{C_1,\ldots,C_n\}$, $S_2 = \{C_1,\ldots,C_k,C_{n+1},\ldots,C_{2n-k}\}$ with $k \geq \lfloor n/2 \rfloor + 1$, and $S_3 = S_{31} \cup S_{32} \cup S_{33} \cup S_{34}$, where $S_{31} \subseteq \{C_1,\ldots,C_k\}$, $S_{32} \subseteq \{C_{k+1},\ldots,C_n\}$, $S_{33} \subseteq \{C_{n+1},\ldots,C_{2n-k}\}$

and $S_{34} \subseteq \{C_{2n-k+1}, \ldots, C_{\lfloor 7n/4 \rfloor -1}\}$, with $|S_{31}| + |S_3| \geq \lfloor n/2 \rfloor + 1$ and $|S_{31}| + |S_{33}| \geq \lfloor n/2 \rfloor + 1$.

At step 2 keep $\lfloor n/2 \rfloor$ $(2,2)$ calls carried by $S_2' \subseteq S_{31} \cup S_{33}$ and hang up all other $(2,2)$ calls. Keep $\lfloor n/2 \rfloor$ $(1,1)$ calls carried by $S_1' \subseteq S_1$ such that $S_1' \cap S' = \emptyset$ (this is always possible) and hang up all other $(1,1)$ calls. Add $\lceil n/2 \rceil$ new $(1,2)$ calls which must be carried by S_{12} disjoint from S_1' and S_2'. Note that $S_1' \cap S_2'$. Note that $S_1' \cap S_3 < \lfloor \frac{n}{2} \rfloor$, Theorem 2 follows from Lemma 1 immediately

Since $C(n,m,r)$ is not WSNB implies $C(n,m,r')$ is not WSNB for all $r' > r$, we have

Corollary 2 *$C(n,m,r)$ is not WSNB for all $r \geq 3$ if $m < \lfloor 7n/4 \rfloor$.*

We now give a sharper bound than Theorem 1 for P.

Lemma 2 *$C(n,m,3)$ is not WSNB under P if $m < \lfloor 15n/8 \rfloor$.*

Proof. Again, we assume 8 divides n for easy presentation.

Construct a matrix M with rows indexed by input switches, columns by output switches. and in which cell m_{ij} contains the set C_{ij} of center switches carrying an (i,j) call. Define $C_i = \bigcup_j C_{ij}$ and $C_{ij} = \bigcup_i C_{ij}$. At step 3 in the proof of Theorem 1, M is shown as below:

$$\begin{array}{ccc} [1, \frac{n}{4}] & [n+1, \frac{3n}{2}] & [\frac{3n}{2}+1, \frac{7n}{4}] \\ & [\frac{n}{2}+1, n] & \\ & & [\frac{n}{4}+1, n] \end{array}$$

Hang up all calls carried by C_{22}. Insert $n/2$ new $(2,1)$ calls and $3n/8$ new $(2,2)$ calls. Since every switch in $[1, 7n/4]$ was used exactly once before these insertions, we can route the $7n/8$ new calls by any set of $7n/8$ switches in $[1, 7n/4]$ without violating the rule P. Assume that the switches are chosen as shown in the following updated M and then the calls carried by $[\frac{n}{4}+1, \frac{3n}{5}]$ in C_{33} are hung up:

$$\begin{array}{ccc} [1, \frac{n}{4}] & [n+1, \frac{3n}{2}] & [\frac{3n}{2}+1, \frac{7n}{4}] \\ [n+1, \frac{3n}{2}] & [1, \frac{3n}{8}] & \\ & & [\frac{3n}{8}+1, n] \end{array}$$

Now when $n/8$ new $(2,3)$ calls arise, we have to use new switches of new switches $[7n/4+1, 15n/8]$.

The reason that this result is SNB under P is that in adding the $n/2$ $(2,1)$ calls, there are choices other than $[n+1, 3n/2]$ under P.

Theorem 3 *$C(n,m,r)$ is not SNB under P for $r \geq 3$ if $m < 2n-1$.*

Proof. It suffices to prove Theorem 3 for $r = 3$. For easier presentation, we ignore the issue of divisibility. For $k = 0, 1, \ldots$, define

$$\begin{aligned} a_k &= (2^k - 1)n/w^{k+3}, \\ b_k &= n/w^{k+3}. \end{aligned}$$

Note that $a_k + b_k = n/8$, $b_k = 2b_{k+1}$ and $2a_{k+1} = 2a_k + b_k = 2(a_k + b_{k+1})$.

To facilitate our induction proof, we convert M to another matrix M'. Two center switches are said to be of the same *pattern* if each carries the same set of calls. The frequency of a pattern is the number of switches of that pattern. The rows of M' are still indexed by input switches, but each column represents a center switch, except that we group all center switches with the same pattern into one *group column* which is indexed by its frequency. Cell m'_{ij} contains the output switch k such that pattern j carries an (i,k) call. To illustrate, consider M at the end of the proof of Lemma 2. After adding the $n/8$ $(2,3)$ calls and hanging up the $n/2$ $(1,2)$ calls, we have

M

$$\begin{array}{lll} [1, \frac{n}{4}] & & [\frac{3n}{2}+1, \frac{7n}{4}] \\ [n+1, \frac{3n}{2}] & [1, \frac{3n}{8}] & [\frac{7n}{4}+1, \frac{15n}{8}] \\ & [\frac{3n}{8}+1, n] & \end{array}$$

M'

	$n/4$	$n/8$	$5n/8$	$n/2$	$n/4$	$n/8$
1	1				3	
2	2	2		1		3
3			3			

For example, the first group column represents the set $[1, n/4]$ which carry both $(1,1)$ and $(2,2)$ calls. The second group column represents the set $[n/4+1, 3n/8]$ which carry $(2,2)$ calls.

For $k = 0, I, \ldots$, define

$$M'_k = \begin{array}{cccccccc} n/4 & a_k & b_k & 5n/8 & n/2 & 2a_k & n/4 & b_k \\ 1 & 3 & & & & & 3 & \\ 2 & & 2 & & 1 & & & 3 \\ & & & 3 & & 1 & & \end{array}$$

Then $M'_0 = M'$. Note that a new call (i, j) can be carried by any switch in a group column not having the entry j at row i. Under the rule P, a column k is chosen only if there exists no other feasible columns having more entries. We now show the existence of a sequence of calls and hang-ups which transform M'_k to M'_{k+1} under P.

	$n/4$	a_k	b_k	$5n/8$	$n/2$	$2a_k$	b_k	b_k	$2a_k$	b_k	
M'_k	$\rightarrow$	1	3					3	3	3	
		2		2		1					3
					3		1	1			2
		$n/4$	a_k	b_k	$5n/8$	$n/2$	$2a_k$	b_k	b_k	$2a_k$	b_k
	$\rightarrow$	1	3					3	3	3	
		2		2		1					
					3		1	1			2
	$\rightarrow$	$n/4$	a_k	b_k	$5n/8$	$n/2$	$2a_k$	b_k	b_k	$2a_k$	b_k
		1	3					3	3	3	3
		2		2		1					
					3		1	1			2
	$\rightarrow$	$n/4$	a_k	b_k	$5n/8$	$n/2$	$2a_k$	b_k	b_k	$2a_k$	b_k
		1	3						3	3	3
		2		2		1					
					3		1	1			
	$\rightarrow$	$n/4$	a_k	b_k	$5n/8$	$n/2$	$2a_{k+1}$		b_{k+1}	$2a_{k+1}$	
		1	3	3					3	3	
		2		2		1					
					3		1				
	$\rightarrow$	$n/4$	a_k	b_k	$5n/8$	$n/2$	$2a_{k+1}$		$n/4$		
		1	3	3					3		
		2		2		1	2				
					3		1				
	$\rightarrow$	$n/4$	a_k	b_{k+1}	b_{k+1}	$5n/8$	$n/2$	$2a_{k+1}$	$n/4$		
		1	3	3					3		
		2			2		1	2			
						3		1			
	$\rightarrow$	$n/4$	a_{k+1}		b_{k+1}	$5n/8$	$n/2$	$2a_{k+1}$	$n/4$	b_{k+1}	
		1	3						3		
		2			2		1	2		3	
						3		1			
	$\rightarrow$	$n/4$	a_{k+1}	b_{k+1}	$5n/8$	$n/2$	$2a_{k+1}$	$n/4$	b_{k+1}		
		1	3					3			
		2		2		1			3	$= M'_{k+1}$	
					3		1				

It is easily verified for each M' matrix during the transformation that every row carries at most n calls, every entry appears in at most n weighted columns, and every call is added following rule P. Since the total number of weighted columns in M_k is $2n - n/2^{k+3}$, after $\log_2 n - 3$ iterations, it reaches $2n - 1$.

The first discrepancy between Theorems 2 and 3 is for $n = 5$. From Theorem 3. $C(5, 8, 3)$ is not WSNB under P, but Theorem 3 does not rule out the general WSNB. We now show that $C(5, 8, 3)$ is not WSNB.

Consider the state $|C_{11}| = |C_{22}| = |C_{33}| = 5$ (all other $|C_{ij}| = 0$). Define $d_1 = |C_{11} \backslash C_{22} = |C_{22} \backslash C_{11}|$, $d_2 = |C_{22} \backslash C_{33}| = |C_{33} \backslash C_{22}|$, $d_3 = |C_{33} \backslash C_{11}| = |C_{11} \setminus C_{33}|$. Without loss of generality, assume $d_1 \geq d_2 \geq d_3$.

Case (i). $d_i \geq 3$. The desired result follows from Lemma 1.

Case (ii). $d_i = 2$. Let $C_{11} = \{1, 2, 3, 4, 5\}$ and $C_{22} = \{1, 2, 3, 6, 7\}$. There are three sub cases:

Subcase 1. C_{33} is $\{1, 2, 3, 4, 6\}$ or $\{1, 2, 4, 5, 6\}$ or $\{1, 2, 4, 6, 8\}$. Hang up $\{1, 2\}$ in C_{11} and $C_{33}\{1, 2, 6\}$ in C_{33}. Add two new $(1, 3)$ calls which must use $\{7, 8\}$. Now C_{22} and the updated C_{11} differ in three elements; Lemma 1 applies.

Subcase 2. $C_{33} = \{1, 4, 5, 6, 7\}$. Hang up $\{1, 2\}$ in On and $\{4, 5\}$ in C_{33}. Add two new $(1, 3)$ calls which must use $\{2, 8\}$. Now C_{22} and the updated C_{33} differ in three elements; Lemma 1 applies.

Subcase 3. $C_{33} = \{1, 2, 3, 4, 8\}$. Hang up $\{1, 2\}$ in C_{22} and $\{3, 8\}$ in C_{33}. Add two new $(2, 3)$ calls which must use 5,8. Now On and the updated 022 differ in three elements; Lemma 1 applies.

Case (iii). $d_1 \leq 1$. Let $C_{11} = \{1, 2, 3, 4, 5\}$ and $C_{22} = \{1, 2, 3, 4, x\}$, where $x \in \{5, 6\}$. Then C_{33} is either $\{1, 2, 3, 4, y\}$ or $\{1, 2, 3, 5, 6\}$ if $x = 6$. Hang up $\{3, 4, 5\}$ from On and $\{1, 2, x\}$ from C_{22}. Add three new $(1, 2)$ calls which must use $\{5, 6, 7\}$, which differs from C_{33} in two elements. Hang up the calls in C_{11} and C_{22} and add two new $(1, 2)$ calls and five new $(2, 1)$ calls. Since $|C_{12} \setminus C_{33}| \geq 2$, we are back to either Case (i) or Case (ii).

3 Asymmetric 3-stage Clos networks

An asymmetric 3-stage Clos network will be denoted by $C(n_1, n_2, m, r_1, r_2)$ where the input stage has r_1 $n_1 \times m$ switches and the output stage has r_2 $m \times n_2$ switches. It is obvious that:

Lemma 3 *If $C(n_1, n_2, m, r_1, r_2)$ is blocking under an algorithm A, then $C(n_1', n_2', m, r_1', r_2')$ is also blocking under A for $n_1' \geq n_1$, $n_2' \geq n_2$, $r_1' \geq r_1$, $r_2' \geq r_2$.*

Lemma 1 can be easily modified to

Lemma 4 *Let $S_i(S_j')$ denote a set of s center switches each carrying a call from I_i (to O_j), but no call from I_i to O_j. Then $|S_i \cup S_j'| \leq x$ implies that at least $\max\{n_1, n_2\} + s - \lceil x/2 \rceil$ center switches are needed.*

Assume without loss of generality that $r_1 \leq r_2 \equiv r$. We study $r_1 = 2$ and then $r_1 = 3$.

Let S denote a set of C_i. Let S^* denote the closure of S under the operation of adding a call, i.e., C_i in $S^* \Rightarrow C_j \in S^*$ when C_j is obtained by adding a call to C_j. For example, let $S(a, b, c)$ denote the C_i of the three following patterns:

$$\begin{array}{lll} a: 001 & b: 000 & c: 000 \\ 000 & 100 & 010 \\ 000 & 010 & 100 \end{array}$$

Then $S^*(a, b, c)$ includes $S(a, b, c)$ and the following patterns

$$\begin{array}{llllll} d: 001 & e: 001 & f: 001 & g: 001 & h: 001 & i: 001 \\ 100 & 010 & 000 & 000 & 100 & 010 \\ 000 & 000 & 100 & 010 & 010 & 100 \end{array}$$

Consider $r_1 = 2$. Let a_j denote the pattern which carries a single call from I_1 to O_j, and let b denote the pattern which carries a single call from I_2 to O_1.

Lemma 5 $|S^*(a_2, \ldots, a_r, b)| \leq \min\{n_1, (r-1)n_2\}$ *under any P or MI for $r \geq 2$.*

Proof. Since $|S^*(a_2, \ldots, a_r, b)|$ is a closure set, we need only consider adding a new call z. If z is a $(2, 1)$ call, then $|a_j| = 0$ for $j = 2, \ldots, r$ by the P or MI rule. Hence

$$|S^*(a_2, \ldots, a_r, b)| = |S^*(b)| \leq \min\{n_1, n_2\}$$

since each pattern in $S^*(b)$ carries a $(2, 1)$ call.

If z is a $(1, j)$ call, then $|b| = 0$. Hence $|S^*(a_2, \ldots, a_r, b)| = |S^*(a_2, \ldots, a_r)| \leq \min\{n_1, (r-1)n_2\}$ since each pattern carries a call from I_1. Lemma 5 follows from the fact that $\min\{n_1, n_2\} \leq \min\{n_1, (r-1)n_2\}$.

Theorem 4 *$C(n_1, n_2, m, 2, r)$, $r \geq 2$, is SNB under P or MI if and only if*

$$\begin{array}{ll} m \geq rn_2 & \text{for } rn_2 \leq n_1, \\ m \geq \lfloor (n_1 + rn_2)/2 \rfloor & \text{for } (r-1)n_2 < n_1 \leq rn_2, \\ m \geq n_1 + \lfloor n_2/2 \rfloor & \text{for } n_2/2 < n_1 \leq (r-1)n_2, \\ m \geq 2n_1 & \text{for } n_1 \leq n_2/2. \end{array}$$

Proof. If $rn_2 \leq n_1$, consider rn_2 calls from the same input switch. Then rn_2 center switches are required. On the other hand, designating a center switch for each of the rn_2 outputs certainly results in nonblocking. The case $2n_1 \leq n_2$ can be similarly argued.

We now consider the other two cases. Let U denote the set of currently used center switches.

Sufficiency. Since hanging up a call cannot increase $|U|$, we need only consider the time at which a new request w occurs which cannot be routed through U. Without loss of generality, assume w is a $(1,1)$ call. Let N denote the set of center switches carrying a call from I_1, and let N' denote the set of center switches carrying a call to O_1. Then

$$\begin{aligned} |N| &\leq n_1 - l, \\ |N'| &\leq n_2 - l. \end{aligned}$$

By Lemma 5,

$$|S^*(a_2, \ldots, a_r, b)| \leq \min\{n_1, (r_1 - 1)n_2\}.$$

Adding up the above three inequalities and dividing by 2, we obtain

$$|U| \leq \lfloor (n_1 + n_2 + \min\{n_1, (r-1)n_2\})/2 \rfloor - 1.$$

Hence $m = \lfloor (n_1 + n_2 + \min\{n_1, (r-l)n_2\}/2 \rfloor$ provides an unused center switch to route w.

Necessity. The following three cases cover the relevant n_i ranges.

(i). $(r-1)n_2 < n_1 < rn_2$. Define $s = rn_2 - n_1$. Let M be the $2 \times r$ matrix

$$\begin{pmatrix} [1, n_2] & [n_2 + 1, 2n_2 - s] & [2n_2 - s + 1, 3n_2 - s] \ldots [(r-1)n_2 - s + 1, rn_2 - s] \\ & [1, s] & \end{pmatrix}$$

Note that

$$rn_2 - s = rn_2 - (rn_2 - n_1) = n_1.$$

Let $S_1 = S_2' = [1, s]$. Then $|S_1 \cap S_2'| = s$. By Lemma 4, at least

$$n_1 + s + \lceil s/2 \rceil = n_1 + \lfloor s/2 \rceil = \lfloor (n_1 + rn_2)/2 \rfloor$$

switches are needed.

(ii). $n_2 \leq n_1 \leq (r-1)n_2$. Let M be the $2 \times r$ matrix

$$\begin{pmatrix} [1, n_2] & & [n_2 + 1, n_2] \cdots [(r-1)n_2 + 1, n_1] \\ & [1, n_2] & \end{pmatrix}$$

Let $S_1 = S_2' = [1, n_2]$. By Lemma 4 at least $n_1 + \lfloor n_2/2 \rfloor$ switches are needed.

(iii). $n_2/2 < n_1 < n_2$. Let M be the 2×2 matrix

$$\begin{pmatrix} [1, n_1] & \\ [n_1 + 1, n_2] & [1, 2n_1 - n_2] \end{pmatrix}$$

Let $S_1 = S_2' = [1, 2n_1 - n_2]$. By Lemma 4, at least

$$n_2 \lfloor (2n_1 - n_2)/2 \rfloor = n_1 + \lfloor n_2/2 \rfloor$$

switches are needed.

Corollary 3 *$C(n, n, m, 2, r)$ is SNB under P or MI if and only if $m \geq \lfloor 3n/2 \rfloor$.*

For $r_1 = 3$, from Theorem 2 and 3 and Lemma 3, we have

Theorem 5 *$C(n, n, m, 3, r)$ is not SNB under P for $r \geq 3$ if $m < 2n - 1$. It is not WSNB for $r \geq 3$ if $m < \lfloor 7n/4 \rfloor$.*

4 Concluding remarks

While fundamental results have been established for the strictly nonblocking and the rearrangeable 3-stage Close networks, not much is known about WSNB. This paper has introduced new techniques to study WSNB. We showed that the packing algorithm does not help for $r \geq 3$. We also completely settled the asymmetric case when $\min\{r_1, r_2\} = 2$.

Since $C(n, 2n-1, r)$ is strictly nonblocking, the room to explore the benefit of WSNB is quite limited. However, we hope that the techniques developed here may also be useful for the multirate environment for which Chung and Ross [3] proved that $C(n, m, r)$ is not strictly nonblocking for any m if the bandwidth of a call has no known positive lower bound.

An earlier version of this paper was submitted to SIAM J. Disc. Math. in 1995. The paper was withdrawn from that journal in November, 2000 as no feedback had ever been received during the 5-year period. Some material has been deleted and the new terms: SNB under A, WSNB under A, are introduced in the current version but the text is otherwise intact.

References

[1] Benes, V. E., Mathematical Theory of Connecting Networks and Telephone Traffic, Academic Press, New York, 1965.

[2] Benes, V. E., Blocking in the NAIU network, AT&T Bell Laboratories Tech. Memo, 1985.

[3] Chung, S.-P and Ross, K. W., On nonblocking multirate interconnection networks, *SIAM J. Comput.* **20** (1991) 726–736.

[4] Clos, C., A study of nonblocking networks, *Bell Syst. Tech. J.* **32** (1953) 406–424.

[5] Duguid, A. M., Structural properties of switching networks, Brown Univ. Progress Report BTL-7, 1959.

[6] Slepian, D., Two theorems on a particular crossbar switching network, AT&T Bell Labs Tech. Memo, 1958.

[7] Smith, D. G., Lower bound on the size of a 3-stage wide-sense nonblocking network, *Elec. Lett.* **13** (1977) 215–216.

[8] Yang, Y. and Wang, W., Wide-sense nonblocking Clos networks under packing strategy, *IEEE Trans. Cmput.* **45** (1999) 265–254.

Switching Networks: Recent Advances
D.-Z. Du and H.Q. Ngo (Eds.) pp. 101-116

Study of Wide-sense Nonblocking Switching Networks from the Approach of Upper Ideals

Siu-Yu Kwok
Department of Information Engineering
The Chinese University of Hong Kong, Hong Kong
E-mail: tony9kwok@hongkong.com

Shuo-Yen Robert Li
Department of Information Engineering
The Chinese University of Hong Kong, Hong Kong
E-mail: bobli@ie.cuhk.edu.hk

Contents

Abstract

A switching network is said to be *wide-sense nonblocking* if there is a non-blocking algorithm for route selection. In 1965, Beneš [1] proved the 3-stage

Clos network $[n \times m,\ 2 \times 2,\ m \times n]$ to be nonblocking when $m \geq \lfloor \frac{3n}{2} \rfloor$. This identified a family of 3-stage networks that are wide-sense nonblocking but not strictly nonblocking. It also raised the question on the existence of any wide-sense nonblocking network $[n \times m,\ r \times r,\ m \times n]$, $r > 2$, that is not strictly nonblocking. We answer this question affirmatively with an algorithm over the network $[6 \times 10,\ 3 \times 3,\ 10 \times 6]$. We also prove that, if a certain packing algorithm over $[n \times m,\ 3 \times 3,\ m \times n]$ is nonblocking, then $m \geq \lfloor \frac{15n}{8} \rfloor$. Consequently a wide-sense nonblocking network does not necessarily admit a packing algorithm.

1 Introduction

Let $[n \times m, r \times r, m \times n]$ denote a 3-stage Clos network where the first-, second- and third-stage nodes are, respectively, $n \times m$, $r \times r$ and $m \times n$. The network is said to be *wide-sense nonblocking* (WSN) if there is a nonblocking algorithm for route selection. Denote by $WS(n,\ r)$ the minimum value of m such that $[n \times m,\ r \times r,\ m \times n]$ is WSN. An algorithm is called a *packing* algorithm if the selected route is always through one of the heaviest loaded second-stage node. Figure 1 illustrates the network $[5 \times 8,\ 3 \times 3,\ 8 \times 5]$.

For $[n \times m,\ 2 \times 2,\ m \times n]$, there is a unique packing algorithm, which is obviously optimal in the sense of requiring the minimnm number of second-stage nodes. In 1965, this packing algorithm was proved by Beneš [1] to be nonblocking when $m \geq \lfloor 3n/2 \rfloor$. Consequently, $WS(n,\ 2) \leq \lfloor 3n/2 \rfloor$. It is not hard to see that $WS(n,\ 2) \geq \lfloor 3n/2 \rfloor$ and hence $WS(n,\ 2) = \lfloor 3n/2 \rfloor$. This result identified a family of 3-stage WSN networks that are not strictly nonblocking. It also raised the question on the existence of any WSN, but not strictly nonblocking, network in the form of $[n \times m,\ r \times r,\ m \times n]$, where $r > 2$. In Section 3, we shall answer this question affirmatively with a nonblocking algorithm over the network $[6 \times 10,\ 3 \times 3,\ 10 \times 6]$. The algorithm relies on the concept of *upper ideals* to be described in Section 2.

Other known facts on values of $WS(n,\ r)$ are as follows. A strictly nonblocking network is automatically WSN, hence $WS(n,\ r) \leq 2n - 1$. E. F. Moore proved that $WS(n,\ r) = 2n - 1$ when $r > (n-1)\binom{2n-2}{n-1}$. This result was recorded in Kurshan-Benes [7]. Du-Fishburn-Gao-Hwang [3] proved that $WS(n\ ,r) \geq \lfloor 7n/4 \rfloor$ when $r \geq 3$. Li [9] proved that

- $WS(n,\ r) \geq \lfloor (7n+1)/4 \rfloor$ when $r \geq 4$ and
- $WS(n,\ r) \geq \lfloor (7n+2)/4 \rfloor$ when $n \geq 3$ and $r > \rho(\lfloor n/4 \rfloor)$, where the

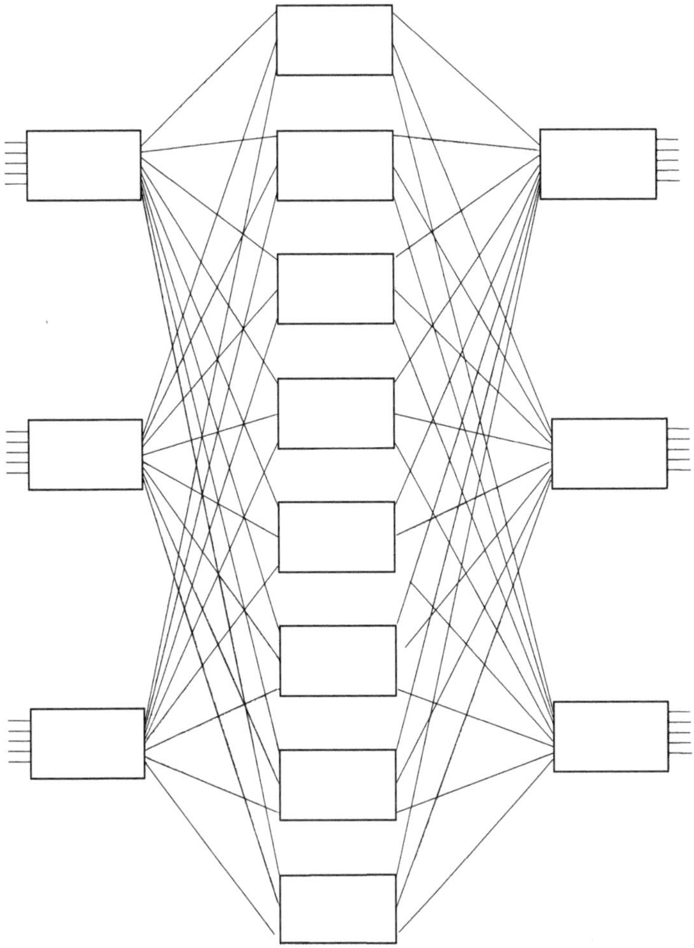

Figure 1: The 3-stage Clos network denoted by [5×8, 3×3, 8×5]

function $\rho(s)$ means the maximum number of $(4s+2)$-element subsets in

a $(7s+3)$-element set such that the pairwise intersection between any two of these subsets contains exactly $2s+1$ elements. For example, since $\rho(1) = 4$ and $\rho(2) = 6$. Therefore, $WS(6,\ r) = 11$ for $n \geq 5$ and $WS(10,\ r) \geq 18$ for $r \geq 7$.

From all these results, Table 1 summarizes the values of $WS(n,\ r)$ for $n \leq 10$. The smallest undetermined value in Table 1 is $WS(5,\ 3)$. In Section 4, we shall prove that the network [5×8, 3×3, 8×5] is not WSN and hence $WS(5,\ 3) = 9$. The proof again relies on the concept of upper ideals.

<table>
<tr><th>n \ r</th><th>2</th><th>3</th><th>4</th><th>5-6</th><th>7-5544</th><th>5545-24024</th><th>24025-102960</th><th>102961-437580</th><th>437581+</th></tr>
<tr><td>2</td><td colspan="9">3</td></tr>
<tr><td>3</td><td>4</td><td colspan="8">5</td></tr>
<tr><td>4</td><td>6</td><td colspan="8">7</td></tr>
<tr><td>5</td><td>7</td><td>8 or 9</td><td colspan="7">9</td></tr>
<tr><td>6</td><td>9</td><td colspan="2">10 or 11</td><td colspan="6">11</td></tr>
<tr><td>7</td><td>10</td><td colspan="4">12 or 13</td><td colspan="4">13</td></tr>
<tr><td>8</td><td>12</td><td colspan="5">14 or 15</td><td colspan="3">15</td></tr>
<tr><td>9</td><td>13</td><td>15-17</td><td colspan="5">16 or 17</td><td colspan="2">17</td></tr>
<tr><td>10</td><td>15</td><td colspan="3">17-19</td><td colspan="4">18 or 19</td><td>19</td></tr>
</table>

Table 1: Known values and bounds of $WS(n,\ r)$ for $n \leq 10$.

Define a network as *strictly nonblocking by packing* if every packing algorithm is nonblocking over the network. Similarly, a network is *WSN by packing* if a certain packing algorithm is nonblocking. Smith [10] proved that the number of second-stage nodes required for a WSN network cannot be less than $\lfloor 2n - n/r \rfloor$ under a large class of packing rules. Du-Fishburn-Gao-Hwang [3] proved that $[n \times m,\ r \times r,\ m \times n]$ is WSN by packing only if $m \geq \lfloor \left(2 - \frac{1}{2^{r-1}}\right) n \rfloor$. Yang [12] sharpened this necessary condition to $m \geq \lfloor \left(2 - \frac{1}{F_{2r-1}}\right) n \rfloor$, where F_{2r-1} is the Fibonacci number. It should be mentioned that Tsai and Wang [11] proved that for r sufficiently large, $[n \times m,\ r \times r,\ m \times n]$ is WSN iff $m \geq 2n - 1$.

Figure 2 shows the obvious relationship among various nonblocking properties of networks. For $[n \times m,\ 2 \times 2,\ m \times n]$ in particular, the two horizontal implications in the figure are reversible since packing obviously minimizes the required number of second-stage nodes. Thus the vertical implication in the figure is not reversible by the aforementioned result of Beneš [1].

A natural question is then: for $[n \times m,\ r \times r,\ m \times n]$ in general, are the two horizontal implications in Figure 2 still reversible? Following the proof of Lemma 2.1.6 and Theorem 2.1.7 in Hwang [5], one can establish the necessity of the inequality $m \geq 2n - 1$ for $[n \times m,\ r \times r,\ m \times n]$ to be strictly nonblocking by packing. Thus the upper horizontal implication in Figure 2 remains reversible for a general r. In Section 5, we shall prove that, if $[n \times m,\ 3 \times 3,\ m \times n]$ is WSN by packing, then $m \geq \lfloor \frac{15n}{8} \rfloor$. Therefore, the WSN network $[6 \times 10,\ 3 \times 3,\ 10 \times 6]$ is not WSN by packing. Figure 3 summarizes these conclusions.

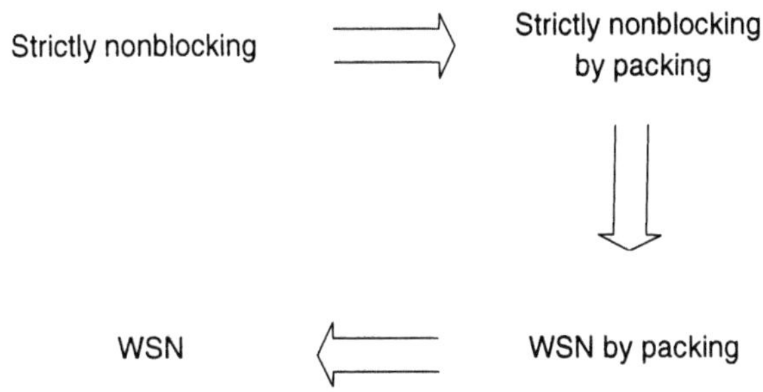

Figure 2: Relationship among various nonblocking properties of networks.

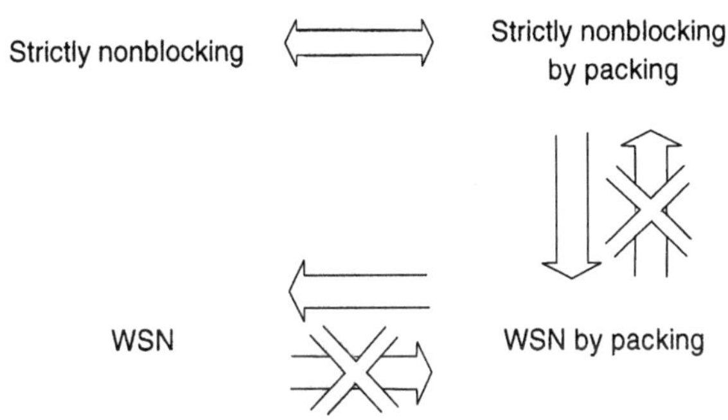

Figure 3: Relationship among various nonblocking properties of the network $[n\times m,\ r\times r,\ m\times n]$.

2 The Concept of the Upper Ideals

The focus of the present paper is on the network $[n\times m,\ 3\times 3,\ m\times n]$. Let the three first-stage nodes be labeled as a, b and c and the three third-stage nodes as x, y and z. As far as a routing algorithm over the 3-stage network is concerned, a connected route can be identified by the node it traverses through at every stage and a connection request can be identified by the first- and third-stage nodes. Label inputs and outputs of a middle-stage node as a, b, c, x, y, and z according to the node each I/O is linked to. This convention is shown in Figure 4. Hereafter a *node* will refer to a middle-stage node unless otherwise specified.

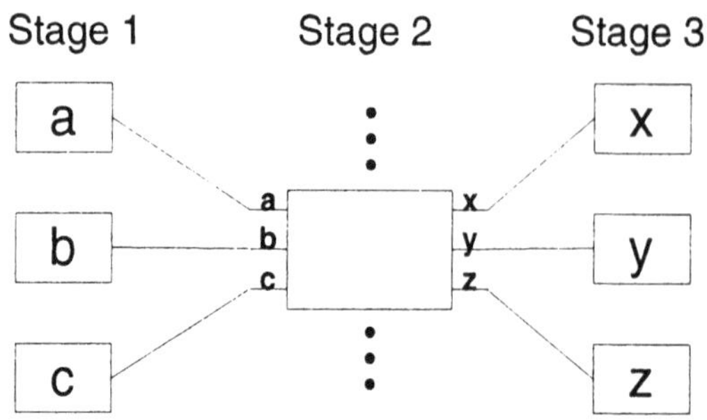

Figure 4: I/O labels of a middle-stage node.

The *state* of a node is the set of connections it is carrying. For example, the set $\{ax, bz\}$ represents the state of carrying two connections, one from a to x and the other from b to z. For the sake of simplicity, the notation for this state will be simplified as $axbz$. The empty state is denoted as 0. Altogether there are 34 possible states displayed in Figure 5. Two states are connected by a line in the figure when one can be obtained by adding a new route to the other.

The set-theoretical containment endows a natural order among the 34 states. The state 0 is regarded as the *smallest*. The *network state* of $[n\times m, r\times r, m\times n]$ is determined by states of the m individual nodes. Following the terminology of lattice theory, we have the definition below.

Definition 1 *With respect to a network state of* $[n\times m,\ 3\times 3,\ m\times n]$, *a set* S *of nodes is called an* **upper ideal** *if, whenever* S *contains a node in the node state* θ, *then* S *contains every node in every node state greater than or equal to* θ. *The upper ideal* generated *by a collection of node states means the smallest upper ideal that includes all nodes in these states.*

Notation *We shall use the notation* $< ... >$ *for upper ideal generation. Thus, for example,* $<axby>$ *represents the set of nodes in the two states* $axby$ *and* $axbycz$, *while* $<ax,\ by>$ *represents the set of all nodes in the twelve states* ax, by, $axby$, $axbz$, $axcy$, $axcz$, $azby$, $bycx$, $bycz$, $axbycz$, $axbzcy$, *and* $azbycx$. *The cardinality of a set* S *of nodes will be denoted as* $|S|$. *For every node state* θ, *let* S_θ *denote the set of nodes in the state* θ.

Definition 2 *Define the upper ideals.*
$H_a = <ax,\ ay,\ az>$, $H_x = <ax,\ bx,\ cx>$,

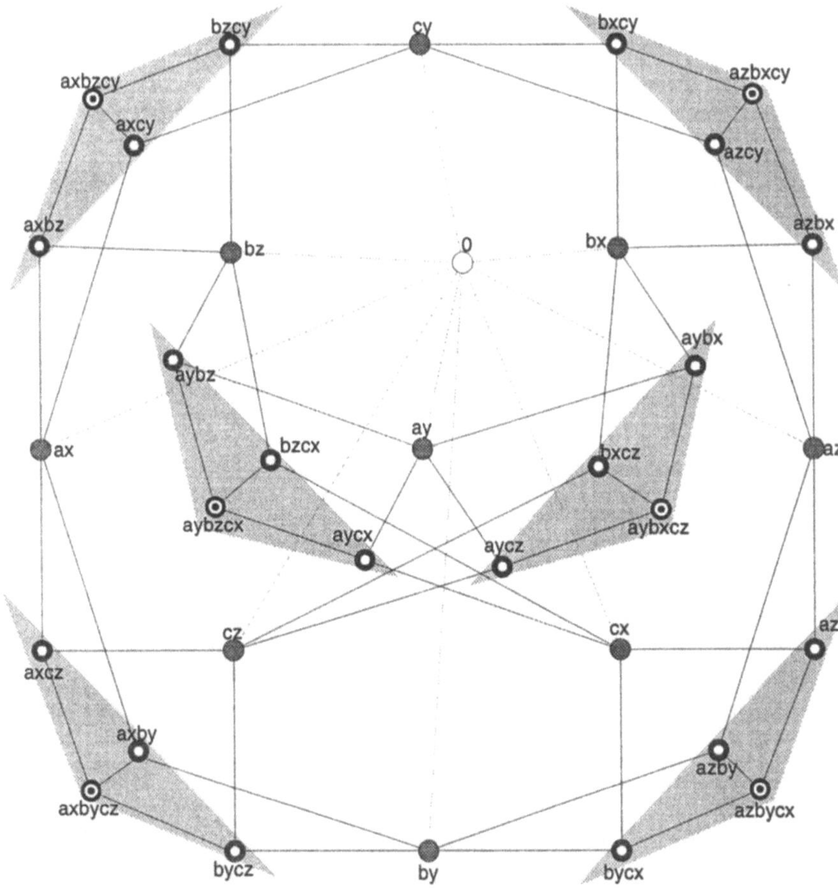

Figure 5: Chessboard for $r = 3$

$I_{ax} =<ay,\ az,\ bx,\ cx>$,
$J_{ax} =<ax,\ bycz,\ bzcy>$, *and*
$K_{ax} =<bycz,\ bzcy>$.
For $s \in \{a,\ b,\ c\}$ *and* $t \in \{x,\ y,\ z\}$, *the upper ideals* H_s, H_t, I_{st}, J_{st}, *and* K_{st} *are defined symmetrically.*

Lemma 2.1 *A network state such that* $|H_a| < n$, $|H_x| < n$, *and* $|H_a \cup H_x| = m$ *leads to blocking regardless of the routing algorithm.*

Proof The two inequalities ensure that the first-stage nodes a and the third-stage x are not saturated. Thus a connection from a to x could be requested. The assumption of $|H_a \cup H_x| = m$ blocks this request. □

Theorem 2.2 *For the network $[n{\times}m,\ 3{\times}3,\ m{\times}n]$ with $m < 2n-1$, a network state such that $|I_{ax}| = m$ leads to blocking regardless of the routing algorithm.*

Proof Delete all routes except those traversing through the first-stage node a or the third-stage node x. Thus every node becomes in the state ay, az, bx, or cx. Let $|S_{ay}|+|S_{az}| = i$ and $|S_{bx}|+|S_{cx}| = m-i$. Thus $n \geq i \geq m-n$.

Case 1: $i < n$ and $m-i < n$.
By Lemma 2.1, the network state leads to blocking.

Case 2: $i = n$. Thus $m-i = m-n < n-1$.
Add a connection from b to x. Then, $|S_{ay}| + |S_{az}| = n-1$, $|S_{aybx}| + |S_{azbx}| = 1$ and $|S_{bx}| + |S_{cx}| = m-n$.
Delete the old route in the node in the state $aybx$ or $azbx$. Then $|S_{ay}| + |S_{az}| = n-1$ and $|S_{bx}| + |S_{cx}| = m-n+1$.
The network state is as in Case 1.

Case 3 : $m-i = n$. The case is symmetric to Case 2. □

The theorem offers a guideline in routing: Avoid the situation when $|I_{ax}| = m$ or, more generally, when $|I_{st}| = m$ for any s and any t.

Theorem 2.3 *For the network $[n{\times}m,\ 3{\times}3,\ m{\times}n]$ with $m = \lfloor \frac{7n}{4} \rfloor$, a network state such that $|J_{ax}| \geq 2\lfloor \frac{3n}{4} \rfloor$ and $\lfloor \frac{3n}{4} \rfloor \leq |K_{ax}| \leq n$ leads to blocking regardless of the routing algorithm.*

Proof Delete some old routes until every node becomes in the state ax, $bycz$ and $bzcy$. Let $|S_{ax}| = i$. Then, $|S_{bycz}| + |S_{bzcy}| = 2\lfloor \frac{3n}{4} \rfloor - i$. Thus $2\lfloor \frac{3n}{4} \rfloor - n \leq i \leq \lfloor \frac{3n}{4} \rfloor$.

Case 1: $i = \lfloor \frac{3n}{4} \rfloor$.
Add $\lceil \frac{n}{4} \rceil$ connections from the first-stage node a to the third-stage node z. Then $|S_{ax}| = \lfloor \frac{3n}{4} \rfloor$, $|S_{bycz}| + |S_{bzcy}| = \lfloor \frac{3n}{4} \rfloor$ and $|S_{az}| = \lceil \frac{n}{4} \rceil$. Thus $|I_{ay}| = |S_{ax}| + |S_{az}| + |S_{bycz}| + |S_{bzcy}| = \lfloor \frac{3n}{4} \rfloor + \lceil \frac{n}{4} \rceil + \lfloor \frac{3n}{4} \rfloor = \lfloor \frac{3n}{4} \rfloor + n = \lfloor \frac{7n}{4} \rfloor$. By Theorem 2.2, the network state leads to blocking.

Case 2: $2\lfloor\frac{3n}{4}\rfloor - n \leq i < \lfloor\frac{3n}{4}\rfloor$.

Add $(\lfloor\frac{3n}{4}\rfloor - i)$ connections from a to x. Then $|S_{ax}| = i+k$, $|S_{axbycz}| + |S_{axbzcy}| = \lfloor\frac{3n}{4}\rfloor - i - k$ and $|S_{bycz}| + |S_{bzcy}| = \lfloor\frac{3n}{4}\rfloor + k$ where $0 \leq k \leq \lfloor\frac{3n}{4}\rfloor - i$.

Delete the two old routes in every node in the state $axbycz$ or $axbzcy$. Then $|S_{ax}| = \lfloor\frac{3n}{4}\rfloor$ and $|S_{bycz}| + |S_{bzcy}| = \lfloor\frac{3n}{4}\rfloor$ +k.

Clear routes in k of the nodes in the state $bycz$ or $bzcy$. The situation becomes as in Case 1. □

Theorem 2.4 *For the network $[n\times m,\ 3\times 3,\ m\times n]$ with $m = \lfloor\frac{7n}{4}\rfloor$, denote by NS_1, a network state with $|S_{ax}| = n$ and $|S_{bycz}| + |S_{bzcy}| = \lfloor\frac{3n}{4}\rfloor - \lceil\frac{n}{4}\rceil$. Then, a network state such that $|J_{ax}| \geq 2\lfloor\frac{3n}{4}\rfloor$ leads to either blocking or NS_1 regardless of the routing algorithm.*

Proof

Delete all routes until every node state becomes ax, $bycz$ and $bzcy$. Let $|S_{ax}| = i$, $|S_{bycz}| + |S_{bzcy}| = 2\lfloor\frac{3n}{4}\rfloor - i$. Thus $2\lfloor\frac{3n}{4}\rfloor - n \leq i \leq n$.

Case 1: $2\lfloor\frac{3n}{4}\rfloor - n \leq i \leq \lfloor\frac{3n}{4}\rfloor$

By Theorem 2.3, the network state leads to blocking state.

Case 2: $\lfloor\frac{3n}{4}\rfloor < i \leq n$

Add $(n-i)$ connections from the first-stage node a to the third-stage node x. Then $|S_{ax}| = i + k$, $|S_{axbycz}| + |S_{axbzcy}| = n - i - k$ and $|S_{bycz}| + |S_{bzcy}| = 2\lfloor\frac{3n}{4}\rfloor - n + k$. Thus $0 \leq k \leq n - i$.

Delete the two old routes in every node in the state $axbycz$ or $axbzcy$. Then the network state becomes $|S_{ax}| = n$, $|S_{bycz}| + |S_{bzcy}| = 2\lfloor\frac{3n}{4}\rfloor - n + k$.

Clear routes in k of the nodes in the state $bycz$ or $bzcy$. The network state becomes in the state NS_1. □

Theorems 2.3 and 2.4, as well as Theorem 2.2, offer guidelines in routing.

3 Routing Algorithm Over The Network [6×10, 3×3, 10×6]

Algorithm 1 *Let a network state of the network $[6\times 10,\ 3\times 3,\ 10\times 6]$ be called* ***admissible*** *if it satisfies the following two conditions*
(1) $|I_{st}| \geq 9$.
(2) Either $|J_{st}| \leq 7$ *or* $|K_{st}| \leq 3$.
for all s and t. Moreover, let an ***admissible*** *state be called* ***preferred*** *when*

$|I_{st}| \leq 9$ and $|J_{st}| \leq 7$ for all s and t. Given a network state of the network, an input node s_0 that is carrying less than six routes, and an output node t_0 that is carrying less than six routes, the algorithm needs to choose a middle-stage node for adding a connection from s_0 to t_0 such that the resulting state of the network is ***admissible****. The criteria of the selection of the middle-stage node are as follows.*

- *First, it is preferred that the resulting network state of the network is a* ***preferred*** *one. Among such choices, priority is given to a middle-stage node that has been carrying 1, 0, or 2 routes, by that order. The tiebreaker is the label of the middle-stage node in the increasing order.*
- *When no choice leads to a* ***preferred*** *state, then one leading to an* ***admissible*** *state is adopted. Priority is given to a middle-stage node already carrying 0, 1, or 2 routes, by that order. The tiebreaker is again the label of the middle-stage node.* □

Theorem 3.1 *The network [6×10, 3×3, 10×6] is WSN. In fact, Algorithm 1 is nonblocking over the network.*

Proof Firstly, the initial state of the network is the empty state. The evolution of the state is through the addition of a new route to a middle node or the deletion of an existent route at a time. In the case of adding a new route, the number N of possible new states may be 0, 1 or more. When $N = 0$, blocking occurs. When $N > 1$, the new state depends upon the algorithm of route selection. Exhaustive computer search identifies a set of **admissible** states that include the empty state and is "closed" under arbitrary route deletion and under route addition by algorithm 1. The theorem is thus proved constructively. □

4 Nonexistence of routing algorithm over the network [5×8, 3×3, 8×5]

The networks $[n\times\lfloor \frac{7n}{4} \rfloor, 3\times3, \lfloor \frac{7n}{4} \rfloor\times n]$ are strict-sense nonblocking for all $n \leq 4$ because $\lfloor\frac{7n}{4}\rfloor = 2n-1$ if $n \leq 4$. Therefore, the first smallest possible WSN network may be the network [5×8, 3×3, 8×5] which is shown in figure 1.

Notation *Label the nodes by 1, 2, ..., m. For $j \leq k$, let $[j, k]$ denote the set of nodes j, $j+1$, ..., k.*

Theorem 4.1 *For the network* $[5 \times 8, 3 \times 3, 8 \times 5]$, *the network state* NS_1 *leads to blocking regardless of the routing algorithm.*

Proof Let $Th2.3(J_{ax})$ represents that the network state leads to blocking by Theorem 2.3 with $|S_{ax}| = \lfloor \frac{3n}{4} \rfloor$ and $|S_{bycz}| + |S_{bzcy}| = \lfloor \frac{3n}{4} \rfloor$.

Let $Th2.2(I_{ax})$ represents that the network state leads to blocking by Theorem 2.2 with $|S_{ay}| + |S_{az}| + |S_{bx}| + |S_{cx}| = m$.

Let the notation "Add one ax" represents "Add one connection from frist-stage node a to the third-stage node x" and so on.

State $T1$: $S_{ay} =[1, 5]$, $S_{bzcx} =[6, 6]$, $S_0 =[7, 8]$.
Add four bz: the state will become $T11$, $T12$ or $T13$ as specified below.
$T11$: $S_{aybz} =[1, 4]$, $S_{ay} =[5, 5]$, $S_{bzcx} =[6, 6]$, $S_0 =[7, 8]$.
Add two cx: the new state become $T111$, $T112$, $T113$, $T114$ or $T115$.
$T111$: $S_{aybzcx} =[1, 2]$, $S_{aybz} =[3, 4]$, $S_{ay} =[5, 5]$, $S_{bzcx} =[6, 6]$, $S_0 =[7, 8]$.
Delete every node beomes in the state ay, $bxcz$ or $bzcx$. Then $Th2.3(J_{ay})$.
$T112$: $S_{aybzcx} =[1, 1]$, $S_{aybz} =[2, 4]$, $S_{aycx} =[5, 5]$, $S_{bzcx} =[6, 6]$, $S_0 =[7, 8]$.
Delete every node beomes in the state cx, $aybz$ or $azby$. Then $Th2.3(J_{cx})$.
$T113$: $S_{aybzcx} =[1, 1]$, $S_{aybz} =[2, 4]$, $S_{ay} =[5, 5]$, $S_{bzcx} =[6, 6]$, $S_{cx} =[7, 7]$, $S_0 =[8, 8]$; ditto
$T114$: $S_{aybz} =[1, 4]$, $S_{aycx} =[5, 5]$, $S_{bzcx} =[6, 6]$, $S_{cx} =[7, 7]$, $S_0 =[8, 8]$; ditto.
$T115$: $S_{aybz} =[1, 4]$, $S_{ay} =[5, 5]$, $S_{bzcx} =[6, 6]$, $S_{cx} =[7, 8]$; ditto.
$T12$: $S_{aybz} =[1, 3]$, $S_{ay} =[4, 5]$, $S_{bzcx} =[6, 6]$, $S_{bz} =[7, 7]$, $S_0 =[8, 8]$.
Add three cx: there are two cases.
Case $T121$: middle-stage nodes 1, 2, 3 carry at least two new cx;
$T121$: $S_{aybzcx} =[1, 2]$, $S_{aybz} =[3, 3]$, $S_{ay} =[4, 5]$, $S_{bzcx} =[6, 6]$, $S_{bzcx} =[7, 7]$, $S_0 =[8, 8]$; Delete every node beomes in the state ay, $bxcz$ or $bzcx$. Then $Th2.3(J_{ay})$.
Case $T122$: middle nodes 4 through 8 carry at least two new cx;
The case will Th $2.3(J_{cx})$ because node 1 to 3 carry three "$aybz$" and node 4 to 8 carry at least three "cx".
$T13$: $S_{aybz} =[1, 2]$, $S_{ay} =[3, 5]$, $S_{bzcx} =[6, 6]$, $S_{bz} =[7, 8]$; $Th2.2(I_{az})$.

All the possible routes are considered beginning from the state NS_1. Therefore, the state NS_1 leads to blocking over the network [5×8, 3×3, 8×5] regardless of the routing algorithms. □

Theorem 4.2 *The network [5×8, 3×3, 8×5] is not WSN.*

Proof Let the notation "5-I-P $|I_{ax}|$" represents that $|I_{ax}| = m = 8$. Then, by Theorem 2.2, the network state leads to blocking if the state is 5-I-P $|I_{st}|$ for some s and some t.

Let the notation "5-J-P $|J_{ax}|$" represents that $|J_{ax}| \geq 2\lfloor \frac{3n}{4} \rfloor$. By Theorem 2.4, the network state leads to blocking if the state is 5-J-P $|J_{st}|$ for some s and some t.

By the 5-I-P and 5-J-P, it suffices to force an empty state leading to blocking.
Add five az: the new state becomes $\mathcal{S}_{az}$ =[1, 5], $\mathcal{S}_0$ =[6,8].
Add five by: the new state becomes T1, T2, T3, or T4 as specified below.
T1: $\mathcal{S}_{azby}$ =[1, 5], $\mathcal{S}_0$ =[6, 8].
  Add five cx: the new state becomes T11, T12, T13 and T14.
    T11: $\mathcal{S}_{azbycx}$ =[1, 5], $\mathcal{S}_0$ =[6, 8].
      Delete routes: $\mathcal{S}_{azby}$ =[1, 3], $\mathcal{S}_{azcx}$ =[4, 5], $\mathcal{S}_0$ =[6, 8].
      Add two bx: $\mathcal{S}_{azby}$ =[1, 3], $\mathcal{S}_{azcx}$ =[4, 5], $\mathcal{S}_{bx}$ =[6, 7], $\mathcal{S}_0$ =[8, 8].
      Add two cy: the new state becomes T111, and T112.
        T111: $\mathcal{S}_{azby}$ =[1, 3], $\mathcal{S}_{azcx}$ =[4, 5], $\mathcal{S}_{bxcy}$ =[6, 7], $\mathcal{S}_0$ =[8, 8]: 0; 5-J-P $|J_{az}|$.
        T112: $\mathcal{S}_{azby}$ =[1, 3], $\mathcal{S}_{azcx}$ =[4, 5], $\mathcal{S}_{bxcy}$ =[6, 6], $\mathcal{S}_{bx}$ =[7, 7], $\mathcal{S}_{cy}$ =[8, 8]; 5-J-P $|J_{az}|$.
    T12: $\mathcal{S}_{azbycx}$ =[1, 4], $\mathcal{S}_{azby}$ =[5, 5], $\mathcal{S}_{cx}$ =[6, 6], $\mathcal{S}_0$ =[7, 8]; 5-J-P $|J_{cx}|$.
    T13: $\mathcal{S}_{azbycx}$ =[1, 3], $\mathcal{S}_{azby}$ =[4, 5], $\mathcal{S}_{cx}$ =[6, 7], $\mathcal{S}_0$ =[8, 8]; 5-J-P $|J_{cx}|$.
    T14: $\mathcal{S}_{azbycx}$ =[1, 2], $\mathcal{S}_{azby}$ =[3, 5], $\mathcal{S}_{cx}$ =[6, 8]; 5-I-P $|I_{bx}|$.
  T2: $\mathcal{S}_{azby}$ =[1, 4], $\mathcal{S}_{az}$ =[5, 5], $\mathcal{S}_{by}$ =[6, 6], $\mathcal{S}_0$ =[7, 8].
    Add three cx: consider four cases T21, T22, T23, T24 and T25.
    T21: the state of node 5 (az) carries a new connection from c to x; 5-J-P $|J_{by}|$ case.
    T22: the state of node 6 (by) carries a new "cx" route; 5-J-P $|J_{az}|$ case.
    T23: both middle-stage noes 7 and 8 carry new connections; 5-J-P $|J_{cx}|$ case.
    T24: $\mathcal{S}_{azbycx}$ =[1, 2], $\mathcal{S}_{azby}$ =[3, 4], $\mathcal{S}_{az}$ =[5, 5], $\mathcal{S}_{by}$ =[6, 6], $\mathcal{S}_{cx}$ =[7, 7], $\mathcal{S}_0$ =[8, 8].
      Delete routes: $\mathcal{S}_{azcx}$ =[1, 2], $\mathcal{S}_{azby}$ =[3, 4], $\mathcal{S}_{az}$ =[5, 5], $\mathcal{S}_{by}$ =[6, 6], $\mathcal{S}_{cx}$ =[7, 7], $\mathcal{S}_0$ =[8, 8].

Add two bx: $\mathcal{S}_{azcx}$ =[1, 2], $\mathcal{S}_{azby}$ =[3, 4], $\mathcal{S}_{azbx}$ =[5, 5], $\mathcal{S}_{by}$ =[6, 6],
$\mathcal{S}_{cx}$ =[7, 7], $\mathcal{S}_{bx}$ =[8, 8].
Add two cy: $\mathcal{S}_{azcx}$ =[1, 2], $\mathcal{S}_{azby}$ =[3, 4], $\mathcal{S}_{azbxcy}$ =[5, 5], $\mathcal{S}_{by}$ =[6, 6],
$\mathcal{S}_{cx}$ =[7, 7], $\mathcal{S}_{bxcy}$ =[8, 8]; 5-J-P $|J_{az}|$.

T25: $\mathcal{S}_{azbycx}$ =[1, 3], $\mathcal{S}_{azby}$ =[4, 4], $\mathcal{S}_{ax}$ =[5, 5], $\mathcal{S}_{by}$ =[6, 6], $\mathcal{S}_0$ =[7, 8].
Add one more cx: the new state becomes T251 or T252.
T251: $\mathcal{S}_{azbycx}$ =[1, 4], $\mathcal{S}_{az}$ =[5, 5], $\mathcal{S}_{by}$ =[6, 6], $\mathcal{S}_0$ =[7, 8]: 0.
Add one more cx. We cannot added the cx into node 5 and node 6 because of $T21$ and $T22$.
the state: $\mathcal{S}_{azbycx}$ =[1, 4], $\mathcal{S}_{az}$ =[5, 5], $\mathcal{S}_{by}$ =[6, 6], $\mathcal{S}_{cx}$ =[7, 7],
$\mathcal{S}_0$ =[8, 8].
Delete routes: $\mathcal{S}_{azcx}$ =[1, 2], $\mathcal{S}_{azby}$ =[3,4], $\mathcal{S}_{az}$ =[5, 5], $\mathcal{S}_{by}$ =[6, 6],
$\mathcal{S}_{cx}$ =[7, 7], $\mathcal{S}_0$ =[8, 8].
Add two bx: $\mathcal{S}_{azcx}$ =[1, 2], $\mathcal{S}_{azby}$ =[3,4], $\mathcal{S}_{azbx}$ =[5, 5], $\mathcal{S}_{by}$ =[6, 6],
$\mathcal{S}_{cx}$ =[7, 7], $\mathcal{S}_{bx}$ =[8, 8].
Add two cy: $\mathcal{S}_{azcx}$ =[1, 2], $\mathcal{S}_{azby}$ =[3,4], $\mathcal{S}_{azbxcy}$ =[5, 5], $\mathcal{S}_{by}$ =[6, 6],
$\mathcal{S}_{cx}$ =[7, 7], $\mathcal{S}_{bxcy}$ =[8, 8]; 5-J-P $|J_{az}|$.
T252: $\mathcal{S}_{azbycx}$ =[1, 3], $\mathcal{S}_{azby}$ =[4, 4], $\mathcal{S}_{az}$ =[5, 5], $\mathcal{S}_{by}$ =[6, 6],
$\mathcal{S}_{cx}$ =[7, 7], $\mathcal{S}_0$ =[8, 8].
Add one cx: $\mathcal{S}_{azbycx}$ =[1, 3], $\mathcal{S}_{azby}$ =[4, 4], $\mathcal{S}_{az}$ =[5, 5]: az,
$\mathcal{S}_{by}$ =[6, 6], $\mathcal{S}_{cx}$ =[7, 7], $\mathcal{S}_{cx}$ =[8, 8]; 5-J-P $|J_{cx}|$.

T3: $\mathcal{S}_{azby}$ =[1, 3], $\mathcal{S}_{az}$ =[4, 5], $\mathcal{S}_{by}$ =[6, 7], $\mathcal{S}_0$ =[8, 8].
Add five cx: consider T31, T32 and T33.
T31: middle node 4 or 5 carries a new route; 5-J-P $|J_{by}|$.
T32: middle node 6 or 7 carries a new route; 5-J-P $|J_{az}|$.
T33: after eliminating the states T31 and T32, we can only add four cx. Then the network state leads to blocking.

T4: $\mathcal{S}_{azby}$ =[1, 2], $\mathcal{S}_{az}$ =[3, 5], $\mathcal{S}_{by}$ =[6, 8]; 5-I-P $|I_{bz}|$ case. □

In view of Table 1, we conclude that $WS(5, 3)$=9.

5 Packing algorithms

By the following Theorem 5.2, we prove if the network $[n\times m,\ 3\times 3,\ m\times n]$ is WSN by packing, then $m \geq \lfloor \frac{15n}{8} \rfloor$.

Lemma 5.1 $\lfloor \frac{7n}{8} \rfloor \geq \lfloor \frac{3n}{2} \rfloor - \lfloor \frac{3n}{4} \rfloor$ *for* $n \geq 3$

Theorem 5.2 *If the network $[n\times m,\ 3\times 3,\ m\times n]$ is nonblocking under some packing algorithms, then $m \geq \lfloor \frac{15n}{8} \rfloor$.*

<u>Proof</u> Firstly, we prove that the theorem is true for $n \geq 3$. The proof starts from an empty state.

Add n ax into the empty state: $\mathcal{S}_{ax} = [1,\ n]$.

Add n by: $\mathcal{S}_{axby} = [1,\ n]$.

Add n cz: $\mathcal{S}_{axbycz} = [1,\ n]$.

Delete routes : $\mathcal{S}_{axby} = [1,\ \lfloor \frac{n}{2} \rfloor]$, $\mathcal{S}_{axcz} = [\lfloor \frac{n}{2} \rfloor + 1,\ n]$.

Add $\lfloor \frac{n}{2} \rfloor$ bz: $\mathcal{S}_{axby} = [1,\ \lfloor \frac{n}{2} \rfloor]$, $\mathcal{S}_{axcz} = [\lfloor \frac{n}{2} \rfloor + 1,\ n]$, $\mathcal{S}_{bz} = [n+1,\ \lfloor \frac{3n}{2} \rfloor]$.

Add $\lfloor \frac{n}{2} \rfloor$ cy: $\mathcal{S}_{axby} = [1,\ \lfloor \frac{n}{2} \rfloor]$, $\mathcal{S}_{axcz} = [\lfloor \frac{n}{2} \rfloor + 1,\ n]$, $\mathcal{S}_{bzcy} = [n+1,\ \lfloor \frac{3n}{2} \rfloor]$.

Delete routes: $\mathcal{S}_{ax} = [1,\ n]$, $\mathcal{S}_{bzcy} = [n+1,\ \lfloor \frac{3n}{2} \rfloor]$.

Add $\lceil \frac{n}{4} \rceil$ bz: $\mathcal{S}_{ax} = [1,\ \lfloor \frac{3n}{4} \rfloor]$, $\mathcal{S}_{axbz} = [\lfloor \frac{3n}{4} \rfloor + 1,\ n]$, $\mathcal{S}_{bzcy} = [n+1,\ \lfloor \frac{3n}{2} \rfloor]$.

Add $\lceil \frac{n}{4} \rceil$ cy: $\mathcal{S}_{ax} = [1,\ \lfloor \frac{3n}{4} \rfloor]$, $\mathcal{S}_{axbzcy} = [\lfloor \frac{3n}{4} \rfloor + 1,\ n]$, $\mathcal{S}_{bzcy} = [n+1,\ \lfloor \frac{3n}{2} \rfloor]$.

Delete routes: $\mathcal{S}_{ax} = [1,\ \lfloor \frac{3n}{4} \rfloor]$, $\mathcal{S}_{bzcy} = [\lfloor \frac{3n}{4} \rfloor + 1,\ \lfloor \frac{3n}{2} \rfloor]$.

Add $(\lfloor \frac{7n}{4} \rfloor - \lfloor \frac{3n}{2} \rfloor)$ az: $\mathcal{S}_{ax} = [1,\ \lfloor \frac{3n}{4} \rfloor]$, $\mathcal{S}_{bzcy} = [\lfloor \frac{3n}{4} \rfloor + 1,\ \lfloor \frac{3n}{2} \rfloor]$, $\mathcal{S}_{az} = [\lfloor \frac{3n}{2} \rfloor + 1,\ \lfloor \frac{7n}{4} \rfloor]$.

delete routes: $\mathcal{S}_{ax} = [1,\ \lfloor \frac{3n}{4} \rfloor]$, $\mathcal{S}_{cy} = [\lfloor \frac{3n}{4} \rfloor + 1,\ \lfloor \frac{3n}{2} \rfloor]$, $\mathcal{S}_{az} = [\lfloor \frac{3n}{2} \rfloor + 1,\ \lfloor \frac{7n}{4} \rfloor]$. ..... (*)

Sort nodes: $\mathcal{S}_{ax} = [1,\ \lfloor \frac{3n}{4} \rfloor]$, $\mathcal{S}_{az} = [\lfloor \frac{3n}{4} \rfloor + 1,\ 2\lfloor \frac{3n}{4} \rfloor - \lfloor \frac{n}{2} \rfloor]$, $\mathcal{S}_{cy} = [2\lfloor \frac{3n}{4} \rfloor - \lfloor \frac{n}{2} \rfloor + 1,\ \lfloor \frac{7n}{4} \rfloor]$.

Since $\lfloor \frac{7n}{4} \rfloor \geq 2\lfloor \frac{7n}{8} \rfloor$, the state can be divided into two parts. The size of each part is greater than or equal to the value $\lfloor \frac{7n}{8} \rfloor$. The desired network state will be like : $\mathcal{S}_{cy} \cup \mathcal{S}_{by} = [1,\ \lfloor \frac{7n}{8} \rfloor]$, $\mathcal{S}_{ax} \cup \mathcal{S}_{az} = [\lfloor \frac{7n}{8} \rfloor + 1,\ \lfloor \frac{7n}{4} \rfloor]$.

$| < cy > | = \lfloor \frac{3n}{2} \rfloor - \lfloor \frac{3n}{4} \rfloor$. Therefore, by lemma 5.1, the $| < cy > | \leq \lfloor \frac{7n}{8} \rfloor$ and $| < az,\ ax > | \geq \lfloor \frac{7n}{8} \rfloor$. In order to achieve the desired state, we need to add $\lfloor \frac{7n}{8} \rfloor - \lfloor \frac{3n}{2} \rfloor + \lfloor \frac{3n}{4} \rfloor$ "by".

Add $(\lfloor \frac{7n}{8} \rfloor - \lfloor \frac{3n}{2} \rfloor + \lfloor \frac{3n}{4} \rfloor)$ by: $\mathcal{S}_{ax} \cup \mathcal{S}_{az} = [1,\ \lfloor \frac{7n}{4} \rfloor - \lfloor \frac{7n}{8} \rfloor]$, $\mathcal{S}_{axby} \cup \mathcal{S}_{azby} = [\lfloor \frac{7n}{4} \rfloor - \lfloor \frac{7n}{8} \rfloor + 1,\ 2\lfloor \frac{3n}{4} \rfloor - \lfloor \frac{n}{2} \rfloor]$, $\mathcal{S}_{cy} = [2\lfloor \frac{3n}{4} \rfloor - \lfloor \frac{n}{2} \rfloor + 1,\ \lfloor \frac{7n}{4} \rfloor]$.

Delete routes: $\mathcal{S}_{ax} \cup \mathcal{S}_{az} = [1,\ \lfloor \frac{7n}{4} \rfloor - \lfloor \frac{7n}{8} \rfloor]$, $\mathcal{S}_{by} = [\lfloor \frac{7n}{4} \rfloor - \lfloor \frac{7n}{8} \rfloor + 1,\ 2\lfloor \frac{3n}{4} \rfloor - \lfloor \frac{n}{2} \rfloor]$, $\mathcal{S}_{cy} = [2\lfloor \frac{3n}{4} \rfloor - \lfloor \frac{n}{2} \rfloor + 1,\ \lfloor \frac{7n}{4} \rfloor]$.

Since $| < by,\ cy > | = \lfloor \frac{7n}{8} \rfloor$ and $| < ax, az > | \geq \lfloor \frac{7n}{8} \rfloor$, the number of free "$ay$" is $n - \lfloor \frac{7n}{4} \rfloor + \lfloor \frac{7n}{8} \rfloor = \lfloor \frac{7n}{8} \rfloor - \lfloor \frac{3n}{4} \rfloor$. By $\lfloor \frac{7n}{8} \rfloor \geq \lfloor \frac{3n}{4} \rfloor$, The free connection request "ay" is always positive.

Add $(\lfloor\frac{7n}{8}\rfloor - \lfloor\frac{3n}{4}\rfloor)$ *ay*: $\mathcal{S}_{ax} \cup \mathcal{S}_{az}$ =[1, $\lfloor\frac{7n}{8}\rfloor$], $\mathcal{S}_{by}$ =[$\lfloor\frac{7n}{8}\rfloor + 1$, $2\lfloor\frac{3n}{4}\rfloor - \lfloor\frac{n}{2}\rfloor$], $\mathcal{S}_{cy}$ =[$2\lfloor\frac{3n}{4}\rfloor - \lfloor\frac{n}{2}\rfloor + 1$, $\lfloor\frac{7n}{4}\rfloor$], $\mathcal{S}_{ay}$ =[$\lfloor\frac{7n}{4}\rfloor + 1$, $\lfloor\frac{15n}{8}\rfloor$].

The final node should be $\lfloor\frac{15n}{8}\rfloor$ because $(\lfloor\frac{7n}{8}\rfloor - \lfloor\frac{3n}{4}\rfloor) + \lfloor\frac{7n}{4}\rfloor = n + \lfloor\frac{7n}{8}\rfloor = \lfloor\frac{15n}{8}\rfloor$.

Since lemma 5.1 only applied for $n \geq 3$, the above proof is only valid for $n \geq 3$. Therefore, $n = 1$ and 2 cases should be considered individually.
Obviously, when $n = 1$ and $m = \lfloor\frac{15}{8}\rfloor = 1$, the theorem is trivial true.
For $n = 2$,
Add two *ax* into the empty state : $\mathcal{S}_{ax}$ =[1, 2].
Add two *by* : $\mathcal{S}_{axby}$ =[1, 2].
Add two *cz* : $\mathcal{S}_{axbycz}$ =[1, 2].
Delete routes : $\mathcal{S}_{axby}$ =[1], $\mathcal{S}_{axcz}$ =[2].
Add one *cy* : $\mathcal{S}_{axby}$ =[1], $\mathcal{S}_{axcz}$ =[2], $\mathcal{S}_{cy}$ =[3].

From the above, three middle-stage switches should be needed for the case $n = 2$. In addition, $\lfloor\frac{(15)(2)}{8}\rfloor = 3$ concludes that the theorem is also true for $n = 2$. Therefore, the theorem is true for all n. □

6 Epilogue

The Beneš theorem [1] identified a family of WSN networks, but not strictly nonblocking. It also raised the question on the existence of any WSN, but not strictly nonblocking for the network [$n{\times}m$, $r{\times}r$, $m{\times}n$] where $r > 2$. Theorem 3.1 answer this question affirmatively with a nonblocking Alogrithm 1 over the network [6×10, 3×3, 10×6]. The smallest value of $WS(n, r)$ that is previously undetermined was $WS(5, 3)$. Theorem 4.2 finds $WS(5, 3) = 9$. Theorem 5.2 establishes the necessity of $m \geq \lfloor\frac{15n}{8}\rfloor$ for [$n{\times}m$, 3×3, $m{\times}n$] to be *WSN by packing*. Thus the example of [6×10, 3×3, 10×6] shows that a WSN network is not necessarily WSN by packing.

For further research, we conjecture that the network [$n{\times}\lfloor\frac{7n}{4}\rfloor$, 3×3, $\lfloor\frac{7n}{4}\rfloor{\times}n$], $n \geq 6$, is WSN through some generalization of Algorithm 1. On the other hand, it could be interesting to find a new proof of Theorem 3.1 without the aid of computer.

References

[1] V. E. Beneš, *Mathematical Theory of Connecting Networks and Telephone Traffic*,New York, Academic Press, 1965.

[2] Charles Clos, *A study of non-blocking switching networks*, Manuscript received October 30, 1952.

[3] D. Z. Du, P. Fishburn, B. Gao, and F. K. Hwang, *Wide-sense nonblocking for 3-stage Clos networks*, Preprint, 1995.

[4] J. Friedman, *A lower bound on strictly non-blocking networks*, Combinatorica 8(2), 1988.

[5] Frank K. Hwang, *The Mathematical Theory of Nonblocking Siwthcing Networks*, Volume 11.

[6] Yu-Ngai Joseph Hui, *Switching and traffic theory for integrated broadband networks*, 1990.

[7] R. P. Kurshan and V. E. Benes, *Wide-sense nonblocking network made of square switches*, Elec. Letters, vol. 17, pp. 697-700, 1980.

[8] Chin-Tau Lea, *Tradeoff of horizontal decomposition versus vertical stacking in rearrangeable nonblocking networks*, IEEE Transactions on Communications, vol. 39, no. 6, June 1991.

[9] Shuo-Yen Robert Li, *Algebraic Switching Theory and Broadband Applications*, Academic Press, 2000.

[10] D. G. Smith, *Lower bound on the size of a 3-stage wide-sense nonblocking network*, Electronics Letters, vol. 13, No. 7, 31st March, 1977.

[11] K. Tsai, D. Wang and F. K. Hwang, *Lower bounds of wide-sense nonblocking Clos networks*, to appear in Theoretical Computer Science.

[12] Yuanyuan Yang, *Wide-sense nonblocking Clos networks under packing strategy*, IEEE Transactions on Computers, vol. 48, no.3, March 1999.

Switching Networks: Recent Advances
D.-Z. Du and H.Q. Ngo (Eds.) pp. 117-141

Nonblocking Routing Properties of Clos Networks

Guo-Hui Lin
Department of Computing and Software
McMaster University, Hamilton, Ontario L8S 4L7, Canada
Email: ghlin@cas.mcmaster.ca

Contents

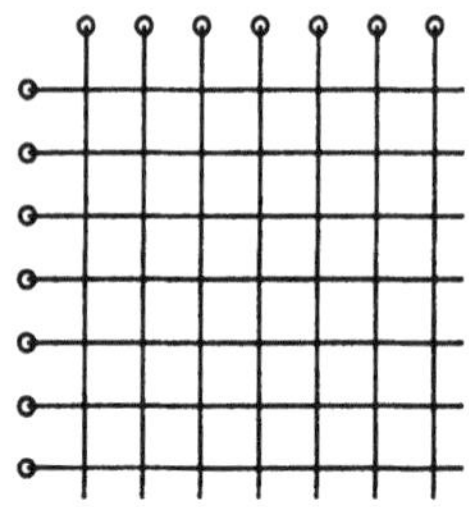

Figure 1: Square array

1 Introduction

By a connecting system we shall mean a physical communication system consisting of (i) a set of terminals, (ii) control units which process requests for connections between pairs of terminals, and (iii) a connecting network through which the connections are effected. The connecting network is an arrangement of arrays of crosspoints (called *crossbars* or *switches*) and transmission links through which certain terminals can be connected together in many combinations.

The terminals of a connecting system usually are classified into two groups, one for input terminals and the other for output terminals. A terminal in the input/output group is called an *inlet/outlet*. A *connection* request is, then, between a pair of an inlet and an outlet. Input group and output group may intersect. In this paper, we only consider the *two-sided* case, in which they are assumed to be disjoint.

In the design of connecting systems, the number of crosspoints, which are assumed of control functions, must be kept at a minimum, and yet the system is able to establish as many simultaneous connections as possible. One common requirement for a connecting network is to be *strictly non-blocking*. In a strictly nonblocking network, it is always possible to establish a connection from an idle inlet to an idle outlet, regardless of the state of the network, *i.e.*, the occupying status of the links. Notice that a simple square array with N inlets, N outlets, and N^2 crosspoints meets this requirement. Thus it can be taken as an upper design limit. Such a square array is depicted in Figure 1.

Since the number of crosspoints in a square array goes up unpleasantly fast with N, it is desirable to consider strictly nonblocking networks in which

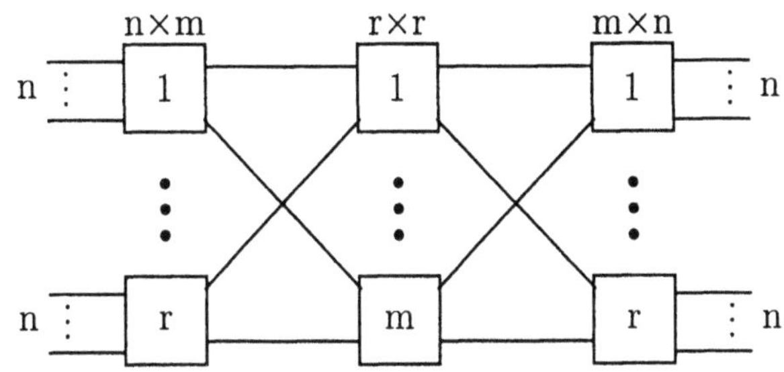

Figure 2: A symmetric 3-stage Clos network $C(n, m, r)$

the number of crosspoints goes up more slowly than N^2. In 1953, C. Clos [3] published the fundamental study of nonblocking connecting networks. In that paper, he elegantly solved the problem of designing a strictly nonblocking network having fewer than N^2 crosspoints. The symmetric three-stage Clos network $C(n, m, r)$, as depicted in Figure 2, is the first improvement, in which there are r crossbars (switches) in the first (or input) stage each of which has n inlets and m outlets, m crossbars in the second (or central) stage each of which has r inlets and r outlets, and r crossbars in the third (or output) stage each of which has m inlets and n outlets. Every crossbar in the first stage has an outlet connected to an inlet of every crossbar in the second stage and every crossbar in the second stage has an outlet connected to an inlet of every crossbar in the third stage. There are a total number rn of inlets in the first stage, which are inlets of the network, and a total number rn of outlets in the third stage, which are outlets of the network. Notice that a crossbar having n inlets and m outlets is a rectangular array with n inlets and m outlets and thus nm crosspoints, which is denoted as an $n \times m$ crossbar, and thus is itself strictly nonblocking.

Clos showed that $C(n, 2n-1, r)$ is strictly nonblocking and thus by taking $n = r$ there are only a number $6N^{3/2} - 3N$ of crosspoints in the network, where $N = rn$ is the number of inlets/outlets. Furthermore, by using the symmetric three-stage Clos networks as construction units, Clos showed that using enough stages one can design strictly nonblocking networks for large N in which the number of crosspoints goes up more slowly than $N^{1+\epsilon}$, for any prescribed $\epsilon > 0$ [1].

Since the emergence of the Clos networks, besides Clos himself, there are a lot of researchers extensively studied them and proved many distinguished routing properties that they possess. These routing properties are strictly nonblocking, *rearrangeably nonblocking* or simply *rearrangeability*, and *wide-sense nonblocking* or simply *WSNB*. To name a few researchers involved, Slepian [20] and Duguid [8] showed a sufficient and necessary condition under which the network is rearrangeable; Paull [19] gave a tight upper bound, for $C(n,n,n)$, on the number of connections to be rearranged such that the new connection can be routed successfully. Beneš [1] generalized the method and obtained a tight upper bound for $C(n,n,r)$.

The development of technology in 1980s enables a wire to route more than one connection, as long as the total *bandwidths* of the connections assigned to the wire does not exceed the wire capacity. The impetus of multirate switching comes from the interest in designing telecommunication switches that handle traffic with a wide range of bandwidth requirements (voice, facsimile, video, etc). In the multirate switching environment, it is usually assumed that the connection links and inlets/outlets have uniform capacity which is normalized to be unity, while every connection requests a bandwidth w, $0 < w \leq 1$, in order to be routed. Nonblocking routing properties of the Clos networks in the multirate switching environment became new research subjects. Representative work in the 1980s includes that done by Jajszczyk [11], Melen and Turner [17], and Chung and Ross [2], etc.

The three-stage Clos network has been widely used now in the design of telecommunication networks, computer networks and Internet. In fact, the multirate interconnection network design and routing is one of the hottest research topics in Asynchronous Transfer Mode (ATM) networks. Recently, a group of researchers, including D.Z. Du, P.C. Fishburn, B. Gao, X.D. Hu, F.K. Hwang, D.S. Kim, J.H. Kim, G.-H. Lin, W. Wu, G.L. Xue, and K. Yoo [5, 6, 7, 9, 10, 12, 13, 14, 15, 16] updated most of the results on the nonblocking routing properties of Clos networks. In this paper, the most significant results on every routing property and the novel techniques involved are surveyed. Some conjectures and open problems, as well as other related problems, are provided for the future study.

1.1 Three Nonblocking Routing Properties

To explain more clearly, let us define the three most interesting routing properties formally. We will give the definitions in the multirate switching environment, and point out the special cases that reduce to that in the classical switching. In the following, when we say a (connecting) network, we are saying a symmetric 3-stage Clos network $C(n,m,r)$. As mentioned earlier, it is assumed that the connection links and inlets/outlets have uniform capacity which is normalized to be unity. A *connection* (or sometimes a *call*) in the network is a triple (i, j, w), where i is an inlet, j is an outlet, and w is the *bandwidth requirement* of the connection which is also called the *weight* of the connection. A *route* r is a path in the network joining an input crossbar to an output crossbar, through a central crossbar. Each route is assigned with a capacity, meaning that every link on the route keeps that amount of capacity for the route. Route r *realizes* a connection (i, j, w) if inlet i and outlet j belong to the input crossbar and the output crossbar joined by r, respectively, and the capacity of r is at least w.

A set of connections is *compatible* if for every inlet/outlet, the sum of bandwidths of the connections involving it is less than or equal to 1. A *connection frame* is a compatible set of connections. A set of routes is *compatible* if for every link in the network, the sum of the capacities of the routes using this link is less than or equal to 1. A *configuration* is a compatible set of routes. A connection frame is said to be *realizable* if there exists a configuration such that the routes therein realize all the connections in the frame. Given a connection frame $\mathcal{F}$, a connection c is said to be *compatible* with $\mathcal{F}$ if $\mathcal{F} \cup \{c\}$ is still compatible. Similarly, a route r is said to be *compatible* with a configuration $\mathcal{C}$ if $\mathcal{C} \cup \{r\}$ is still compatible.

Note that if all the connections have bandwidth requirement 1, then the multirate switching environment reduces to the classical switching environment, whereby every route realizing a connection must have capacity 1. A network is said *multirate/classical rearrangeable* if in the multirate/classical switching environment every connection frame is realizable. In the following we abbreviate multirate/classical rearrangeable to rearrangeable, whenever the routing environment is clear from the context. A network is said to be *strictly nonblocking* if for every configuration $\mathcal{C}$ realizing a connection frame $\mathcal{F}$ and every connection c compatible with $\mathcal{F}$, there exists a route r such that r realizes c and r is compatible with $\mathcal{C}$. The third important routing property is *wide-sense nonblocking* or *WSNB* for short. When talking about the WSNB property, a concrete routing algorithm is involved. A network is WSNB if for the configuration $\mathcal{C}$ realizing a connection frame $\mathcal{F}$ and a connection c compatible with $\mathcal{F}$, the routing algorithm can generate a route r such that r realizes c and r is compatible with $\mathcal{C}$, provided that the configuration $\mathcal{C}$ is generated by the routing algorithm.

1.2 Layout of the Paper

In Section 2, we retrospect the results on the three nonblocking routing properties in the classical switching environment. We do not intend to be systematic, but the results quoted here are either historic or the latest. Section 3 is devoted to the three nonblocking routing properties in the multirate switching environment. As there are so many interesting techniques involved and so many historic results, the section is divided into three subsections dealing with strictly nonblocking, WSNB, and rearrangeability, respectively. Particularly, in the second subsection, we will have a deep look at a very useful notion of *1-rate WSNB*, where results on the similarity between the 1-rate WSNB and the classical WSNB are reviewed. However, we note that proof of the genuine similarity answers the conjecture made by Fishburn *et al.* [9]. In the last subsection, we retrospect the most efforts paid to the proof of truth of the famous Chung-Ross's Conjecture, particularly, in the *restricted discrete bandwidth case* and the *k-rate cases* for small k. Section 4 introduces two kinds of generalizations, one is the generalization of the network structure from symmetric to asymmetric, and the other is the generalization of the type of connections from one-to-one connection requests to one-to-many broadcast requests and many-to-many multicast requests. The last section introduces two important related problems. Extensive study in them may lead to significant progress on the nonblocking routing properties and designing better routing algorithms.

2 Classical Switching

Recall that in the classical switching environment, all connections have bandwidth requirement 1. Therefore, whenever a link is used to route some connection, it is fully occupied.

In 1953, in the paper [3] that Clos suggested and studied the 3-stage Clos network $C(n, m, r)$, he showed

Theorem 2.1 *$C(n, m, r)$ is strictly nonblocking if and only if $m \geq 2n - 1$.*

The following result on rearrangeability is attributed to Slepian [20] and Duguid [8], which was proved by induction on n, the number of inlets putting on an input crossbar.

Theorem 2.2 *$C(n, m, r)$ is rearrangeable if and only if $m \geq n$.*

Strictly nonblocking property enables arbitrary routing, by sacrificing $n - 1$ central crossbars compared to that needed in a rearrangeable network. However, in a rearrangeable network $C(n, n, r)$, when a new connection arrives, one has to make a lot of rearrangements (as many as $r - 1$ [19, 1]) in order to route it. This makes the routing much more complicated. One natural question is to ask *If it is possible that, by routing the connections carefully, the number of central crossbars required in a nonblocking network would decrease while we don't need to reroute existing connections when routing a new connection.* The answer to this question is *affirmative* and the concept of wide-sense nonblocking was introduced for this purpose. In the following, when we say that a network is WSNB, it is WSNB under some routing algorithm; and whenever possible, we will point out the algorithm explicitly. (However, in some cases, the routing algorithm is a modification of some other algorithm, or a combination of several other algorithms. The interested reader may refer to the original articles for the detail description.)

Let STU denote the class of algorithms called SAVE-THE-UNUSED. Each algorithm in this class do not route a new connection through an unused central crossbar unless necessary. Beneš [1] proved

Theorem 2.3 *$C(n,m,2)$ is WSNB under* STU *if $m \geq \lfloor \frac{3n}{2} \rfloor$, and is not WSNB (under any routing algorithm) if $m < \lfloor \frac{3n}{2} \rfloor$.*

While this $r = 2$ case had been completely solved in 1965, when $r \geq 3$, it was unknown if there is a routing algorithm under which $C(n,m,r)$ is WSNB for some $m < 2n - 1$. One easy observation is that if $C(n,m,r)$ is not WSNB (under any routing algorithm), then neither is $C(n,m,r+1)$. Therefore, Theorem 2.3 implies that $C(n, \lfloor \frac{3n}{2} \rfloor - 1, r)$ is not WSNB when $r \geq 3$. Du *et al.* [5] was able to push this lower bound to $\lfloor \frac{7n}{4} \rfloor$.

Theorem 2.4 *$C(n,m,3)$ is not WSNB if $m < \lfloor \frac{7n}{4} \rfloor$.*

The algorithm P (PACKING) assigns the new connection to the most heavily-loaded central switch (with any tie-breaking rule). While algorithm MI (MINIMUM-INDEX) assigns the new connection to the available central switch with the minimum index. Notice that both P and MI are STU algorithms, and they are considered two of the most heralded algorithms. It was expected that P would perform well when $r \geq 3$. Nonetheless, Du *et al.* [5] showed that

Theorem 2.5 *When $r \geq 3$, $C(n,m,r)$ is not WSNB under* P *if $m < 2n-1$; $C(n,m,r)$ is not WSNB under* MI *if $m < \lfloor 2n - \frac{n}{2^{r-1}} \rfloor$.*

The difficulty lying in the study of WSNB involves not only the designing of the algorithm, but also its analysis. Du *et al.* [5] proposed an improved packing algorithm called MCP (MOST-CANDIDATE-PACKING), and showed that routing under MCP requires less central switches.

Theorem 2.6 *$C(n, m, 3)$ is WSNB under* MCP *for* $m \geq \lfloor \frac{11n+19}{6} \rfloor$.

It is conjectured that algorithm MCP, or a suitable modification of MCP, would perform also well when $r \geq 4$. That is, the WSNB network under MCP requires less than $2n - 1$ central switches.

3 Multirate Switching

There are two kinds of restrictions imposed on the bandwidth of the connections in the multirate switching environment. One is the *continuous bandwidth case* and the other is the *discrete bandwidth case.* In the continuous bandwidth case, the bandwidth of all connections belongs to an interval $[b, B]$ (or sometimes $(b, B]$), where $1 \geq B > b \geq 0$. In the discrete bandwidth case, the bandwidth of all connections belongs to a given finite set $\{b_1, b_2, \cdots, b_k\}$, where $1 \geq b_1 > b_2 > \cdots > b_k > 0$ and b_i is an integral multiple of b_k, denoted by $b_k \mid b_i$, for every i. Letting $B = b_1$ and $b = b_k$ in the discrete bandwidth case.

The *restricted discrete bandwidth case* is a discrete bandwidth case in which $b_{i+1} \mid b_i$, for $i = 1, 2, \cdots, k-1$.

If there are k possible rates in the discrete bandwidth case and there is no integral multiplicity restriction imposed on them, namely, b_i need not necessarily be an integral multiple of b_k, then we simply call it the *k-rate case.* The *k-rate case* might be regarded as the *relaxed* discrete bandwidth case. $C(n, m, r)$ is rearrangeable/strictly nonblocking/WSNB in the k-rate case is called *k-rate rearrangeable/strictly nonblocking/ WSNB.*

For any connection frame $\mathcal{F}$ of the network $C(n, m, r)$, construct a *weighted bipartite connection graph* G with the input switches as one part, the output switches as the other part, and an edge of weight w between vertices I and J for each connection $(i, j, w) \in \mathcal{F}$, where i (j) is an inlet (outlet, respectively) on input switch I (output switch J, respectively). We call G the connection graph *induced* by $\mathcal{F}$. Notice that G is generally not a simple graph. G can be *m-colored* if there exists an edge-coloring of G using m distinct colors such that the weight of like-colored edges incident to each vertex sum up to a value of at most 1. Given such an edge-coloring of G, let all the connections corresponding to like-colored edges pass through one central switch. This gives a configuration C which realizes $\mathcal{F}$.

3.1 Multirate Strictly Nonblocking

Several necessary and sufficient conditions for strictly nonblocking for multirate 3-stage Clos networks have been established by Niestegge [18], Melen and Turner [17], and Chung and Ross [2]. The following two are due to Chung and Ross.

Theorem 3.1 *$C(n,m,r)$ is strictly nonblocking in the discrete bandwidth case if and only if $m \geq 2\left\lfloor \frac{n-B}{1-B+b} \right\rfloor + 1$.*

Theorem 3.2 *When $B \in (1-b, 1]$, $C(n,m,r)$ is strictly nonblocking in the continuous bandwidth case if and only if $m \geq 2\lfloor \frac{1}{b} \rfloor (n-1) + 1$.*

Notice that when the bandwidth interval is $(0, B]$ in the continuous bandwidth case, Theorem 3.2 is no longer valid. The following general sufficient condition is due to Melen and Turner [17].

Theorem 3.3 *When $1 - B + b > 0$, $C(n,m,r)$ is strictly nonblocking in the continuous bandwidth case if*

$$m \geq \begin{cases} 2\lfloor \frac{n-B}{1-B} \rfloor + 1, & \text{when } B \leq 1-b, \\ 2\lfloor \frac{n-1+b}{b} \rfloor + 1, & \text{when } B \geq 1-b. \end{cases}$$

Note that in either of the above three theorems, the number of central switches goes to infinity as b approaches to zero and/or B approaches to 1.

It is easy to see that when the bandwidth of all connections belongs to interval $(\frac{1}{p+1}, \frac{1}{p}]$ for some integer $p \geq 1$, then we may in fact enlarge, whenever necessary, the bandwidth of all connections to $\frac{1}{p}$. Therefore, this special continuous bandwidth case reduces to the 1-rate case, where the rate is $\frac{1}{p}$. From Theorem 3.1, we have the following result on 1-rate strictly nonblocking:

Corollary 3.4 *$C(n,m,r)$ is 1-rate strictly nonblocking if and only if $m \geq 2n - 1$.*

3.2 Multirate Wide-Sense Nonblocking

From Theorem 3.3, we know that all connections having bandwidth less than or equal to $\frac{1}{2}$ could be routed through $4n - 1$ central switches, as long as these $4n - 1$ switches are reserved for routing this kind of *light* connections. By Theorem 3.2 or Corollary 3.4, all connections having bandwidth greater than $\frac{1}{2}$ can be routed through $2n-1$ central switches, which are reserved for routing this kind of *heavy* connections. It follows that, under the simple algorithm which reserves $2n-1$ central switches for routing heavy connections and $4n-1$ central switches for routing light connections, $C(n, 6n-2, r)$ is multirate WSNB in the continuous bandwidth case, where the bandwidth interval is $(0, 1]$. This is a significant decrement in the number of central switches compared to that needed to be strictly nonblocking.

Suppose the bandwidth interval is $(0, B]$, and let ρ be a value such that $0 < \rho \leq B$. A connection is a *heavy* connection if its bandwidth requirement is greater than ρ, and a *light* connection otherwise. In the above paragraph, $B = 1$ and $\rho = \frac{1}{2}$. The QUOTA scheme [10] $P(x, y)$ denotes the algorithm that reserves x central switches each of them is allowed to carry no more than y light connections but as many heavy connections as possible. Note that the simple algorithm in the above paragraph is actually $P(2n-1, 0)$. QUOTA scheme can be generalized to route more than two kinds of connections.

Let $p = \lfloor \frac{1}{B} \rfloor$. This time, a connection is a *heavy* connection if its bandwidth requirement is greater than $\frac{1}{p+1}$, is a *median* connection if its bandwidth requirement is greater than $\frac{1}{p+2}$ but less than or equal to $\frac{1}{p+1}$, and a *light* connection otherwise. The QUOTA scheme $P(x_1, y_1; x_2, y_2)$ denotes the algorithm that reserves x_1 central switches each of them is allowed to carry no more than y_1 light and median connections but as many heavy connections as possible, and reserves x_2 central switches each of them is allowed to carry no heavy connections, no more than y_2 light connections but as many median connections as possible. Gao and Hwang [10] showed that QUOTA scheme requires less central switches.

Theorem 3.5 *In the continuous bandwidth case with bandwidth interval* $(0, B]$*,* $C(n, m, r)$ *is multirate WSNB under* QUOTA $P(x_1, 0; x_2; 0)$*, where*

$$(x_1, x_2) = \begin{cases} \left(\frac{2(p+1)(Bp+B-1)n}{p^2}, 0\right), & \text{for } B \leq \frac{23}{32} = 0.71875, \\ (2n-1, 0.75n), & \text{for } B \geq \frac{23}{32}, \end{cases}$$

and $m \geq m^* = \min\left\{5.75n, \frac{2(p+1)(Bp+B+p-1)n}{p^2}\right\}$.

3.2.1 k-Rate Wide-Sense Nonblocking

Better upper bounds on the number of central switches needed to be WSNB in the k-rate case, for small k, were also obtained in [10]. Let the k rates be $b_1, b_2, \cdots, b_k$, where $1 \geq b_1 > b_2 > \cdots > b_k > 0$.

Theorem 3.6 *$C(n, 3n, r)$ is 2-rate WSNB (under a combination of* QUOTA *schemes) when $b_1 \le \frac{1}{2}$.*

Theorem 3.7 *When $b_1 > \frac{1}{2} \ge b_2$ and $b_1 + l_1 b_2 \le 1 < b_1 + (l_1 + 1)b_2$ and $l_2 b_2 \le 1 < (l_2 + 1)b_2$, $C(n, m, r)$ is 2-rate WSNB (under a combination of* QUOTA *schemes), where*

$$m > m^* = \begin{cases} 2 + \frac{2(l_2-l_1)(l_2-l_1-1)}{l_2^2-l_2l_1-l_1^2-l_1} & \text{for } l_2 \ge 2l_1 + 1, \\ 2 + \frac{2l_1}{l_1+1} & \text{for } l_2 \le 2l_1. \end{cases}$$

Theorem 3.8 *$C(n, 5n, r)$ is 3-rate WSNB (under a combination of* QUOTA *schemes).*

For the 1-rate case, whether or not $C(n, m, r)$ is WSNB depends also on what the rate is, apart from the number m of central switches and the routing algorithm. Suppose the rate is w and let $p = \lfloor \frac{1}{w} \rfloor$. It is trivial to examine that the minimum number of central switches needed in a WSNB network depends on p, rather than the exact value of w. Therefore, if $C(n, m, r)$ is WSNB under some algorithm, we say that $C(n, m, r)$ is 1-rate-p WSNB under that algorithm, while ignoring w. Du *et al.* [7] showed the following relationship between 1-rate-p WSNB and 1-rate-cp WSNB.

Theorem 3.9 *Suppose c is an integer, then $C(n, m, r)$ is 1-rate-cp WSNB if and only if $C(cn, cm, r)$ is 1-rate-p WSNB.*

Proof. Suppose that $C(n, m, r)$ is 1-rate-cp WSNB. Treating each link (inlet, outlet) as a union of c sub-links (sub-inlets, sub-outlets, respectively) each with capacity $\frac{1}{c}$, and replacing each central switch by c copies each having a sub-link connecting to an input/output switch, this transforms $C(n, m, r)$, where the allowed connection bandwidth is $\frac{1}{cp}$, into $C(cn, cm, r)$, where the allowed connection bandwidth is $\frac{1}{p}$.

Without loss of generality, assume that central switch C_i in $C(n, m, r)$ is replaced by $C_{i_1}, C_{i_2}, \cdots$, and C_{i_c} in $C(cn, cm, r)$. Let A denote the routing algorithm involved in $C(n, m, r)$. Let $\mathcal{F}$ denote a connection frame for $C(cn, cm, r)$. Notice that by ignoring the detail inlet/outlet while only focusing on input/output switch, and decreasing the bandwidth from $\frac{1}{p}$ to $\frac{1}{cp}$, $\mathcal{F}$ is also a connection frame for $C(n, m, r)$. Let A$'$ denote the algorithm which, if algorithm A assigns a connection to central switch C_i in $C(n, m, r)$, assigns that connection to arbitrary available one of $C_{i_1}, C_{i_2}, \cdots$, and C_{i_c} in $C(cn, cm, r)$. It is clear that by using the routing algorithm A as guiding algorithm, $C(cn, cm, r)$ is 1-rate-p WSNB under A$'$.

Conversely, suppose that $C(cn, cm, r)$ is 1-rate-p WSNB under algorithm A$'$. Let $\mathcal{F}$ denote a connection frame for $C(n, m, r)$. Similarly, by ignoring the detail inlet/outlet while focusing on input/output switch, and increasing the bandwidth from $\frac{1}{cp}$ to $\frac{1}{p}$, $\mathcal{F}$ is also a connection frame for $C(cn, cm, r)$. Let A denote the algorithm which, if algorithm A$'$ assigns a connection to any one of $C_{i_1}, C_{i_2}, \cdots$, and C_{i_c} in $C(cn, cm, r)$, assigns that connection to central switch C_i in $C(n, m, r)$. Therefore, by using A$'$ as guiding algorithm, $C(n, m, r)$ is 1-rate-cp WSNB under algorithm A. □

Theorem 3.10 [9] *Suppose that $C(n, f(n, r), r)$ is 1-rate-p WSNB for all n. Then $C(n, f(n, r), r)$ is 1-rate-cp WSNB for all n if $f(cn, r) \leq cf(n, r)$.*

Proof. Since $C(cn, f(cn, r), r)$ is 1-rate-p WSNB and $f(cn, r) \leq cf(n, r)$, $C(cn, cf(n, r), r)$ is also 1-rate-p WSNB. By Theorem 3.9, $C(n, f(n, r), r)$ is 1-rate-cp WSNB. And clearly this holds for all n. □

Theorem 3.11 [7] *Suppose that $C(n, f(n, r), r)$ is not 1-rate-p WSNB for any n. Then neither is $C(n, f(n, r), r)$ 1-rate-cp WSNB for any n, if $f(cn, r) \geq cf(n, r)$.*

Proof. Since $C(cn, f(cn, r), r)$ is not 1-rate-p WSNB and $f(cn, r) \geq cf(n, r)$, $C(cn, cf(n, r), r)$ is neither 1-rate-p WSNB. It follows from Theorem 3.9 that nor is $C(n, f(n, r), r)$ 1-rate-cp WSNB. □

Theorems 3.10 and 3.11 are so powerful that almost all results on classical WSNB have their counterparts in the 1-rate WSNB. Notice that classical WSNB is 1-rate-1 WSNB. Suppose that $C(n, m, r)$ is classical WSNB under routing algorithm A, then according to the proof of Theorem 3.9, we may easily design the routing algorithm under which $C(n, m, r)$ is 1-rate-p WSNB, by using A as guiding algorithm. In this case, we will let A$'$ denote this new routing algorithm. For example, Theorems 2.3–2.6 imply the following.

Corollary 3.12 *$C(n, m, 2)$ is 1-rate WSNB under* STU$'$ *if $m \geq \lfloor \frac{3n}{2} \rfloor$; and is not 1-rate WSNB (under any routing algorithm) if $m < \lfloor \frac{3n}{2} \rfloor$.*

Corollary 3.13 *$C(n, m, 3)$ is not 1-rate WSNB if $m < \lfloor \frac{7n}{4} \rfloor$.*

Corollary 3.14 *When $r \geq 3$, $C(n, m, r)$ is not WSNB under* P$'$ *if $m < 2n - 1$; $C(n, m, r)$ is not WSNB under* MI$'$ *if $m < \lfloor 2n - \frac{n}{2^{r-1}} \rfloor$.*

Corollary 3.15 *$C(n, m, 3)$ is WSNB under* MCP$'$ *if $m \geq \lfloor \frac{11n+19}{6} \rfloor$.*

In the proofs of Theorems 3.10 and 3.11, we need the monotonousness of function $f(n,r)$ in n. Nevertheless, does 1-rate WSNB imply the classical WSNB? This is stated in the following Conjecture, which was first proposed by Fishburn *et al.* [9].

Conjecture 3.16 *If $C(n,m,r)$ is 1-rate-p WSNB for any integer $p \geq 2$, then it is also WSNB in the classical switching.*

3.3 Multirate Rearrangeability

Melen and Turner [17], and Chung and Ross [2] also studied multirate rearrangeability. In particular, Chung and Ross [2] showed that

Theorem 3.17 *$C(n,m,r)$ is 1-rate rearrangeable if and only if $m \geq n$.*

They also made the following conjecture:

Conjecture 3.18 *$C(n, 2n-1, r)$ is rearrangeable in the discrete bandwidth case.*

And were able to show

Lemma 3.19 *$C(n, 2n-1, r)$ is rearrangeable if all connections have bandwidth either b or 1.*

Let $\mathcal{F}$ be a connection frame and $G = (V_1, V_2, E)$ its induced connection graph. For each vertex $v \in V_1 \cup V_2$, E_v denotes the set of edges incident at v. Number the edges in E_v from 0, in non-increasing order of their weight, and let E_v^i comprise the edges indexed in the range $\{ik, ik+1, \cdots, (i+1)k-1\}$ for all $i \geq 0$. The CAP (Constrained Alternating Path) algorithm assigns distinct colors to edges in each subset E_v^i. In particular, given a partial (edge) coloring of G, the algorithm colors an uncolored edge (u,v), which is in E_u^i and E_v^j, as follows: If there is a color $c \in \{0, 1, \cdots, k-1\}$ that is not already in use within E_u^i and E_v^j, then the algorithm uses it to color (u,v). Otherwise, suppose c_1 is the unused color within E_u^i and c_2 is the unused color within E_v^j. The algorithm finds a maximal *constrained alternating path* from u, that is, a longest path starting at u and with edges colored c_2 and c_1 alternatively, and such that for every interior vertex z on the path, the path edges incident at z belong to a common set E_z^l. Since G is bipartite, the last edge in the path couldn't be a member of either E_u^i or E_v^j. It follows that by interchanging the colors c_1 and c_2 for all edges on the path, the algorithm successfully colors edge (u,v) by c_2.

This CAP algorithm is proposed by Melen and Turner [17]. Notice that by using k colors, in the coloring obtained by CAP, the sum of the weight of like-colored edges incident at any vertex is upper bounded by $B + \frac{n-B}{k}$. It follows that if $B + \frac{n-B}{k} \leq 1$, then G could be k-colored. In the continuous bandwidth case with bandwidth interval $[b, B]$, let $M_{[b,B]}(n, r)$ denote the minimum value of m such that $C(n, m, r)$ is rearrangeable. In the k-rate case, let $M_k(n, r)$ denote the minimum value of m such that $C(n, m, r)$ is rearrangeable. The above argument implies the following theorem:

Theorem 3.20 *$M_{(0,B]}(n, r) \leq \lceil \frac{n-B}{1-B} \rceil$, when $B < 1$.*

Corollary 3.21 $M_{(0,\frac{1}{2}]}(n, r) \leq 2n - 1$.

3.3.1 Continuous Bandwidth Case

When $B = 1$ and $b \in (\frac{1}{p+1}, \frac{1}{p}]$ for some integer $p \geq 1$, by enlarging the bandwidth in $(\frac{1}{q+1}, \frac{1}{q}]$ to $\frac{1}{q}$, we reduce the continuous bandwidth case to a p-rate case where the p rates are $1, \frac{1}{2}, \cdots, \frac{1}{p}$. Using n central switches to route those connections having bandwidth $\frac{1}{q}$, $C(n, pn, r)$ is p-rate rearrangeable and thus rearrangeable for the continuous bandwidth case where the bandwidth interval is $(\frac{1}{p+1}, 1]$. The next theorem is an improvement for the case $p = 2$, and its proof is an interesting application of the CAP algorithm.

Theorem 3.22 $M_{(\frac{1}{3},1]}(n, r) \leq 2n - 1$.

Proof. Since every connection has bandwidth greater than $\frac{1}{3}$, each inlet/outlet can be involved in at most two connections. That means each input/output switch is involved in at most $2n$ connections. For every input/output switch, if it is involved in less than $2n$ connections, then those connections are put into one group; if it is involved in exactly $2n$ connections, then $2n - 1$ of them are put into one group, and the one requiring the least bandwidth is put into another group. Noticing that this least bandwidth and any one of the $2n - 1$ bandwidths would sum up to a value at most 1. Using the CAP routing algorithm by Melen and Turner [17], we can route any connection frame successfully through $2n - 1$ central switches. □

From Theorem 3.5, we conclude that $C(n, 5.75n, r)$ is rearrangeable in the continuous bandwidth case where the bandwidth interval is $(0, 1]$. Notice that $M_{(\frac{1}{2},1]}(n, r) = n$, which, together with Corollary 3.21, implies the following better result.

Theorem 3.23 $M_{(0,1]}(n,r) \leq 3n - 1$.

Eventually, by employing graph theory on edge coloring, Du *et al.* [6] was able to prove that $M_{(0,1]} \leq 2.5625n$. Before describing this result, let us have a look at the lower bounds ever established for $M_k(n,r)$ and $M_{(0,1]}(n,r)$.

Theorem 3.24 [6] *$M_1(n,r) = n$; $M_2(n,r) \geq n+1$ when $r \geq 3$; $M_3(n,r) \geq \lceil \frac{11}{9}n \rceil$ when $r \geq 3$.*

Corollary 3.25 *$M_{(0,1]}(n,r) \geq \lceil \frac{11}{9}n \rceil$ when $r \geq 3$.*

Lemma 3.26 [6] *Given a bipartite connection graph G (induced by some $\mathcal{F}$), suppose that all edges having weight greater than $\frac{1}{f}$ are colored by a set C of c colors, where f is an integer and $c \geq 2n$. Then at most $\lceil \frac{(c-2)}{f} - c + 2n \rceil$ additional colors are needed which, together with C, color all the edges having weight less than or equal to $\frac{1}{f}$.*

Suppose that G is an unweighted bipartite graph. Let $d_G(v)$ denote the degree of vertex v in G. A *spanning subgraph* S of G is a subgraph with the same vertex set as G. The following lemma partitioning the edge set of a bipartite graph is due to de Werra [4].

Lemma 3.27 *Let G be a (unit weighted) bipartite graph. Then for any integer $k \geq 1$, G can be partitioned into k edge-disjoint spanning subgraphs $G_1, G_2, \cdots, G_k$ such that for every vertex v in the vertex set,*

$$\left\lfloor \frac{d_G(v)}{k} \right\rfloor \leq d_{G_i}(v) \leq \left\lceil \frac{d_G(v)}{k} \right\rceil .$$

Lemma 3.28 $M_{(\frac{1}{4},1]}(n,r) \leq \lceil \frac{9}{4}n \rceil$.

Proof. Let G be the bipartite connection graph induced by connection frame $\mathcal{F}$. An edge in G with weight greater than $\frac{1}{2}$ is called *heavy*, and *light* otherwise. Let h_v denote the number of heavy edges incident at vertex v; and l_v the number of light edges incident at vertex v. It follows from the compatibility of $\mathcal{F}$ that $h_v \leq n$ and $l_v \leq h_v + 3(n - h_v)$ since each inlet/outlet involved in a heavy connection could be involved in one more light connection, and could be involved in up to three light connections. For ease of exposition, assume that 4 divides n.

Let G_h and G_l denote the spanning subgraphs of G including all heavy edges and all light edges, respectively. By Lemma 3.27, G_l could be partitioned into two spanning subgraphs G_l^1 and G_l^2, in each of which the degree of vertex v is less than or equal to $\frac{3}{2}n - h_v$. Therefore, the degree of vertex v in $G_l^1 \cup G_h$ is at most $\frac{3}{2}n$. It follows that graph $G_l^1 \cup G_h$ can be $\frac{3}{2}n$-colored. Notice that graph G_l^2 can be $\frac{3}{4}n$-colored. Thus, graph G can be $\frac{9}{4}n$-colored, which means that connection frame $\mathcal{F}$ can be routed through $C(n, \frac{9}{4}n, r)$. □

Lemmas 3.26 and 3.28 lead to

Theorem 3.29 $M_{(0,1]}(n,r) \leq \lceil \frac{41n - E_n}{16} \rceil \approx 2.5625n$, *where* $E_n = 8, 5, 6, 3$ *if* $n \equiv 0,\ 1,\ 2,\ 3\ (mod\ 4)$.

3.3.2 Discrete Bandwidth Case

Chung and Ross [2] conjectured that $C(n, 2n-1, r)$ is rearrangeable in the discrete bandwidth case. It could be easily checked that this conjecture is equivalent to that $C(n, 2n-1, r)$ is k-rate rearrangeable for any k and any set of k rates. In other words, in the discrete bandwidth case, whether or not b_i is an integral multiple of b_k, for every i, is not an important issue to the conjecture. Nonetheless, integral multiplicity does play an important role. Lin *et al.* [14] have shown that $C(n, 2n-1, r)$ is rearrangeable in the restricted discrete bandwidth case. More generally,

Theorem 3.30 $C(n, 2n-1, r)$ *is multirate rearrangeable when the bandwidth of all connections belongs to a given finite set* $\{b_1, b_2, \cdots, b_k\}$, *where* $1 \geq b_1 > b_2 > \cdots > b_k > 0$ *and* $b_{i+1} \mid b_i$, *for* $i = 2, 3, \cdots, k-1$.

Proof. Notice that b_1 is not necessarily an integral multiple of b_2 and thus this particular k-rate case is a little relaxed restricted discrete bandwidth case.

By Corollary 3.21, we know that the theorem holds when $b_1 \leq \frac{1}{2}$. Therefore, we may assume that $b_1 > \frac{1}{2}$. The proof is done by induction on k, the number of distinct rates. Note that the base case $k = 1$ is true by Theorem 3.17.

Let l_1 and l_2 be the integers such that $b_1 + l_1 b_k \leq 1 < b_1 + (l_1 + 1)b_k$ and $l_2 b_k \leq 1 < (l_2 + 1)b_k$. For any given connection frame $\mathcal{F}$, route first the connections with bandwidth $b_1, b_2, \cdots, b_{k-2}$, and b_{k-1}. By the induction hypothesis, these connections can be routed successfully through $C(n, 2n-1, r)$. After that, consider the routing of the connections with bandwidth b_k, which are called b_k-connections, in an arbitrary ordering. Suppose to the contrary that we fail to route a connection (i, j, b_k) at some time. Let I be the input switch containing inlet i and J the output switch containing outlet j. Define the I-load (the J-load) of a central switch as the sum of the bandwidth of all connections from input switch I to the central switch (from the central switch to the output switch J, respectively). Then every central switch has either its I-load or its J-load greater than $1 - b_k$. Note that there are $2n-1$ central switches in the network. Therefore, we can find either n central switches each of which has I-load greater than $1 - b_k$ or n central switches each of which has J-load greater than $1-b_k$. Without loss of generality, we assume that the former setting occurs. This means that each of those n central switches has I-load equal to either $b_1 + l_1 b_k$ or $l_2 b_k$. On the other hand, as every inlet can be involved in at most one b_1-connection, it follows that every inlet on input switch I has load of either $b_1 + l_1 b_k$ or $l_2 b_k$, too. Therefore, the connection (i, j, b_k) cannot exist, a contradiction. □

The following even more general theorem could be similarly proved.

Theorem 3.31 *$C(n, 2n-1, r)$ is rearrangeable in the following k-rate case: $1 \geq b_1 > b_2 > \cdots > b_{i-1} > \frac{1}{2} \geq b_i > b_{i+1} > \cdots > b_k > 0$ and $b_{l+1} \mid b_l$, for $l = i, i+1, \cdots, k-1$.*

Corollary 3.32 *$C(n, 2n-1, r)$ is 2-rate rearrangeable.*

Suppose that $\{b_1, b_2, \cdots, b_k\}$ and $\{b'_1, b'_2, \cdots, b'_k\}$, where $1 \geq b_1 > b_2 > \cdots > b_k > 0$ and $1 \geq b'_1 > b'_2 > \cdots > b'_k > 0$, are two sets of k rates. They are said to be *isomorphic* if for every set of k natural numbers $n_1, n_2, \cdots, n_k$, $\sum_{i=1}^{k} n_i b_i \leq 1$ if and only if $\sum_{i=1}^{k} n_i b'_i \leq 1$. Clearly, if two sets of k rates are isomorphic, then $C(n, m, r)$ is k-rate rearrangeable with one set if and only if it is k-rate rearrangeable with the other set. As an application using this concept, consider a 3-rate set $\{b_1, b_2, b_3\}$ where $1 \geq b_1 > \frac{1}{2} \geq b_2 > \frac{1}{3}$ and $\frac{1}{4} > b_3 > \frac{1}{5}$, and they satisfy that $b_1 + b_3 > 1$ and $2b_2 + b_3 > 1$. Then, $\{b_1, b_2, b_3\}$ is isomorphic to $\{1, 0.42, 0.21\}$. Theorem 3.30/3.31 says that $C(n, 2n-1, r)$ is 3-rate rearrangeable with $\{1, 0.42, 0.21\}$, and thus it is also 3-rate rearrangeable with $\{b_1, b_2, b_3\}$.

3.3.3 3-Rate Rearrangeability

To attack Chung and Ross's Conjecture, one way is to prove the conjecture in the k-rate case, for any k. Theorem 3.17 and Corollary 3.32 tell us that the conjecture is true when $k = 1$ and $k = 2$. But how many central switches are required for the network to be 3-rate rearrangeable? This subsection summarizes several results achieved. Assume that the three rates are b_1, b_2 and b_3, satisfying $1 \geq b_1 > b_2 > b_3 > 0$. For simplicity, an edge in the induced connection graph is called a b_i-*edge* if it represents a b_i-connection.

Given a connection frame $\mathcal{F}$, besides the induced connection graph G, we can construct another edge-weighted bipartite connection graph H, as follows: H takes the rn inlets as one part, the rn outlets as the other part, and an edge of weight w between vertices i and j for each connection $(i,j,w) \in \mathcal{F}$. Suppose T is a spanning subgraph of graph H, by contracting the n inlets/outlets at each input/output switch into one vertex, we obtain a spanning subgraph S for graph G. Subgraph S is called the *translation* of subgraph T.

Lemma 3.33 *$C(n,2n,r)$ is 3-rate rearrangeable when* $1 \geq b_1 > b_2 > b_3 > \frac{1}{5}$.

Proof. Notice that $C(n,2n,r)$ is 3-rate rearrangeable if one of the following two conditions is satisfied:

(a) $b_1 \leq \frac{1}{2}$;

(b) Two of b_1, b_2 and b_3 lie in the same interval $(\frac{1}{p+1}, \frac{1}{p}]$, for some positive integer p.

Therefore, we need only to consider the following three cases. Let $\mathcal{F}$ be a connection frame and G and H its two induced connection graphs.

Case 1. $1 \geq b_1 > \frac{1}{2} \geq b_2 > \frac{1}{3} \geq b_3 > \frac{1}{4}$. Since an input/output switch can generate at most $2n$ calls having bandwidth greater than $\frac{1}{3}$ and at most $3n$ calls having bandwidth greater than $\frac{1}{4}$, we apply the CAP algorithm to G using $2n$ colors. At each vertex, a b_1-edge and a b_2-edge can never be colored the same. Thus, when $b_1 + b_3 \leq 1$, the sum of the weight of like-color edges at any vertex never exceeds unity. This means G can be $2n$-colored.

When $b_1 + b_3 > 1$, again letting h_v denote the number of (heavy) b_1-edges incident at vertex v and l_v denote the number of (light) b_2- and b_3-edges incident at v, then $l_v \leq 3n - 3h_v$. By Lemma 3.27, we can partition the spanning subgraph G_l (containing all light edges) into three spanning subgraphs G_l^1, G_l^2, and G_l^3, in each of which the degree of vertex v is at most $n - h_v$. It follows that $G_l^1 \cup G_h$ can be n-colored, so can $G_l^2 \cup G_l^3$. Thus, G can be $2n$-colored too.

Case 2. $1 \geq b_1 > \frac{1}{2} \geq b_2 > \frac{1}{3}$, *and* $\frac{1}{4} \geq b_3 > \frac{1}{5}$. In G, there might be up to $4n$ edges incident at any vertex and, among them, at most $2n$ edges having weight b_1 and b_2. When $b_1 + b_3 \leq 1$, a same argument as in Case 1 leads to a $2n$-coloring of G.

When $b_1 + b_3 > 1$, for any vertex v, $h_v \leq n$ and $l_v \leq 4n - 4h_v$. By Lemma 3.27, we can partition the spanning subgraph G_l into 4 spanning subgraphs G_l^1, G_l^2, G_l^3, and G_l^4, in each of which vertex v has degree at most $n - h_v$. It follows that $G_l^1 \cup G_h$ can be n-colored. Since there are at most two b_2-edges incident at any vertex, if $2b_2 + b_3 \leq 1$ then $G_l^2 \cup G_l^3 \cup G_l^4$ can be n-colored too. Otherwise, $\{b_1, b_2, b_3\}$ is isomorphic to $\{1, 0.42, 0.21\}$ and thus by Theorem 3.30 $2n - 1$ central switches suffice to route $\mathcal{F}$.

Case 3. $1 \geq b_1 > \frac{1}{2}$, *and* $\frac{1}{3} \geq b_2 > \frac{1}{4} \geq b_3 > \frac{1}{5}$. Let us consider first the setting $b_1 + b_2 > 1$, which implies that $b_1 + 2b_3 > 1$. When $b_1 + b_3 > 1$, for every vertex v in G, $l_v \leq 4n - 4h_v$. By Lemma 3.27, the spanning subgraph G_l can be partitioned into 4 spanning subgraphs G_l^1, G_l^2, G_l^3, and G_l^4, in each of which vertex v has degree less than or equal to $n - h_v$. Therefore, $G_l^1 \cup G_h$ can be n-colored and so can $G_l^2 \cup G_l^3 \cup G_l^4$. When $b_1 + b_3 \leq 1$, we turn to consider graph H. Similarly, let h_v denote the number of (heavy) b_1-edges incident at vertex v, and l_v the number of (light) b_2- and b_3-edges incident at vertex v. It follows that $h_v \leq 1$, $l_v \leq 4$, and $h_v = 1$ implies $l_v \leq 1$. Let H_h denote the spanning subgraph of H consisting of edges having weight b_1 and H_l the spanning subgraph of H consisting of edges having weight b_2 and b_3. By Lemma 3.27, H_l can be partitioned into 4 spanning subgraphs H_l^1, H_l^2, H_l^3, and H_l^4, each of which is a (spanning) matching. Consider the spanning subgraph $H_h \cup H_l^1$. In the translation of this spanning subgraph (recall that the translation is a spanning subgraph of graph G), at every vertex v, there are h_v b_1-edges, at most $n - h_v$ b_2-edges and at most n b_3-edges incident. It follows that this translation can be n-colored by the CAP algorithm. The translation of $H_l^2 \cup H_l^3 \cup H_l^4$ can also be n-colored since $b_3 < b_2 \leq \frac{1}{3}$. Therefore, G can be $2n$-colored.

In the other setting, that is $b_1 + b_2 \leq 1$, there are at most $4n$ edges incident at any vertex v in graph G and, among them, at most n edges are b_1-edges. Using $2n$ colors in the CAP algorithm gives a feasible coloring for G. □

Lemma 3.34 $C(n, \lceil \frac{9}{4}n \rceil, r)$ *is 3-rate rearrangeable when* $b_2 \leq \frac{1}{5}$.

Proof. Setting $B = \frac{1}{5}$ in Theorem 3.20, we conclude that $\lceil \frac{5}{4}n \rceil$ central switches are sufficient to route all b_2- and b_3-calls in any connection frame $\mathcal{F}$. Those b_1-calls can be routed though the other n central switches. Therefore, $\mathcal{F}$ is realizable in $C(n, \lceil \frac{9}{4}n \rceil, r)$. □

Lemma 3.35 $C(n, \lceil \frac{9}{4}n \rceil, r)$ *is 3-rate rearrangeable when* $b_3 \leq \frac{1}{8}$.

Proof. Notice that all b_1-calls and b_2-calls in any connection frame $\mathcal{F}$ can be routed through $2n$ central switches. The lemma follows immediately from Lemma 3.26. □

By using similar techniques as those in the proofs of Lemmas 3.33, 3.34 and 3.35, the following three lemmas are proved in [15].

Lemma 3.36 *$C(n, \lceil (2+\frac{1}{p})n \rceil, r)$ is 3-rate rearrangeable when $b_1 > \frac{1}{2} \geq b_2 > \frac{1}{3}$, and $\frac{1}{p} \geq b_3 > \frac{1}{p+1}$, where $5 \leq p \leq 7$.*

Lemma 3.37 *Suppose that $1 \geq b_1 > \frac{1}{2}$, $\frac{1}{3} \geq b_2 > \frac{1}{4}$, and $\frac{1}{p} \geq b_3 > \frac{1}{p+1}$, where $5 \leq p \leq 7$. Then, $C(n, \lceil \frac{9}{4}n \rceil, r)$ is 3-rate rearrangeable when $p = 5$; $C(n, \lceil \frac{20}{9}n \rceil, r)$ is 3-rate rearrangeable when $p = 6$; $C(n, \lceil \frac{15}{7}n \rceil, r)$ is 3-rate rearrangeable when $p = 7$.*

Lemma 3.38 *$C(n, \lceil \frac{9}{4}n \rceil, r)$ is 3-rate rearrangeable when $b_1 > \frac{1}{2}$, $\frac{1}{4} \geq b_2 > \frac{1}{5}$, and $\frac{1}{p} \geq b_3 > \frac{1}{p+1}$, where $5 \leq p \leq 7$.*

Corollary 3.21, Lemmas 3.33–3.38 and the fact that if two of b_1, b_2 and b_3 lie in a same interval $(\frac{1}{p+1}, \frac{1}{p}]$ for some positive integer p then $C(n, 2n, r)$ is rearrangeable, imply the following theorem.

Theorem 3.39 *$C(n, \lceil \frac{9}{4}n \rceil, r)$ is 3-rate rearrangeable.*

How to further reduce the $\lceil \frac{9}{4}n \rceil$-bound is a challenging problem. Obviously, it stays in the way to Chung and Ross's Conjecture. Solving this problem might serve as a critical practice for proving/disproving the conjecture.

4 Extensions

In the literature, there are several kinds of extensions and generalizations of the symmetric 3-stage Clos networks studied in the above. These include more general assumptions on the link/inlet/outlet capacity, more general connecting structures of the networks, different types of crossbars/switches, and the number of senders and the number of receivers involved in a connection. Regarding the link/inlet/outlet capacity, it is assumed that connection links have unit capacity while inlet/outlet may have capacity β. Some papers study using arrays which have n inlets and m outlets but less than nm crosspoints. Some others study the asymmetric networks where there are r_1 $n_1 \times m$ input switches and r_2 $m \times n_2$ output switches. Apart from the one-to-one connections (calls), there are efforts put on routing one-to-many broadcasts and many-to-many multicasts. In this section, we briefly review some results on routing broadcasts in the symmetric 3-stage Clos networks and the routing properties of the asymmetric 3-stage Clos networks. Nonetheless, we remark that basically the techniques involved are those used in the above.

4.1 Broadcasting

A *broadcast* requests a message to be sent from an inlet i to several outlets, say $j_1, j_2, \cdots, j_l$. It is usually assumed that every switch has the ability to fan-out the message coming from one of its inlet to arbitrary set of outlets, as long as the outlet has enough capacity. Typically, if there are two outlets j_1 and j_2 on a same output switch J and they are receivers of some broadcast request, then when J receives the message it will fan-out the massage to both j_1 and j_2, simultaneously. For this reason, we may assume that the receivers of a broadcast request are on different output switches. We use $(i; j_1, j_2, \cdots, j_l; w)$ to denote a broadcast request, where w again denotes the bandwidth requirement and $1 \le l \le r$.

Although it is reasonable to let the output switches have fan-out n (such that all its n outlets are receivers of a same broadcast request), it would be better to restrict the fan-out of input switches and central switches during routing. Let $f(r) = \min_{1 \le x \le r}(x + r^{\frac{1}{x}})$. It is not hard to show that $f(r) = O(\frac{\log r}{\log\log r})$ and the minimum is achieved at $x^* = \frac{1}{\delta}\frac{\log r}{\log\log r}$, where $0 < \delta < 1$. Restricting the maximum fan-out of input switches and central switches to be x^* would save a lot of central switches. Two typical results are cited in the following.

Theorem 4.1 [21] *$C(n, m, r)$ is strictly nonblocking in the classical switching, for general broadcast requests, if $m > (n-1)f(r)$ and the input switches and central switches have fan-out x^*.*

Kim and Du [12] proposed a suitable modification of the QUOTA scheme [10], and proved the following theorem.

Theorem 4.2 *Suppose that the input switches and central switches have fan-out x^*. Then, in the continuous bandwidth case where the bandwidth interval is $(0, B]$, $C(n, m, r)$ is multirate WSNB if $m > 2.875nf(r)$, or more precisely, letting $p = \lfloor \frac{1}{B} \rfloor$,*

$$m > \begin{cases} \frac{(p+1)(Bp+B+p-1)n}{p^2} f(r) & \text{for } B < \frac{23}{32}, \\ (\frac{15n}{8} + n - 1)f(r) & \text{for } B \ge \frac{23}{32}. \end{cases}$$

4.2 Asymmetric 3-Stage Clos Networks

Let $C(n_1, r_1, m, n_2, r_2)$ denote the asymmetric 3-stage Clos network in which there are r_1 input switches each having n_1 inlets, r_2 output switches each having n_2 outlets, and m central switches. Again, every central switch has a link to every input/output switch.

Results obtained for the symmetric 3-stage Clos networks may be generalized to the asymmetric 3-stage Clos networks [3, 6, 12, 16, 21]. Let $n = \max\{n_1, n_2\}$ and assume that $n = n_1$. By adding $n - n_2$ imaginary outlets (having capacity 0) to each output switch, $C(n_1, r_1, m, n_2, r_2)$ can be treated as $C(n, r_1, m, n, r_2)$. The following three theorems correspond to Theorems 2.1, 2.2 and 3.29, respectively.

Theorem 4.3 $C(n_1, r_1, m, n_2, r_2)$ *is strictly nonblocking in the classical switching if and only if* $m \geq n_1 + n_2 - 1$.

Theorem 4.4 $C(n_1, r_1, m, n_2, r_2)$ *is rearrangeable in the classical switching if and only if* $m \geq n$*, where* $n = \max\{n_1, n_2\}$.

Theorem 4.5 $C(n_1, r_1, m, n_2, r_2)$ *is multirate rearrangeable if* $m \geq \lceil \frac{41n - E_n}{16} \rceil$*, where* $n = \max\{n_1, n_2\}$.

5 Related Problems

5.1 Weighted Edge-Coloring

Let $G = (V_1, V_2, E, w(\cdot))$ denote an edge-weighted bipartite multigraph, where each edge $e \in E$ is associated with weight $w(e)$, $b \leq w(e) \leq B$ and b and B are constants satisfying $0 < b \leq B \leq 1$. The edges are to be colored such that for each vertex v and each color c, the sum of weight of all edges incident at v with color c is at most unity. The goal is to minimize the number of colors. Notice that for $b = B = 1$, the problem reduces to the usual edge-coloring problem.

In the proofs of many theorems, most particularly those in Section 3.3.3, an edge-weighted bipartite multigraph G is constructed for a connection frame $\mathcal{F}$. Moreover, the number of colors used in a feasible edge-coloring of G corresponds to the number of central switches needed in the network in order to route $\mathcal{F}$. We note that in such an induced graph G, there is an additional condition that the sum of weight of all edges incident at a vertex is at most unity. For the problem without this additional condition, let $M(n)$ denote the minimum number of colors required in a feasible edge-coloring, where $n = |V_1| = |V_2|$.

Theorem 5.1 [6] $2n - 1 \leq M(n) \leq \lfloor \frac{17n-5}{6} \rfloor$.

Extensive investigation into this weighted edge-coloring problem (particularly, for bipartite graphs) is expected and new insights may be helpful in getting better nonblocking results for Clos networks.

5.2 Graph Factor and Factorization

Given a graph $G = (V, E)$, F is called a *factor* of G if F is a spanning subgraph of G. Furthermore, if for each vertex v, the degree of v in F is 1, then F is called a 1-*factor* (or a *matching*) of G. Similarly, if the degree of vertex v in F is p for some constant p, then F is called a p-*factor* of G. If the degree of vertex v is in the interval $[a(v), b(v)]$ for some integer-value functions $a(\cdot) \leq b(\cdot)$ defined on V, then F is called an $[a(\cdot), b(\cdot)]$-*factor* of G.

For a pair of functions $a(\cdot)$ and $b(\cdot)$, if the edge set E of graph G can be decomposed into (disjoint) subsets E_1, E_2, $\cdots$, E_l, such that each spanning subgraph $F_i = (V, E_i)$, $i = 1, 2, \cdots, l$, is an $[a(\cdot), b(\cdot)]$-factor, then $\{F_1, F_2, \cdots, F_l\}$ forms an $[a(\cdot), b(\cdot)]$-*factorization* of G. In the special case, those two functions are constants.

Lemma 3.27 due to de Werra tells that a bipartite graph always has a $\left[\lfloor \frac{d(v)}{k} \rfloor, \lceil \frac{d(v)}{k} \rceil\right]$-factorization, for any $k \geq 1$, where $d(v)$ denotes the degree of vertex v in the graph. Any improvement in this line of research, in particular for those cases studied in the proofs of lemmas and theorems in Section 3.3.3, would implies better results on (multirate) rearrangeability.

References

[1] V.E. Beneš. *Mathematical Theory of Connecting Networks and Telephone Traffic.* Mathematics in Science and Engineering, Vol. 17. Academic Press, New York, 1965.

[2] S.-P. Chung and K.W. Ross. On nonblocking multirate interconnection networks. *SIAM Journal on Computing*, 20:726–736, 1991.

[3] C. Clos. A study of non-blocking switching networks. *The Bell System Technical Journal*, 32:406–424, 1953.

[4] D. de Werra. Balanced schedules. *INFOR. Canadian Journal of Operational Research and Information Processing*, 9:230–237, 1971.

[5] D.Z. Du, P.C. Fishburn, B. Gao, and F.K. Hwang. Wide-sense nonblocking for 3-stage Clos networks. Technical report, Department of Computer Science, University of Minnesota, 1995.

[6] D.Z. Du, B. Gao, F.K. Hwang, and J.H. Kim. On multirate rearrangeable Clos networks. *SIAM Journal of Computing*, 28:463–470, 1999.

[7] D.Z. Du, X.D. Hu, G.-H. Lin, H.Z. Shi, and S.X. Gao. On wide-sense nonblocking in 1-rate environment for 3-stage Clos networks. Technical report, Institute of Applied Mathematics, Chinese Academy of Sciences, 1997.

[8] A.M. Duguid. Structural properties of switching networks. Progr. rept. BTL-7, Brown Univ., 1959.

[9] P.C. Fishburn, F.K. Hwang, D.Z. Du, and B. Gao. On 1-rate wide-sense nonblocking for 3-stage Clos networks. *Discrete Applied Mathematics*, 78:75–87, 1997.

[10] B. Gao and F.K. Hwang. Wide-sense nonblocking for multirate 3-stage Clos networks. *Theoretical Computer Sciences*, 182:171–182, 1997.

[11] A. Jajszczyk. On nonblocking switching networks composed of digital symmetrical matrices. *IEEE Transactions on Communications*, COM-31:2–9, 1983.

[12] D.S. Kim and D.Z. Du. Multirate broadcast switching networks nonblocking in a wide sense. In *Advances in Switching Networks*, DIMACS Series in Discrete Mathematics and Theoretical Computer Science 42, pages 59–74. American Mathematics Society, Providence, RI, 1998.

[13] D.S. Kim and D.Z. Du. Multirate multicast switching networks. In *Proceedings of the Fourth Annual International Computing and Combinatorics Conference (COCOON'98)*, LNCS, pages 219–228, 1998.

[14] G.-H. Lin, D.-Z. Du, X.-D. Hu, and G. Xue. On rearrangeability of multirate Clos networks. *SIAM Journal on Computing*, 28:1225–1231, 1999.

[15] G.-H. Lin, D.Z. Du, W. Wu, and K. Yoo. On 3-rate rearrangeability of Clos networks. In *Advances in Switching Networks*, DIMACS Series in Discrete Mathematics and Theoretical Computer Science 42, pages 315–333. American Mathematics Society, Providence, RI, 1998.

[16] G.-H. Lin, D.S. Kim, and D.Z. Du. Strictly nonblocking multirate multicast Clos networks. In *Proceedings of the 10th International Conference on Parallel and Distributed Computing and Systems*, pages 417–420, 1998.

[17] R. Melen and J.S. Turner. Nonblocking multirate networks. *SIAM Journal on Computing*, 18:301–313, 1989.

[18] G. Niestegge. Nonblocking multirate switching networks. In M. Bonatti and M. Decina, editors, *Traffic Engineering for ISDN Designing and Planning*, Elsevier, Amsterdam, 1988.

[19] M.C. Paull. Reswitching of connection networks. *The Bell System Technical Journal*, 41:833–855, 1962.

[20] D. Slepian. Two problems on a particular crossbar switching network. Unpublished manuscript, 1952.

[21] Y. Yang and G.M. Masson. Nonblocking broadcast switching networks. *IEEE Transactions on Computers*, 40:1005–1015, 1991.

SWITCHING NETWORKS: RECENT ADVANCES
D.-Z. Du and H.Q. Ngo (Eds.) pp. 143-178

Supporting Multicast Communication in Clos-Type Switching Networks

Yuanyuan Yang
Department of Electrical and Computer Engineering
State University of New York at Stony Brook, Stony Brook, NY 11794
E-mail: yang@ece.sunysb.edu

Contents

References

Abstract

Multicast (or one-to-many) is a vital operation in both broadband integrated services digital networks (BISDN) and scalable parallel and distributed computing systems. There are broad ranges of applications for multistage networks which can realize multicast connection requests in a nonblocking manner in the sense that a request from an idle network input port to be connected to some set of idle network output ports can always be satisfied without any disturbance of existing multicast connections in the network. In this chapter, we overview the recent development in supporting multicast in a well-known multistage network called Clos network. Both nonblocking and almost nonblocking multicast networks are considered. We first present results that lead to the currently best available nonblocking multicast network, and describe a linear time routing algorithm for satisfying multicast connection requests in the network. Next, we will show that the sufficient nonblocking condition used in the network design is also a necessary condition under several typical routing strategies for a multicast network to be nonblocking, discuss the complexity of the control strategies used, and briefly describe the hardware implementation of the best control strategy, which further speeds up the routing process. Finally, we discuss the blocking behavior of the multicast network that does not satisfy the nonblocking condition. We present two analytical models for the blocking probability of the multicast network, and show the simulation results on the blocking behavior of the network under several routing strategies. The analytical and simulation results demonstrate that a Clos-type multicast network with a comparable cost to a permutation is almost nonblocking for multicast connections and can provide cost-effective support for multicast communication.

1 Introduction

Multicast or one-to-many communication is highly demanded in broadband integrated services digital networks (BISDN) and scalable parallel and distributed computing systems. Some examples are video conference calls and video-on-demand services in BISDN networks, and barrier synchronization and write update/invalidate in directory-based cache coherence protocols in parallel computers. In general, providing multicast support at hardware/network level is the most efficient way supporting such communication operations. In this chapter, we look into the issue of supporting multicast in the well-known three-stage Clos network [1]. Clos-type networks have been widely used in various interconnection problems. Some recent applications include the NEC ATOM switch designed for BISDN, the IBM GF11 multiprocessor and ANSI Fibre Channel Standard for interconnection of processors to the I/O system. More recently, it was shown that the network in the IBM SP2 is functionally equivalent to the Clos network.

Clos-type networks have been extensively studies for both unicast communication and multicast communication in the literature. In this chapter, we overview the recent development in supporting multicast in this type of network. We will first present results that lead to the currently best available nonblocking multicast network, and describe a linear time routing algorithm for satisfying multicast connection requests in the network. Next, we will show that the sufficient nonblocking condition used in the network design is also a necessary condition under several typical routing strategies for a multicast network to be nonblocking, discuss the complexity of the control strategies used, and briefly describe the hardware implementation of the best control strategy, which further speeds up the routing process. Finally, we will discuss the blocking behavior of the multicast network that does not satisfy the nonblocking condition. We will present two analytical models for the blocking probability of the multicast network, and show the simulation results on the blocking behavior of the network under several routing strategies. The analytical and simulation results demonstrate that a multicast network with a comparable cost to a permutation is almost nonblocking for multicast connections and can provide cost-effective support for multicast communication.

2 Preliminaries and Previous Work

In general, a three-stage Clos network or a $v(m,n,r)$ network has r $(n \times m)$ switches in the first stage (or input stage), m $(r \times r)$ switches in the middle stage, and r $(m \times n)$ switches in the third stage (or output stage). The network has exactly one link between every two switches in its consecutive stages. Figure 1 illustrates a general schematic of a $v(m,n,r)$ network. Since two of the $v(m,n,r)$ network parameters, n and r, are restricted by the number of network input/output ports, the main focus of the study is to determine the minimum value of the network parameter m for a certain type of connecting capability to achieve the minimum network cost.

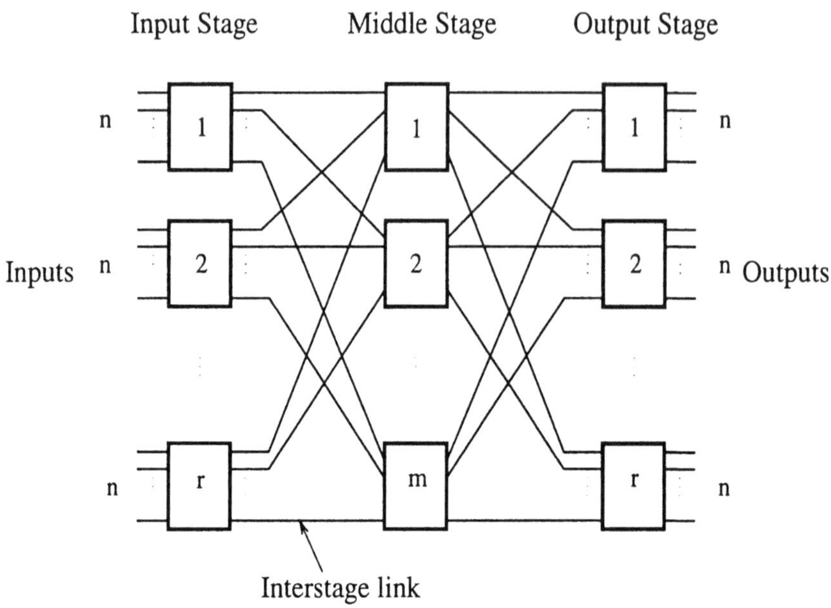

Figure 1: A general schematic of an $N \times N$ $v(m,n,r)$ network, where $N = nr$.

When the $v(m,n,r)$ network is considered for supporting multicast, it is reasonable to assume that every switch in the network has multicast capability. Then since output stage switches have multicast capability, a *multicast connection* from an input port can be simply expressed in terms of output stage switches it connects to. The number of output stage switches in a multicast connection is referred to as the *fanout* of the multicast connection. A maximum set of multicast connections is referred to as a *multicast assignment.*

Earlier work in the literature [2, 3] showed that a $v(m,n,r)$ network is *nonblocking* for arbitrary multicast assignments if the number of middle stage switches $m \geq cnr$, where $c \approx 1$. By nonblocking we mean that any arbitrary multicast connection between an idle network input port and a set of idle network output ports can always be realized without any *rearrangement* of the existing connection in the network. Clearly, the number of middle stage switches required for a nonblocking multicast network is much larger than that for a nonblocking permutation network which requires only $m \geq 2n - 1$ [1]. For the Clos-type network to be used for multicast applications, we need to greatly reduce the number of middle stage switches required for nonblocking. In the next section, we will describe such a design which achieves this goal.

3 An Efficient Design for Nonblocking Multicast Networks

As we saw earlier, the number of middle stage switches required for a nonblocking multicast network is much larger than that for a nonblocking permutation network. This is mainly due to the non-uniform nature of multicast connections. We must find some way to make the multicast connections more regular. Otherwise, in the worst case, to guarantee the network nonblocking, the number of middle stage switches can become very large. Aiming at the non-uniformity of multicast connections, in a recent design [4], a routing control strategy was adopted, which effectively reduces such non-uniformity and leads to the currently best available nonblocking multicast network.

In this section, we present the main idea of this design and the details of the design can be found in [4]. We will first show the sufficient conditions for nonblocking multicast networks and then present a linear time routing algorithm for satisfying multicast connection requests in the network. The hardware implementation of this routing algorithm will be discussed at the end of the next section.

3.1 Sufficient nonblocking conditions

First of all, we need to introduce some terminologies. For each input port $i \in \{1, \ldots, rn\}$, let $I_i \subseteq O = \{1, \ldots, r\}$ denote the subset of the switches in the output stage to which input i is to be connected in a multicast assignment. I_i is called a *connection request from input* i. Let $M_j \subseteq O = \{1, \ldots, r\}, j = 1, 2, \ldots, m$, denote the subset of the switch modules in the output stage to which the M_j is providing connection paths from the input ports. We will refer to the sets $M_j \subseteq O = \{1, \ldots, r\}, j = 1, 2, \ldots, m$, as the *destination sets* of the middle stage switches. For a connection request from input i, we will refer to the set of middle stage switches with currently unused links to the input stage switch associated with input i as the *available* middle switches for this connection request.

We characterize the condition on the status of the middle stage switches for realizing a multicast connection in the following theorem:

Theorem 3.1 *We can satisfy a new connection request I_i, $i \in \{1, 2, \ldots, nr\}$, in a $v(m, n, r)$ network using some x ($x \geq 1$) middle stage switches, say,*

$j_1, \ldots, j_x$, from among the available middle switches if and only if I_i and the current destination sets of these x middle stage switches satisfy

$$I_i \cap (\bigcap_{k=1}^{x} M_{j_k}) = \phi. \tag{1}$$

Now, we will see how to satisfy a multicast connection in a nonblocking manner. To reduce the non-uniformity of multicast connections, an "intelligent" routing control strategy is employed for satisfying each multicast connection request: choose no more than a certain number of middle stage switches, say, x, whose destination set intersections are empty from available middle switches. In other words, a limit is imposed on the number of middle stage switches which are used for satisfying a multicast connection request in order to reduce the total number of middle stage switches required for any multicast assignment. We then determine how many available middle switches can guarantee that these x middle stage switches can always be chosen. We are interested in finding as few as possible available middle switches with this property.

The following theorem reveals an inherent property of destination sets which answers this question.

Theorem 3.2 *For all n', $1 \leq n' \leq n$, and for all x, $1 \leq x \leq \min\{n', r\}$, let m' be the maximum number of middle stage switches whose destination sets satisfying the condition that there are at most n' 1's, n' 2's, ..., n' r's distributed among the destination sets and the intersection of any x of the destination sets is nonempty. Then we have that*

$$m' \leq n' r^{\frac{1}{x}}$$

Since this theorem is critical in determining the nonblocking condition, we give a brief proof here.

Proof sketch: Without loss of generality, suppose these m' middle stage switches are $1, 2, \ldots, m'$ with destination sets $M_1, M_2, \ldots, M_{m'}$. Clearly,

$$\sum_{i=1}^{m'} |M_i| \leq n' r$$

Let

$$c_1 = \min_i\{|M_i|\}.$$

Then, we obtain that

$$m' \leq \frac{n'r}{c_1} \tag{2}$$

Without loss of generality, suppose that the destination set of middle stage switch 1 has cardinality c_1, and $M_1 = \{1, 2, \ldots, c_1\}$.

We intersect each of the destination sets $M_1, M_2, \ldots, M_{m'}$ with M_1 and obtain m' sets $M_1^1, M_2^1, \ldots, M_{m'}^1$ which consist of only elements in $\{1, 2, \ldots, c_1\}$, and distributed among the M_i^1 's are at most n' 1's, n' 2's, $\ldots$, n' c_1's.

Let

$$c_2 = \min_i\{|M_i^1|\}$$

Again, without loss of generality, suppose that M_2^1 has cardinality c_2, and $M_2^1 = \{1, 2, \ldots, c_2\}$. We then get that

$$m' \leq \frac{n'c_1}{c_2} \tag{3}$$

Then intersect each of $M_1^1, M_2^1, \ldots, M_{m'}^1$ with M_2^1 and obtain m' sets $M_1^2, M_2^2, \ldots, M_{m'}^2$, which consist of only elements in $\{1, 2, \ldots, c_2\}$.

After repeating the above process $x - 1$ times, we have

$$m' \leq \min\{\frac{n'r}{c_1}, \frac{n'c_1}{c_2}, \ldots, \frac{n'c_{x-2}}{c_{x-1}}\} \tag{4}$$

We now have a set of m' intersected destination sets $M_1^{x-1}, M_2^{x-1}, \ldots, M_{m'}^{x-1}$. Moreover, each M_k^{x-1} consists of only elements in $\{1, 2, \ldots, c_{x-1}\}$, and there are at most n' 1's, n' 2's, $\ldots$, and n' c_{x-1}'s distributed among these m' sets. Thus, we have

$$m' \leq n'c_{x-1} \tag{5}$$

Therefore, from (3) and (4), we get that

$$m' \leq \min\{\frac{n'r}{c_1}, \frac{n'c_1}{c_2}, \ldots, \frac{n'c_{x-2}}{c_{x-1}}, n'c_{x-1}\}$$

It can be shown that

$$\max_{c_1, c_2, \ldots, c_{x-1}} \min\{\frac{n'r}{c_1}, \frac{n'c_1}{c_2}, \ldots, \frac{n'c_{x-2}}{c_{x-1}}, n'c_{x-1}\} = n'r^{\frac{1}{x}}$$

Therefore,

$$m' \leq n' r^{\frac{1}{x}}$$

□

By considering the worst case: satisfying a connection request with fanout r from an input port on the input stage switch whose all other $n-1$ input ports already had some connections, we can obtain a sufficient condition for realizing all possible multicast connections in a nonblocking manner:

Theorem 3.3 *A $v(m,n,r)$ network is nonblocking for multicast assignments if*

$$m > \min_{1 \leq x \leq \min\{n-1,r\}} \{(n-1)(x + r^{\frac{1}{x}})\}$$

Note that according to the routing control strategy, the existing connections from the $n-1$ other input ports can pass through at most $(n-1)x$ middle stage switches. By Theorem 3.2, if there are still more that $(n-1)r^{1/x}$ available middle switches for the new connection request, this connection request can be realized, which means $m > (n-1)x + (n-1)r^{1/x} = (n-1)(x + r^{1/x})$. We then take some x which minimizes $(x + r^{1/x})$. Note that in the previous work [2, 3] the nonblocking condition is approximately $m > rn$. Now we have $m > \min(x + r^{1/x})n$.

Let's take a closer look at function $y = x + r^{\frac{1}{x}}$. In Figure 2, we plot the curve of this function for $r = 64$, 256, and 1024, respectively. We can see that the function achieves its minimum value approximately at $x = 3.5$, 4 and 5, respectively. In Table 1, we list the minimum values of this function for some typical r values. By comparing the left column with the right column in the table, we can see how much the number of middle stage switches required for nonblocking has been reduced.

The nonblocking condition in Theorem 3.3 is expressed by a combinatorial formula. However, sometimes it is also desirable to represent m as an explicit function of n and r. The following theorem gives a slightly looser, but an explicit bound on m as a function of n and r.

Theorem 3.4 *A $v(m,n,r)$ network is nonblocking for multicast assignments if*

$$m > 3(n-1)\frac{\log r}{\log \log r}$$

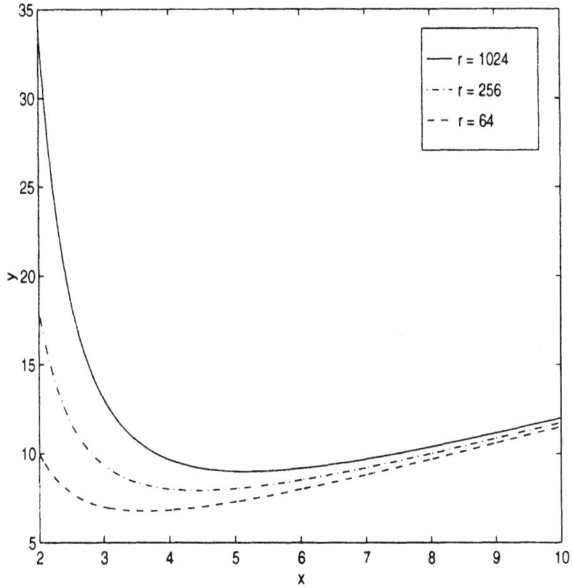

Figure 2: The curve of function $y = x + r^{\frac{1}{x}}$.

Table 1: Relationship between r and the coefficient of m: $\min(x + r^{\frac{1}{x}})$

r	x	$x + r^{\frac{1}{x}}$
1	1	2
2	1	3
$4 = 2^2$	2	4
$9 = 3^2$	2	5
$27 = 3^3$	3	6
$81 = 3^4$	4	7
$256 = 4^4$	4	8
$1024 = 4^5$	5	9
$4096 = 4^6$	6	10
$16384 = 4^7$	7	11
$78125 = 5^7$	7	12
$390625 = 5^8$	8	13
$1953125 = 5^9$	9	14
$10077696 = 6^9$	9	15

Theorem 3.4 is obtained by letting $x = \frac{2\log r}{\log\log r}$ in $(x + r^{\frac{1}{x}})$.

Furthermore, the above results for $v(m, n, r)$ networks can be generalized to restricted fanout multicast assignments in which each input port can have connection paths to at most d, $1 \leq d < r$, output stage switches:

Corollary 3.5 *A $v(m, n, r)$ network is nonblocking for restricted multicast assignments, in which each input port can be connected to at most d $(1 \leq d < r)$ output stage switches, if*

$$m > \min_{1 \leq x \leq \min\{n-1, d\}} \{(n-1)(x + d^{\frac{1}{x}})\}$$

In particular, we have

$$m > (n-1)\left(\frac{2\log d}{\log\log d} + (\log d)^{\frac{1}{2}}\right)$$

From the above Corollary, we can easily derive the special case of the classical nonblocking condition for permutation assignments:

Corollary 3.6 *Setting $d = 1$ in Corollary 1 yields $m \geq 2n - 1$, which is the bound on m associated with the classical Clos nonblocking permutation networks.*

So far we have discussed the nonblocking condition for the three-stage Clos-type multicast network. By recursively applying the design criteria on each middle stage switch, the results can be generalized to any $(2k+1)$-stage networks for $k > 1$.

A traditional means of evaluating and comparing switching network designs is to determine the asymptotic growth of crosspoint switching elements as the number of input ports and output ports increase. The following theorem gives an upper bound on the minimum number of crosspoints of a $(2k+1)$ stage multicast network based on the above design.

Theorem 3.7 *For each fixed integer $k \geq 1$, the minimum number of crosspoints of the $(2k+1)$ stage $(N \times N)$ multicast network is*

$$O\left(N^{1+\frac{1}{k+1}}(\log N/\log\log N)^{\frac{k+2}{2}-\frac{1}{k+1}}\right)$$

In Table 2, we compare the crosspoints of the network with other known constructive multistage multicast networks. Clearly, the network has the best crosspoints among all constructive designs.

Table 2: Crosspoints comparison of $N \times N$ multistage multicast networks

Networks	Stages	Crosspoints
Masson [2]	3	$O(N^{\frac{5}{3}})$
Hwang and Jajszczyk [3]	3	$O(N^{\frac{5}{3}})$
Feldman, Friedman and Pippenger [5]	2	$O(N^{\frac{5}{3}})$
Feldman, Friedman and Pippenger [5]	3	$O(N^{\frac{11}{7}})$
Yang and Masson [4]	3	$O(N^{\frac{3}{2}}(\frac{\log N}{\log\log N}))$

3.2 A linear time routing algorithm

In the previous subsection, we have determined the sufficient condition for a nonblocking multicast network. In this subsection, we look at how to actually find paths for a multicast connection in the network. Given a $v(m, n, r)$ network satisfying the nonblocking condition on m in Theorem 3.3, suppose we have determined some x, $1 \le x \le min\{n-1, r\}$ for this network. We now have a new connection request I_i with $|I_i| = r' \le r$. According to Theorem 3.3, we know that there are at least $m' = (n-1)r'^{\frac{1}{x}}+1$ available middle switches with destination sets, say, $M_1, M_2, \ldots, M_{m'}$, for this connection request. The following algorithm can be used for finding no more than x middle stage switches for satisfying I_i [4].

Algorithm:

Step 1: $mid\_switch \leftarrow \phi$;
  for $j = 1$ **to** m' **do**
    $S_j \leftarrow M_j \bigcap I_i$;

Step 2: **repeat**
    find S_k $(1 \le k \le m')$ such that
    $|S_k| = \min\{|S_1|, |S_2|, \ldots, |S_{m'}|\}$;
    $min\_set \leftarrow S_k$;
    $mid\_switch \leftarrow mid\_switch \bigcup \{k\}$;
    if $min\_set \neq \phi$ **then**
      for $j = 1$ **to** m' **do**
        $S_j \leftarrow S_j \bigcap min\_set$;
  until $min\_set = \phi$;

Step 3: connect I_i through the middle stage switches in *mid_switch* and update the destination sets of these middle stage switches.

End

It was shown that the time complexity of this algorithm is linear to the network size N [4].

It should be mentioned that this complexity analysis is based on the software, sequential implementation of the algorithm. This routing algorithm can also be realized with a relatively simple hardware implementation to satisfy those applications demanding very short network set-up times. We will briefly discuss this issue after we explore the necessary nonblocking conditions under several typical routing control strategies (including the strategy this algorithm is based on) in the next section.

4 Necessary Nonblocking Conditions for Multicast Networks

In the previous section, we presented an efficient design for nonblocking multicast networks. At this point, we may ask such questions: Can the sufficient nonblocking condition used in the design be further reduced? What is the optimal design for this type of multicast network?

To answer these questions, we now look at the necessary conditions for supporting arbitrary multicast assignments in this type of network. We will consider this issue under several typical control strategies by constructing worst case network states which force us to use a certain number of middle stage switches. We will state the main results without proofs. The proofs can be found in [6].

4.1 Network routing control strategies

The following three control strategies are used in deriving the necessary conditions.

Strategy 1 *For each input connection request, I_i, $i \in \{1, 2, \ldots, nr\}$, in the network, always choose the middle stage switch with the minimum cardinality of destination sets with regard to the unsatisfied portion of I_i from available middle switches, until I_i is satisfied, that is, until all middle stage switches chosen satisfy condition (1).*

Notice that this is the strategy used in deriving sufficient nonblocking condition and designing the routing algorithm in the previous section.

Strategy 2 *For each input connection request I_i, $i \in \{1, 2, \ldots, nr\}$, choose the minimum number of middle stage switches that satisfy condition (1) for the current network state from the available middle switches.*

Notice the difference between Strategy 2 and Strategy 1. Instead of choosing middle stage switches in the order of smaller destination set cardinality first for an input connection request in Strategy 1, Strategy 2 always finds the minimum number of available middle switches for the current input connection request. In other words, Strategy 2 achieves "local optimization" within a single input stage switch by using as few as possible output links in the input stage switch and keeping the maximum number of available middle switches for future connection requests from that input stage switch. In general, for a given $v(m, n, r)$ network, it is clear that if we can satisfy all connection requests under Strategy 1, we can definitely satisfy all connection requests under Strategy 2. Of course, since Strategy 2 needs to exam all possible subsets of available middle switches, its complexity is much higher than that of Strategy 1. Even so, it is interesting to see whether Strategy 2 can lead to a tighter nonblocking bound for a $v(m, n, r)$ multicast network.

Strategy 3 *For each input connection request I_i, $i \in \{1, 2, \ldots, nr\}$, use an empty available middle switch (i.e., middle stage switch with no connections) only when there is no any subset of non-empty available middle switches can satisfy condition (1).*

Strategy 3 keeps the total number of non-empty middle stage switches in the network as small as possible at any time with the intent of reducing the total number of middle stage switches required for nonblocking. Unlike Strategy 2, it achieves some degree of "global optimization" for all input stage switches.

The following is a fundamental lemma for constructing worst case network states:

Lemma 4.1 *For sufficiently large n, r and $\frac{m}{n}(m > n)$, there exist $m + n$ subsets of set $\{1, 2, \ldots, r\}$, $I_1, I_2, \ldots, I_n, M_1, M_2, \ldots, M_m$, which satisfy the following conditions:*

1. *the flattened set of $\{I_1, I_2, \ldots, I_n, M_1, M_2, \ldots, M_m\}$ is a multiset chosen from set $\{1, 2, \ldots, r\}$ with multiplicity of each element no more than n;*

2. *for some* $x = \Theta\left(\frac{\log r}{\log m - \log n}\right)$ *and for any* I_i $(1 \le i \le n)$ *and any* $M_{j_1}, M_{j_2}, \ldots, M_{j_x}$ $(1 \le j_1 < j_2 < \cdots < j_x \le m)$ $I_i \cap (\bigcap_{k=1}^{x} M_{j_k}) \neq \phi$.

4.2 Necessary nonblocking conditions

For each control strategy considered, we start from a network without any connections, generate a sequence of worst case connection/disconnection requests based on Lemma 4.1 and then realize these connection requests according to the control strategy. After these connection requests are realized, we look at the number of middle stage switches in use, which is the necessary number of middle stage switches required for the network to be nonblocking under this control strategy. We have following theorems regarding the necessary conditions under no strategy (i.e., strictly nonblocking) and three control strategies described in the previous subsection.

Theorem 4.2 *The necessary condition for a* $v(m,n,r)$ *network to be strictly nonblocking for multicast assignments is* $m \ge \Theta\left(n\frac{\log r}{\log\log r}\right)$.

Theorem 4.3 *The necessary and sufficient condition for a* $v(m,n,r)$ *multicast network to be nonblocking under Strategy 1 is* $m \ge \Theta\left(n\frac{\log r}{\log\log r}\right)$.

Theorem 4.4 *The necessary and sufficient condition for a* $v(m,n,r)$ *multicast network to be nonblocking under Strategy 2 is* $m \ge \Theta\left(n\frac{\log r}{\log\log r}\right)$.

Theorem 4.5 *The necessary condition for a* $v(m,n,r)$ *multicast network to be nonblocking under Strategy 3 is* $m \ge \Theta\left(n\frac{\log r}{\log\log r}\right)$.

Based on the above theorems, we can further conjecture that

Conjecture 1 *The necessary condition for a* $v(m,n,r)$ *multicast network to be nonblocking under any strategy is* $m \ge \Theta\left(n\frac{\log r}{\log\log r}\right)$.

From the necessary conditions we derived, we know that $m = O\left(n\frac{\log r}{\log\log r}\right)$ is optimal for a nonblocking $v(m,n,r)$ multicast network under Strategy 1 and Strategy 2. Furthermore, if the above conjecture is true, then $m = O\left(n\frac{\log r}{\log\log r}\right)$ should be optimal for nonblocking $v(m,n,r)$ multicast network under any control strategy.

4.3 Complexity of control strategies and implementation issues

In the last subsection, we have shown the necessary number of middle stage switches required for a multicast network to be nonblocking under different control strategies. We now further compare the complexity of these strategies in order to find a practical control strategy which leads to the minimum network cost and has the lowest time complexity.

From the last subsection we know that the necessary and sufficient conditions on the number of middle stage switches required for the network to be nonblocking under both Strategy 1 and Strategy 2 are $m \geq \Theta(n\frac{\log r}{\log\log r})$. Now, let's look at the time complexity of the routing algorithm based on these two control strategies. In [4], it was shown that the routing algorithm based on Strategy 1 has a linear time complexity. However, for Strategy 2, the problem of choosing the minimum number of middle stage switches for satisfying a multicast connection request is NP-complete. In fact, consider the sets of all idle outputs of each available middle switch. The problem of satisfying a multicast connection under Strategy 2 is equivalent to the problem of finding the minimum number of such sets which cover all destinations of the multicast connection request. Therefore, this problem is exactly the *minimum cover* problem [15], which is NP-complete. This suggests that Strategy 1 is a better strategy than Strategy 2 in terms of time complexity.

Now let's consider the routing algorithm using Strategy 3. It can be implemented by trying all non-empty middle stage switches one by one to see whether the multicast connection request can be satisfied by these non-empty middle stage switches. If not, a single empty middle stage switch will be used for this connection. Thus, the time complexity is still linear. However, it is unknown whether the condition $m \geq \Theta(n\frac{\log r}{\log\log r})$ is sufficient for the network to be nonblocking. Nevertheless, the necessary condition obtained in the last subsection suggests that Strategy 3 cannot have a lower order sufficient condition than that of Strategy 1.

Furthermore, adopting Strategy 1 has another advantage over other two strategies. Due to the regularity of the algorithm, it can be efficiently implemented in hardware to further reduce the time complexity to $O\left(\frac{\log^2 r}{\log\log r}\right)$ gate propagations. We briefly discuss this issue next, and the detailed implementation can be found in [7].

When considering hardware implementation, we may easily parallelize some frequently performed operations in the routing algorithm. For example, we can represent the destination set of a middle stage switch by an

r-bit binary vector with a binary "1" in the ith bit representing that the ith output of the middle stage switch is occupied. Then the cardinality of a destination set is simply the number of 1's in the vector and can be evaluated either by a sequential circuit, say, a counter, or by a combinational circuit, say, an adder. Furthermore, the evaluation can be performed either sequentially or in parallel among all middle stage switches. Finally, the step of finding the minimum cardinality among all the destination sets can be done by a tree-structure comparator array when the cardinalities are evaluated in parallel. Figure 3 shows an overview structure of such an implementation, where *middle stage switch state registers (M)* are used to store the destination set vectors of middle stage switches, the *counting circuit (CC)* is used to count the number of 1's in the destination sets, the *comparator array (CA)* is used to determine the middle stage switch with the minimum cardinality destination set, the *masking register (MASK)* accomplishes the bitwise intersection operations, the *minimum register (MIN)* stores the content of the current minimum cardinality middle stage switch state register, and finally, the *selected register set (SEL)* is used to store the middle stage switches selected for satisfying the connection request.

Based on the different hardware cost and the operational time trade-off, four basic types of controller designs were considered in [7]: sequential-with-counter, sequential-with-adder, parallel-with-counters and parallel-with-adders. We summarize these designs of the network controller in Table 3. From Table 3, we can see that each of hardware controller designs requires only minor additional circuitry beyond that used in the switch modules, and that each hardware controller design offers significant improvements in the time required to determine connection path routes when compared with the previous software control algorithm. In particular, in the parallel-with-adders controller design, routing paths for a connection request can be determined in $O(\frac{\log^2 r}{\log\log r})$ gate propagations compared with the $O(nr)$ (or $O(N)$) steps required in the software control algorithm. Such a controller design renders the nonblocking multicast networks useful in the applications which require high-speed network connection path set-ups.

From the above discussion, we can conclude that Strategy 1 is the best strategy among all strategies we have considered and $m \geq \Theta\left(n\frac{\log r}{\log\log r}\right)$ is optimal for this type of nonblocking multicast network under Strategy 1. As the conclusion for this section, we give several design examples in Table 4 for some practical size nonblocking multicast networks under Strategy 1. The values of r, n and m shown in the table yield the minimum cost network

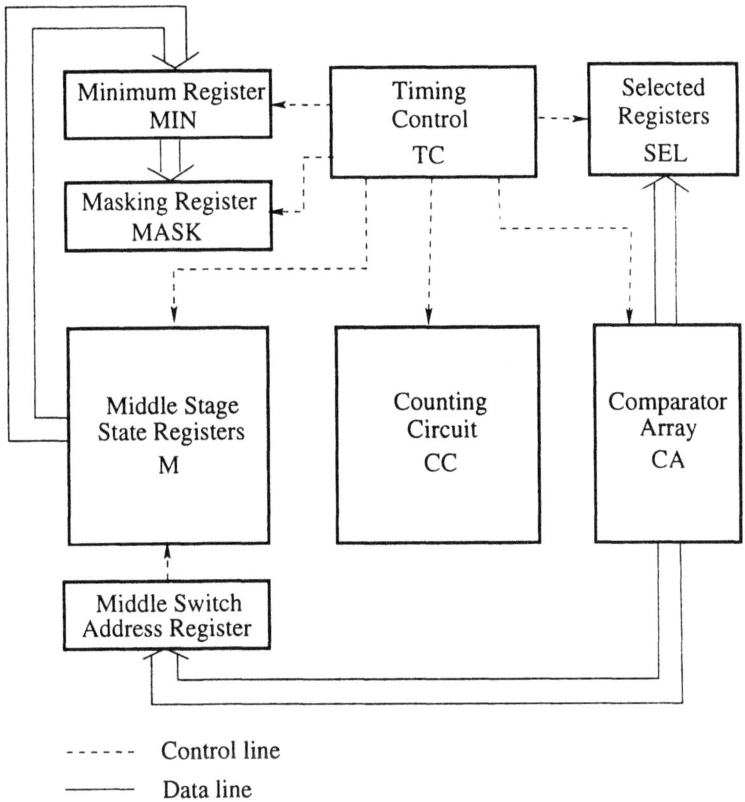

Figure 3: Overview structure of the hardware network controller.

Table 3: Summary of the various designs of the network controller.

Design	Gates	Clocks	Gate Delays
Sequential-with-counter	$O(\log r + \log n)$	$O(\frac{mr \log r}{\log \log r})$	–
Sequential-with-adder	$O(r)$	$O(\frac{m \log r}{\log \log r})$	–
Parallel-with-counters	$O(m(\log r + \log n))$	$O(\frac{r \log r}{\log \log r})$	–
Parallel-with-adders	$O(mr)$	$O(\frac{\log r}{\log \log r})$	$O(\frac{\log^2 r}{\log \log r})$

Table 4: Examples of optimal $v(m, n, r)$ nonblocking multicast networks under Strategy 1, where $N = nr$.

Network size N	r	n	m
100	20	5	23
200	20	10	52
400	25	16	89
1000	40	25	155
2000	50	40	260
4000	80	50	343
10000	100	100	710

for the corresponding network size.

5 Blocking Probability of Multicast Networks

The necessary and sufficient nonblocking conditions in the previous sections suggest that there is little room for further improvement on the multicast nonblocking condition. We now turn our attention to another type of problem. Recall that a nonblocking multicast network still requires a higher network cost than a permutation network. In some multicast applications, we may not need the nonblocking capability and can tolerate a small portion of blocked connections. In this case, we may want to use a lower cost network, for example, a network with a comparable network cost to a permutation network. However, little has been known on the blocking behavior of such networks in realizing multicast connections. In this section, we look into this issue by developing analytical models for the blocking probability of the multicast network and studying the blocking behavior of the network under various routing control strategies through simulations. A complete discussion on this issue can be found in [9, 10].

5.1 The first analytical model

Let's start by looking at a simple model, Lee Approximation [8], for a $v(m, n, r)$ unicast network, in which the events that individual links are busy are assumed to be independent. Consider the paths between a given input and output pair in Figure 4. Let the probability that a typical input port is busy be a and the probability that a typical output port is busy be also a, and assume that the incoming traffic is uniformly distributed over the m interstage links. Then the probability that an interstage link is busy is given by $p = \frac{an}{m}$, and the probability that an interstage link is idle is given by $q = 1 - p$. Thus the probability that one path cannot be used is $(1 - q^2)$, and the blocking probability (i.e., all m paths cannot be used) is $P_B = [1 - q^2]^m$.

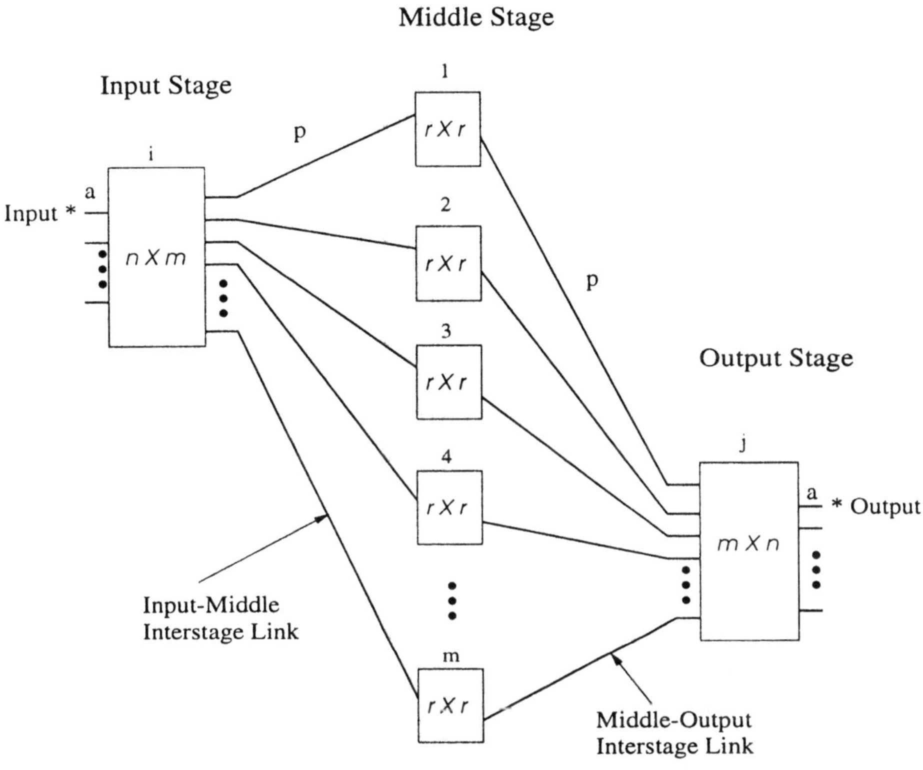

Figure 4: The paths between a given input and output pair in the $v(m, n, r)$ network.

It is interesting to see whether Lee's approach can be generalized to $v(m, n, r)$ multicast networks. In general, for a given multicast connection with fanout f, we can have different ways to realize it. As shown in Figure 5, we can have the connection fanout to one middle stage switch, or fanout to two middle stage switches, etc.

Then the total number of ways to realize a multicast connection with

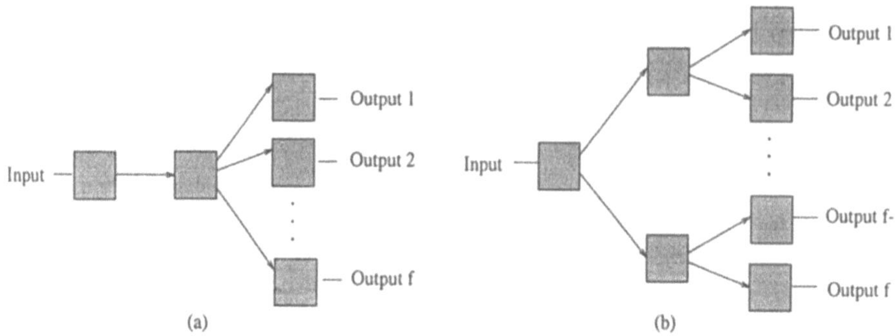

Figure 5: Different ways to realize a multicast connection.

fanout f $(1 \leq f \leq r)$ is

$$\sum_{j=1}^{f} \binom{m}{j} S(f,j) j!,$$

where $S(f,j)$ is the Stirling number of the second kind. Although we can obtain the total number of ways to realize a multicast connection with fanout f, unlike unicast connections, the dependencies among various multicast trees make the problem almost intractable. Consequently, Lee's approach cannot be directly extended to multicast networks.

In [9], a different approach was employed to derive the blocking probability for the $v(m,n,r)$ multicast networks. We briefly discuss this approach here and still follow Lee's assumption that the events that individual links are busy are independent.

Consider the subnetwork in Figure 6. Let ε be the event that the connection request with fanout f cannot be realized in the subnetwork. Notice that any interstage link is either busy or idle. Denote the event that the link a_i is busy as $\mathbf{a_i}$ and the event that the link a_i is idle as $\bar{\mathbf{a}}_{\mathbf{i}}$ for $1 \leq i \leq m$. Let σ represent the state of the input-middle interstage links $a_1, a_2, \ldots, a_m$, $P(\varepsilon|\sigma)$ be the conditional blocking probability in this state, and $P(\sigma)$ be the probability of being in state σ. If in state σ, k a_i's are idle and the rest of a_i's are busy, by the link independence assumption we have $P(\sigma) = q^k p^{m-k}$. Considering all states of input-middle interstage links $a_1, a_2, \ldots, a_m$, and using symmetry of the states, it is apparent that the *blocking probability for*

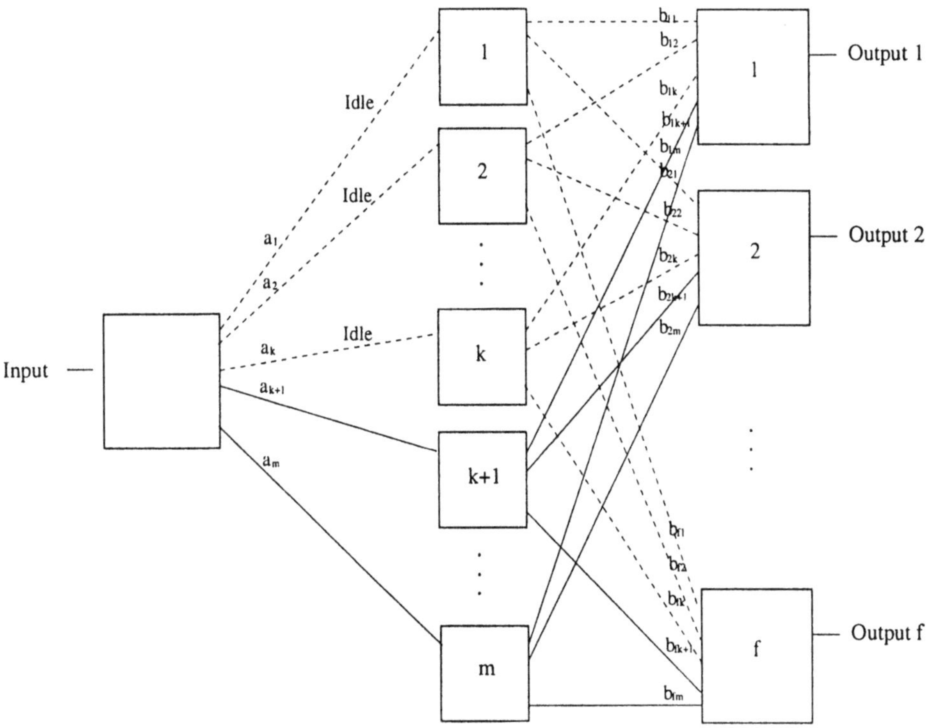

Figure 6: The subnetwork associated with a multicast connection with fanout f. The dashed lines indicate the idle link subnetwork.

a multicast connection with fanout f is given by

$$\begin{aligned} P_B(f) &= P(\varepsilon) = \sum_{\sigma} P(\sigma)P(\varepsilon|\sigma) \qquad (6) \\ &= \sum_{k=0}^{m} \binom{m}{k} q^k p^{m-k} P(\varepsilon|\bar{\mathbf{a}}_{\mathbf{1}}, \ldots, \bar{\mathbf{a}}_{\mathbf{k}}, \mathbf{a}_{\mathbf{k+1}}, \ldots, \mathbf{a}_{\mathbf{m}}). \end{aligned}$$

Under the condition that $a_{k+1}, \ldots, a_m$ are busy and rest of a_i's are idle, finding the blocking probability of the network is equivalent to finding the blocking probability of a smaller subnetwork which is obtained by removing the busy input-middle interstage links, the middle stage switches connected to these links, and the middle-output interstage links associated with these middle stage switches in the original network. Such a subnetwork is shown in Figure 6 by dashed lines. We have the following lemma concerning the blocking property of this subnetwork.

Lemma 5.1 *Assume that the interstage links $a_1, a_2, \ldots, a_k$ in the subnetwork are idle. A multicast connection from an input of the input stage switch to f distinct output stage switches cannot be realized if and only if there exists an output stage switch whose first k inputs are busy.*

Based on this lemma, we can obtain the blocking probability for a multicast connection with fanout f:

$$P_B(f) = \sum_{k=0}^{m} \binom{m}{k} q^k p^{m-k} [1 - (1 - p^k)^f]. \tag{7}$$

In particular, letting $f = 1$, we have

$$\begin{aligned} P_B(1) &= \sum_{k=0}^{m} \binom{m}{k} q^k p^{m-k} [1 - (1 - p^k)] = p^m \sum_{k=0}^{m} \binom{m}{k} q^k \\ &= p^m (1 + q)^m = (1 - q)^m (1 + q)^m = (1 - q^2)^m. \end{aligned}$$

This is exactly Lee's blocking probability for the $v(m, n, r)$ permutation network [8].

Figure 7 gives some numerical examples of $P_B(f)$. From Figure 7 we can see that, for a fixed m the blocking probability increases as the fanout gets larger, and for a fixed fanout, the blocking probability decreases sharply as m gets larger.

In general, we may be more interested in the typical behavior of the blocking probability and ask about its "average" value over all fanouts. Suppose the probability distribution for different fanouts in a multicast connection is

$$\{w_f | 0 \le w_f \le 1, 1 \le f \le r, \sum_{i=1}^{r} w_i = 1\}.$$

Then the "average" value of the blocking probability can be written as

$$P_B = \sum_{f=1}^{r} P_B(f) \cdot w_f. \tag{8}$$

Now, suppose the fanout is uniformly distributed over 1 to r. Then (8) becomes

$$P_B = \frac{1}{r} \sum_{f=1}^{r} P_B(f) = \frac{1}{r} \sum_{f=1}^{r} \sum_{k=0}^{m} \binom{m}{j} q^k p^{m-k} [1 - (1 - p^k)^f]. \tag{9}$$

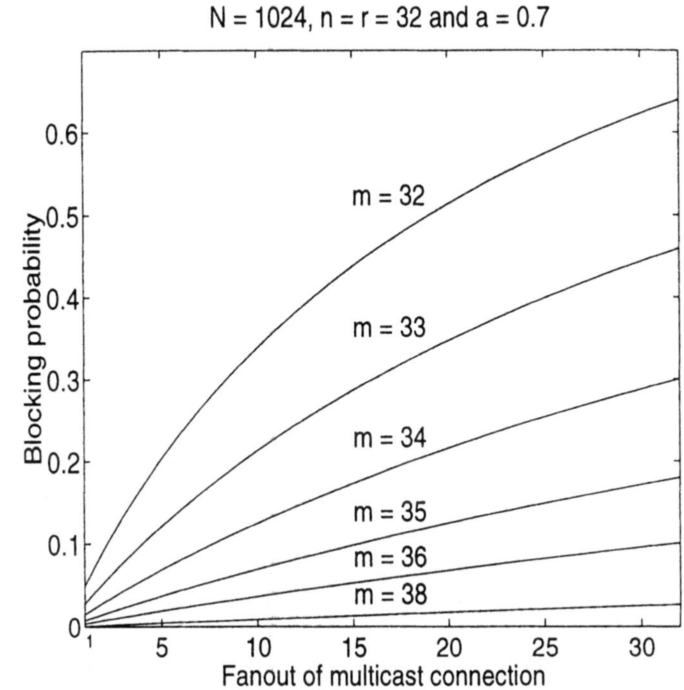

Figure 7: Blocking probabilities for $v(m, 32, 32)$ network with fanouts between 1 and 32.

In the rest of the chapter, we simply refer to P_B as the *blocking probability of the* $v(m,n,r)$ *multicast network.*

Finally, since there is (apparently) no closed form for the blocking probability P_B in (9), it is appropriate that we derive a closed form for the asymptotic bound on it. We are interested in the networks with small m values and consider the following two cases:

Case 1 $m = n + c$, for some constant integer $c > 1$.

Case 2 $m = dn$, for some constant $d > 1$.

In both cases, if $r = O(n)$, which is generally true in practical networks, we can obtain $P_B = O(e^{-\epsilon n})$, where ϵ is a constant > 0. This suggests that the blocking probability tends to zero very quickly as n increases. In other words, for a sufficiently large n, the network is almost nonblocking for multicast connections. This means that, in practice, even when the network parameter m is as small as dn or $n + c$, the network performance is still fairly good. Such m values are much smaller than the theoretical bound $\Theta(n\frac{\log r}{\log\log r})$ given in [4, 6].

5.2 Experimental simulations

In this subsection, we look into the blocking probability of the multicast networks through the simulation of real networks. In the simulation, the following seven different routing control strategies were employed for choosing a middle stage switch. Each strategy specifies a criteria on the cardinality of the destination set of the chosen middle stage switch.

1. *Smallest Absolute Cardinality Strategy:* Choose a middle stage switch whose destination set has the smallest cardinality.

2. *Largest Absolute Cardinality Strategy:* Choose a middle stage switch whose destination set has the largest cardinality.

3. *Average Absolute Cardinality Strategy:* Choose a middle stage switch such that the cardinality of its destination set is equal to the average cardinality of all available middle switches.

4. *Smallest Relative Cardinality Strategy:* Choose a middle stage switch whose destination set has the smallest cardinality with respect to the connection request (that is, first intersect the connection request with the destination sets and then choose the smallest cardinality).

5. *Largest Relative Cardinality Strategy:* Choose a middle stage switch whose destination set has the largest cardinality with respect to the connection request.

6. *Average Relative Cardinality Strategy:* Choose a middle stage switch such that the cardinality of its destination set with respect to the connection request is equal to the average cardinality of all available middle switches with respect to the connection request.

7. *Random Strategy:* Choose a middle stage switch at random.

Three types of traffic distributions are considered: uniform traffic, uniform/constant traffic, and Poisson traffic. In the uniform traffic model, both the inter-arrival time of connection requests and the connecting time of each multicast connection follow the uniform distribution. In the uniform/constant traffic model, the inter-arrival time of connection requests follows the uniform distribution and the connecting time of each multicast connection is a constant. In the Poisson traffic model, the arrival process of connection requests is a Poisson process and the connecting time of each multicast connection follows the exponential distribution.

The blocking probability in the simulation is computed by

$$P_B = \frac{\text{The total number of connection requests blocked}}{\text{The total number of connection requests generated}}.$$

Extensive simulations were carried out on the $v(m,n,r)$ multicast networks for different m values under seven routing control strategies. We present and discuss the simulation results for the following two configurations of the $v(m,n,r)$ multicast networks.

Configuration 1: $N = 1024, n = r = 32$, and $32 \leq m \leq 48$.

Configuration 2: $N = 4096, n = r = 64$, and $64 \leq m \leq 84$.

In Figure 8, we plotted the blocking probability corresponding to $32 \leq m \leq 48$ for configuration 1, and the blocking probability corresponding to $64 \leq m \leq 84$ for configuration 2. The results were obtained under seven routing control strategies for uniform traffic, uniform/constant traffic and Poisson traffic with initial network utilization = 90%.

Although network sizes and traffic models differ, the simulation results demonstrate a similar trend. We observe that, for all seven control strategies, when $m = n$ the blocking probability is relatively high, and as the number of middle stage switches increases the blocking probability decreases quickly.

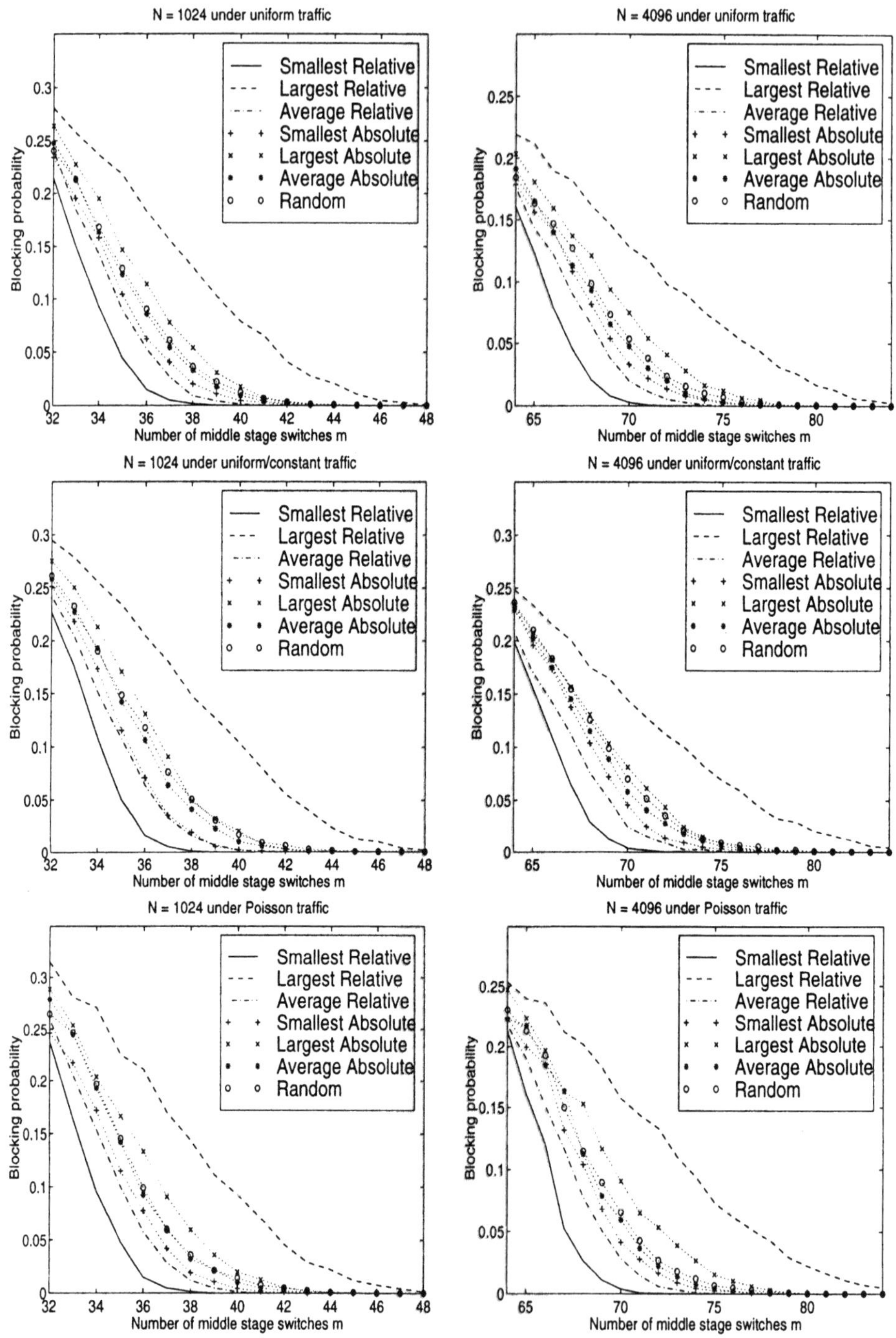

Figure 8: The blocking probability of the $v(m, 32, 32)$ and $v(m, 64, 64)$ multicast networks under seven routing control strategies.

In particular, in Figure 8 when $m \geq 48 = n + 16 = 1.50n$ for network size 1024, and when $m \geq 84 = n + 20 \approx 1.31n$ for network size 4096, the blocking probabilities approach to zero. We also see that, for any of the three traffic models, *smallest relative* strategy leads to the lowest blocking probability. This observation indicates that *smallest relative* strategy, which was employed in the routing algorithm to achieve the currently best available sufficient nonblocking condition in Section 3, is also the best among the seven strategies for a $v(m, n, r)$ multicast network with a much smaller m than the nonblocking condition.

In addition, we compare the analytical model with the simulation results. Figure 9 depicts the comparisons between the analytical blocking probability P_B and the simulation results under *smallest relative* and *random* strategies for configuration 1 and configuration 2. Notice that, although the analytical model and the simulation results were obtained under quite different assumptions, they reveal the same trend in the blocking behavior of the $v(m, n, r)$ multicast network: when m gets slightly larger than n, the network becomes almost nonblocking.

5.3 An improved analytical model

Although the analytical model in [9] shows the same trend as the simulation for the blocking behavior of the network, and it matches well with the simulation when $m > cn$ for $c > 1$, there is still some gap in blocking probabilities between the analytical model and simulation results, especially when m is very close to n. In [10], the analytical model was further improved by introducing more reasonable assumptions based on the properties of multicast communication and the Clos network itself. In the following we briefly discuss the improved model.

First we can make some correction to the blocking probability (7) based on the limitation of output stage switches. As only one-to-many or one-to-one connections are considered and there are n outputs on each output stage switch, if an output stage switch is chosen as one of destinations in a multicast connection, this output stage switch must have at least one idle output and have at most $n - 1$ busy inputs. From Lemma 5.1 and Figure 6, in the case of $k > n - 1$, there must exist some idle input on each of f output stage switches. Thus the conditional blocking probability in this

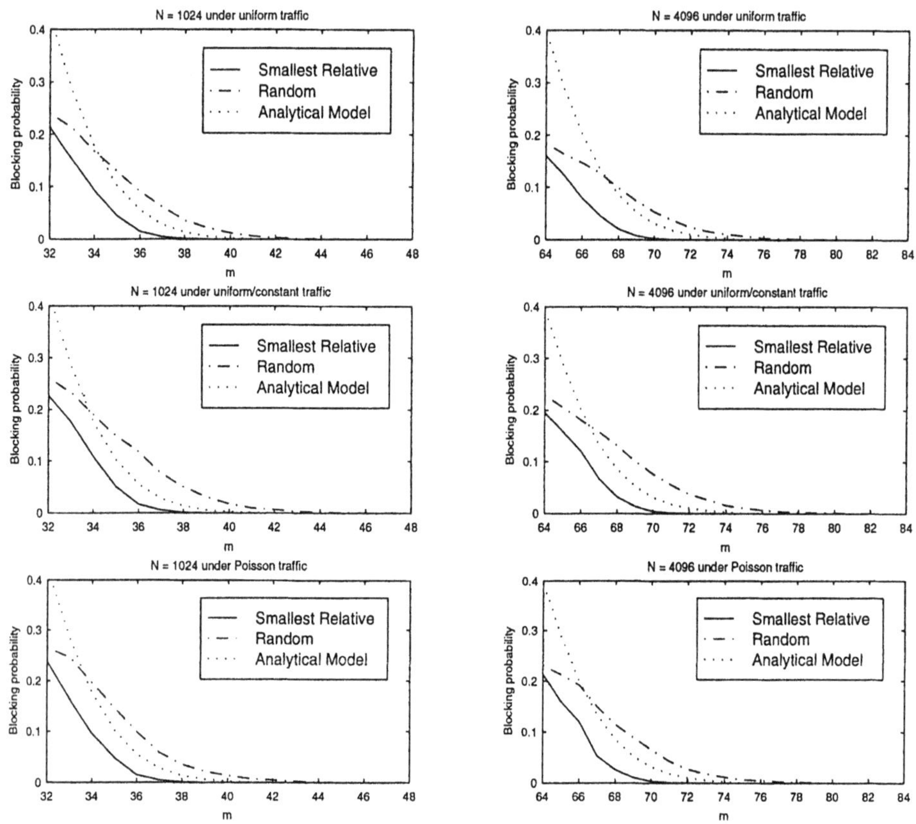

Figure 9: The comparison between the analytical model and the simulation results for the $v(m, 32, 32)$ and $v(m, 64, 64)$ multicast networks.

case becomes 0, and it can be modified to

$$P(\varepsilon|\bar{\mathbf{a}}_1, \ldots, \bar{\mathbf{a}}_k, \mathbf{a}_{k+1}, \ldots, \mathbf{a}_m) = \begin{cases} 1 - (1-p^k)^f & \text{if } 1 \leq k \leq n-1 \\ 0 & \text{if } k \geq n. \end{cases} \quad (10)$$

Then the blocking probability for a multicast connection with fanout f in (7) should be corrected to

$$P_B(f) = \sum_{k=0}^{n-1} \binom{m}{k} q^k p^{m-k} [1 - (1-p^k)^f]. \quad (11)$$

Now we consider another improvement which is based on the busy interstage link ratio. Consider the number of busy input-middle interstage links and the number busy middle-output interstage links at any time. As only one-to-many or one-to-one connections are considered, in general we should have fewer busy input-middle interstage links than busy middle-output interstage links. Thus the assumption that all interstage links have the same probability of being busy used in the previous model may not hold. Let the probability that a middle-output interstage link is busy and idle be

$$p_b = \frac{an}{m}, \quad q_b = 1 - p_b \quad (12)$$

respectively as before, and the probability that an input-middle interstage link is busy and idle be corrected to

$$p_a = \alpha p_b \text{ and } q_a = 1 - p_a, \quad (13)$$

respectively, where $\alpha \leq 1$. Then the blocking probability for a multicast connection with fanout f in (11) can be rewritten as

$$P_B(f) = \sum_{k=0}^{n-1} \binom{m}{k} q_a^k p_a^{m-k} [1 - (1-p_b^k)^f]. \quad (14)$$

Hence, the overall blocking probability of $v(m, n, r)$ *multicast network* in the modified model can be expressed as

$$P_B = \frac{1}{r} \sum_{f=1}^{r} \sum_{k=0}^{n-1} \binom{m}{k} q_a^k p_a^{m-k} [1 - (1-p_b^k)^f]. \quad (15)$$

Now it remains to determine the value of α which actually represents the average ratio of the number of busy input-middle interstage links to the number of busy middle-output interstage links.

Let's first consider the multicast connection in Figure 6. For the given input stage switch, k middle stage switches and f output stage switches, there are many ways to realize the multicast connection from the input stage switch to the f output stage switches via up to $\min(k, f)$ middle stage switches. Clearly, this multicast connection contributes f busy links among all middle-output interstage links. In order to find the average number of busy input-middle interstage links contributed by such multicast connection, we consider all possible multicast connections to the f output stage switches, and assume that each such a multicast connection is chosen equally likely. In general, a multicast connection can be fanouted through j middle stage switches to the f output stage switches, where $1 \leq j \leq \min(k, f)$. Thus, this multicast connection contributes j busy links among all input-middle interstage links. We now calculate how many ways such a multicast connection uses j busy input-middle interstage links in the subnetwork shown in Figure 6. We have $\binom{k}{j}$ ways to choose j middle stage switches, and then $S(f, j) \cdot j!$ ways to partition the f output stage switches to j disjoint sets so that each of the j middle stage switches is routed to a different set of output stage switches, where $S(f, j)$ is the Stirling number of the second kind [16]. Denote $\binom{k}{j} \cdot j!$ as $(k)_j$, which equals $k(k-1)\cdots(k-j+1)$. Hence there are a total of $(k)_j S(f, j)$ ways. We can obtain the following lemma.

Lemma 5.2 *Under the assumption that each multicast connection from the input stage switch to the f output stage switches via up to $\min(k, f)$ middle stage switches (in Figure 6) is chosen equally likely, the average number of busy input-middle links contributed by such a multicast connection is*

$$k\left[1-\left(1-\frac{1}{k}\right)^f\right] \tag{16}$$

and the average ratio of the number of busy input-middle interstage links to the number of busy middle-output interstage links is

$$\frac{k}{f}\left[1-\left(1-\frac{1}{k}\right)^f\right] \tag{17}$$

We now discuss the ratio shown in (17). We observe that for any $k, f \geq 1$, the ratio

$$\frac{k}{f}\left[1-\left(1-\frac{1}{k}\right)^f\right] \leq \frac{k}{f}\left[1-\left(1-\frac{1}{k}\cdot f\right)\right] = 1.$$

In general, it is difficult to determine the actual ratio of the busy input-middle interstage links to the busy middle-output interstage links. In this chapter, we make a reasonable assumption that the ratio is obtained by randomly picking up a multicast connection with fanout f which uses up to f middle stage switches among m middle stage switches. By applying Lemma 5.2 and noticing that f is chosen equally likely among $1, 2, \cdots, r$, we can let the α in (13) take the average value of (17) for $k = m$ and for all possible values of f. That is,

$$\alpha = \frac{1}{r} \sum_{f=1}^{r} \frac{m}{f} \left[1 - \left(1 - \frac{1}{m} \right)^f \right] \tag{18}$$

In Figure 10, we compare the blocking probabilities of the new model in (14) and the previous model in (7) for a $v(m, 32, 32)$ network with different fanouts. We can see that the blocking probabilities under the new model are smaller than those under the previous model.

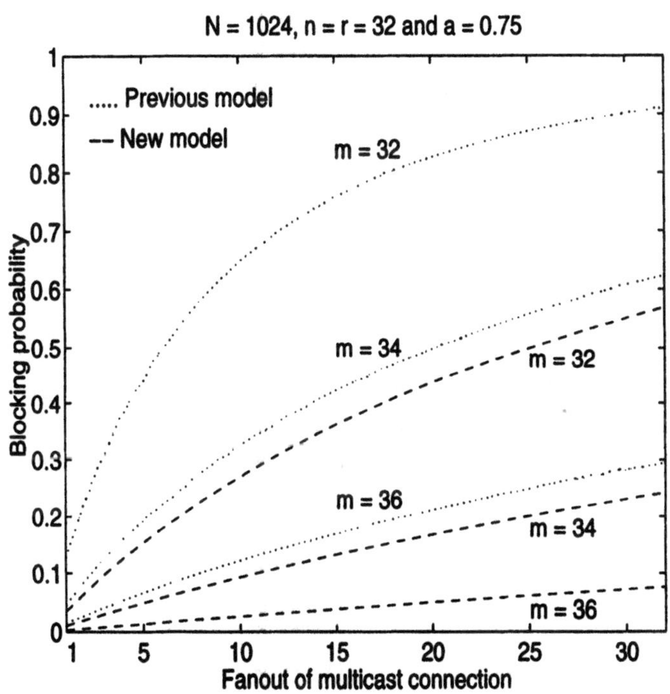

Figure 10: Comparison of blocking probabilities in two models for a $v(m, 32, 32)$ multicast network with fanouts between 1 and 32.

5.4 Comparison between the improved model and simulation results

In this subsection, we compare the analytical models with the simulation results. For simplicity, we employ the following three typical routing strategies in the simulation for choosing middle stage switches from the available middle stage switches in a $v(m,n,r)$ multicast network: *smallest relative cardinality, largest relative cardinality* and *random strategies.* Now, we present the simulation results for two network sizes: (1) $N = 1024, n = r = 32$, and $32 \leq m \leq 48$; (2) $N = 4096, n = r = 64$, and $64 \leq m \leq 84$.

For each network size, control strategy and traffic model, the network is simulated for five runs with different initial network states, and the final results are averaged over these five runs. In each run, 5000 connection requests are handled for network size 1024 with the average network utilization = 75%, and 10000 connection requests are handled for network size 4096 with network utilization = 80%. In both cases, 95% confidence interval is achieved.

In Figure 11, we plot the blocking probability corresponding to $32 \leq m \leq 48$ for network size 1024, and the blocking probability corresponding to $64 \leq m \leq 84$ for network size 4096. The results were obtained under three routing control strategies for uniform traffic and Poisson traffic. We also plot the analytical curves of the previous model (9) and the improved model (15) in Figure 11. We can see that the curve of the new model matches better with the simulation results, especially in the case of m is very close to n.

Finally, notice that, although the analytical models and the simulation results were obtained under quite different assumptions, they reveal the same trend in the blocking behavior of the $v(m,n,r)$ multicast network: when m gets slightly larger than n, the network becomes almost nonblocking.

6 Conclusions

In this chapter, we have overviewed the recent development in supporting multicast in Clos-type switching networks. In particular, we have presented the currently best available explicit construction of nonblocking multicast networks and given the blocking probability analysis of almost nonblocking multicast networks. The main results are summarized below:

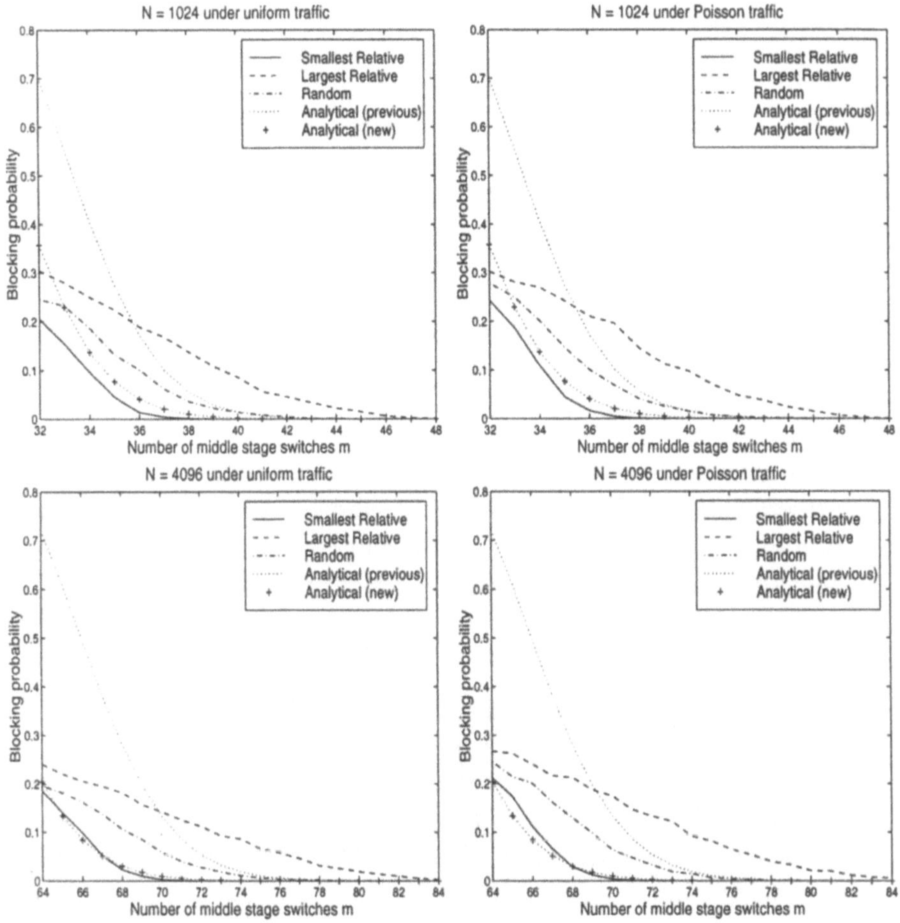

Figure 11: The blocking probability comparison between the analytical models and the simulation results under three routing control strategies for the $v(m, 32, 32)$ and $v(m, 64, 64)$ multicast networks.

- The number of middle stage switches required for nonblocking is reduced from $O(nr)$ to $O(n\frac{\log r}{\log\log r})$.

- A linear time routing algorithm is designed for satisfying connection requests.

- The necessary conditions for the nonblocking multicast networks under several typical control strategies are

$$m \geq \Theta\left(n\frac{\log r}{\log\log r}\right)$$

- The necessary conditions obtained match the sufficient nonblocking conditions under Strategy 1 and Strategy 2.

- The hardware implementations of the controller provide very fast path routings and require only a small amount of hardware compared with the switching hardware.

- Two analytical models for the blocking probability of the multicast networks are developed. The second model improves the first one by introducing more reasonable assumptions based on the properties of multicast communication and the Clos network itself.

- Extensive simulations were conducted to validate the models.

- The analytical and simulation results demonstrate that a network with a small m, such as $m = n + c$ or dn, is almost nonblocking for multicast connections, and indicate that a Clos-type multicast network with a comparable cost to a permutation is almost nonblocking for multicast connections and can provide cost-effective support for multicast communication.

Finally, besides the topics discussed in this chapter, readers may find other related topics on Clos-type multicast networks in [11]-[14].

Acknowledgments

This work was supported in part by the U.S. Army Research Office under Grant No. DAAH04-96-1-0234 and by the National Science Foundation under Grant No. CCR-0073085, MIP-9522532 and OSR-9350540.

References

[1] C. Clos, "A study of non-blocking switching networks," *The Bell System Technical Journal,* Vol. 32, pp. 406-424, 1953.

[2] G.M. Masson and B.W. Jordan, "Generalized multi-stage connection networks," *Networks,* Vol. 2, pp. 191-209, 1972.

[3] F.K. Hwang and A. Jajszczyk, "On nonblocking multiconnection networks," *IEEE Transactions on Communications,* Vol. 34, pp. 1038-1041, 1986.

[4] Y. Yang and G.M. Masson, "Nonblocking broadcast switching networks," *IEEE Transactions on Computers,* Vol. 40, No. 9, pp. 1005-1015, 1991.

[5] P. Feldman, J. Friedman, and N. Pippenger, "Wide-sense nonblocking networks," *SIAM Journal of Discrete Mathematics,* Vol. 1, No. 2, pp. 158-173, May 1988.

[6] Y. Yang and G.M. Masson, "The necessary conditions for Clos-type nonblocking multicast networks," *IEEE Transactions on Computers,* Vol. 48, No. 11, pp. 1214-1227, 1999.

[7] Y. Yang and G.M. Masson, "A fast network controller for nonblocking multicast networks," *International Journal of Parallel and Distributed Systems and Networks,* Vol. 1, No. 3, 1998, pp. 149-156.

[8] C.Y. Lee, "Analysis of switching networks," *The Bell System Technical Journal,* Vol. 34, No. 6, Nov. 1955, pp. 1287-1315.

[9] Y. Yang and J. Wang, "On blocking probability of multicast networks," *IEEE Transactions on Communications,* Vol. 46, No. 7, July 1998, pp. 957-968.

[10] Y. Yang and J. Wang, "A more accurate analytical model on blocking probability of multicast networks," *IEEE Transactions on Communications,* Vol. 48, No. 11, November 2000.

[11] Y. Yang and G.M. Masson, "Non-blocking Broadcast Network," United States Patent Number 5,451,936, issued 1995.

[12] Y. Yang and G.M. Masson, "Controller for a Non-Blocking Broadcast Network," United States Patent Number 5,801,641, issued September 1998.

[13] Y. Yang, "A class of interconnection networks for multicasting," *IEEE Transactions on Computers*, Vol. 47, No. 8, August 1998, pp. 899-906.

[14] Y. Yang, "Nonblocking multirate broadcast networks for parallel and distributed computing systems," *International Journal of Computers and Applications*, Vol. 19, No. 1, 1997, pp. 45-53.

[15] M.R. Garey and D.S. Johnson, *Computers and Intractability: A Guide to the Theory of NP-Completeness*, W.H. Freeman and Company, New York, 1979.

[16] K.P. Bogart, *Introductory Combinatorics, 2nd Edition*, Harcourt Brace Jovanovich, 1990.

SWITCHING NETWORKS: RECENT ADVANCES
D.-Z. Du and H.Q. Ngo (Eds.) pp. 179-189

The Number of Rearrangements in a 3-stage Clos Network Using an Auxiliary Switch

Frank K. Hwang & Wen-Dar Lin
Department of Applied Mathematics,
National Chiao Tung University,
HsinChu 30050, Taiwan, ROC

Contents

Abstract

We consider the problem raised by Bassalygo: "What is the maximum number of rearrangements required by a rearrangeable 3-stage Clos network when there is an auxiliary middle switch carrying a light load?" For a 3-stage Clos network with an auxiliary middle switch carrying s connections, he claimed that the maximum number of rearrangements $\varphi_1(n, n, r; s)$ is less than $s + \sqrt{2s} + 1$. In this paper, we give a lower bound $3 \times \lfloor s/2 \rfloor$ and an upper bound $2s + 1$, where the lower bound shows that the upper bound given by Bassalygo does not hold in general.

1 Introduction

The 3-stage Clos network $\nu(n,m,r)$ is the most basic multistage interconnection network and has been widely studied. The first stage of $\nu(n,m,r)$ consists of r $n \times m$ crossbars, the second stage m $r \times r$ crossbars, and the third stage r $m \times n$ crossbars, and the connections between adjacent stages are complete bipartite graphs (see Fig. 1).

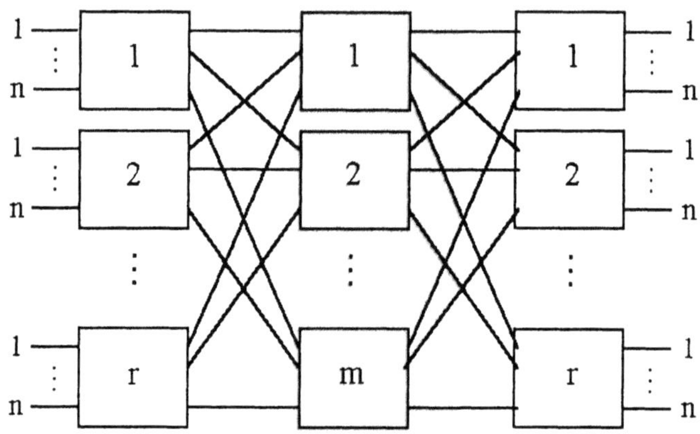

Figure 1: $\nu(n,m,r)$

A *request* is a pair of input and output requesting a connection. The requests come in sequentially; once connected, they turn into *connections* which can be released any time. A network is *rearrangeable* if under the condition that existing connections are allowed to be rearranged (rerouted), then a request can always be routed, meaning there exist link-disjoint paths one for each connection (including the request). It is well known [2] that $\nu(n,n,r)$ is rearrangeable, while $\nu(n,n-1,r)$ is not.

Let $\varphi(n,n,r)$ denote the maximum number of rearrangements required to guarantee the routing of any request for any network states. So far, all rearrangement algorithms use Paull's matrix [3] (see Sec 2.) with all connections rearranged lying on one path in Paull's matrix. We call this a one-path algorithm and let $\varphi_1(n,n,r;s)$ denote the minimum number of rearrangements required under a *one-path algorithm*. Paull proved $\varphi_1(n,n,r) \leq 2(r-1)$ and Benes [2] improved to $r-1$.

Bassalygo [1] considered the case that there exists a third middle switch c with a light load s. By taking advantage of the existence of c, he claimed $\varphi_1(n,n,r;s) \leq s + \sqrt{2s} + 1$. He gave an outline of proof but no detail. He also commented that $\varphi_1(n,n,r;s) = s+1$ is probable. In this paper we give a lower bound $\varphi_1(n,n,r;s) \geq 3 \times \lfloor s/2 \rfloor$. Note that $3 \times \lfloor s/2 \rfloor \geq s + \sqrt{2s} + 1$ for $s \geq 16$, thus invalidating the upper bound of Bassalygo. We also give an example that $\varphi(n,n,r;s) \geq s+2$. Finally, we give an upper bound $\varphi_1(n,n,r;s) \leq 2s$.

2 One-path algorithm

The Paull's matrix P is an $r \times r$ matrix, where rows are indexed by input switches I_i, columns by output switches O_j, and cell p_{ij} by the set of middle switches each carrying a connection (I_i, O_j). By convention, the request to be routed is (I_1, O_1) and we may assume that there exist two middle switches a and b, one appearing in the first row and the other the first column, but not both. Paull's method is to switch the connection carried by a in the first row to be carried by b; then the request can be routed through a. But switching that a to b means that if there is already a b in the same column as that a, then we must switch that b to a. Again, we need to check whether there is already an a in the same row with that b, and so on. Such a path, called an ab-path, alternates in vertical and horizontal turns, with an a at every vertical turn and b at every horizontal turn, and stops if and only if it hits a line not containing the other symbol. Similarly, we can start the ba-path from the b in the first column (Fig. 2). Note that both ab-path and ba-path go from a to b in a column and b to a in a row. Therefore when we trace back such a path, it is always from b to a in a column and from a to b in a row. However, tracing back of a path from a symbol either a or b is unique since there exist at most one a and one b in each row and column. Hence the ab-path and the ba-path cannot meet. Since there are at most $2r-2$ a's and b's. One of the two paths contains at most $r-1$ a's and b's.

Bassalygo's idea is to alter an ab-path or a ba-path at some point to an ac-path or a bc-path, and use the scarcity of c to show the existence of a shorter path. He gave the following rules for an one-path algorithm:

1. A path must start with an ab-path or a ba-path.

2. Once it turns into an ac-path or a bc-path, it cannot turn back to ab-path or ba-path.

3. When the first c appears in a path, the line containing two letters preceding it does not contain letter c.

4. Every matrix letter is visited at most once.

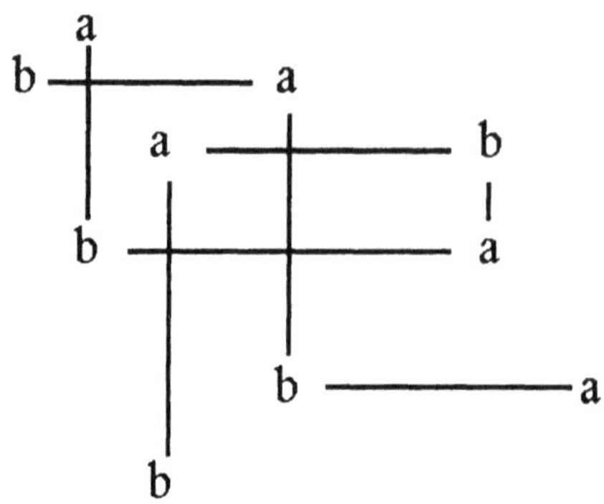

Figure 2: The ab-path and the ba-path

5. A path stops if and only if either it hits a line not containing the other symbol, or an ab-path (or ba-path) hits a symbol whose row and column contain no c.

Definition 1 *We call a path a rerouting path if the path follows the above rules. A rerouting path can be divided into two parts, where the first part is an ab-path (or ba-path) and the second part is an ac-path (or bc-path). Note that a rerouting path can have null second part.*

For example, there are three rerouting paths in the Paull's matrix of Fig. 3. They are *ababab*, *abcb*, and *babababababab*. Note that the path *acaca* starting from cell $(1, 12)$ is not a rerouting path, because by definition a rerouting path must start with an ab-path or a ba-path. And the a's in cell $(9, 4)$ and $(10, 5)$ can't turn to an ac-path because it will revisit an a already on the path (violating rule 4). After we choose a rerouting path, we can reroute the existing connections lying on the rerouting path, the scheme is to exchange a's and b's in the first part, which then starts an exchange of c's and the other letter of the second part. If a rerouting path has null second part and the first part (an ab-path or ba-path) hits a symbol whose row and column contain no c, we change the last letter to c.

3 The lower bound and upper bound

Before giving the construction for $\varphi_1(n, n, r; s) \geq 3 \times \lfloor s/2 \rfloor$, we define the function φ_1 more clearly.

Definition 2 *A network state is a legitimate arrangement of a Paull's matrix (each row and each column has at most n entries, all disjoint).*

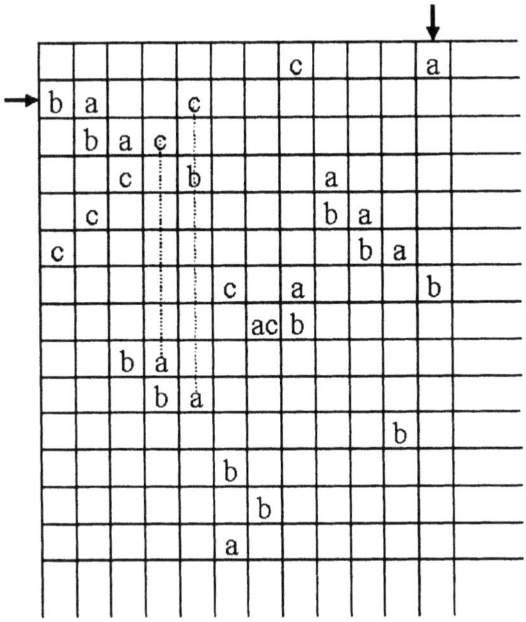

Figure 3: A Paull's matrix

Definition 3 $\phi(x) = \min(\{\textit{lengths of rerouting paths of network state } x\})$

Definition 4 $\varphi_1(n,n,r;s) = \max\{\phi(x) \mid x \in N(n,n,r;s)\}$, *where* $N(n,n,r;s)$ *denotes the set of all states of* $\nu(n,n,r)$ *in which the minimum-load middle switch carries* s *connections.*

Now we give a series of constructions that $\varphi_1(n,n,r;2k) \geq 3k$. The general pattern of construction should be clear from the examples given in Fig. 4-8.

Three points should be noted for understanding these examples:

1. We have permuted the rows and columns such that the first s rows and the first s columns contain all c's. The heavy lines denote this $s \times s$ submatrix C. By rule 3, a rerouting path can turn to c only when it is out of C.

2. Sometimes, a path is out of C but cannot turn to c because the c lies in the same line with a symbol already on the path. Turning to that c would violate rule 5 if the path stops at c, and violate rule 4 if the path continues.

3. The shortest paths in these examples are the ab-paths or the ba-paths.

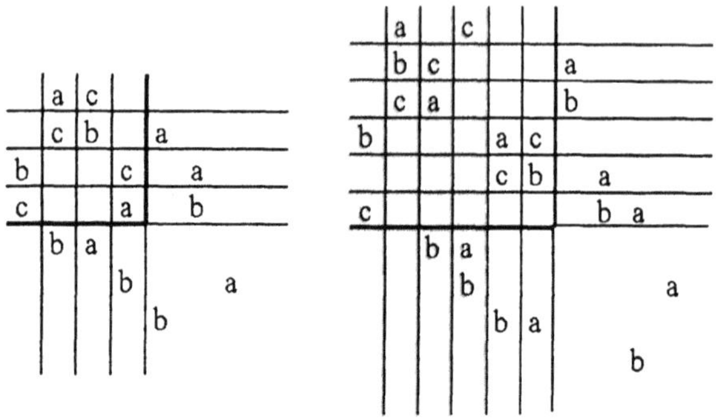

Figure 4: $k = 2$ and $k = 3$

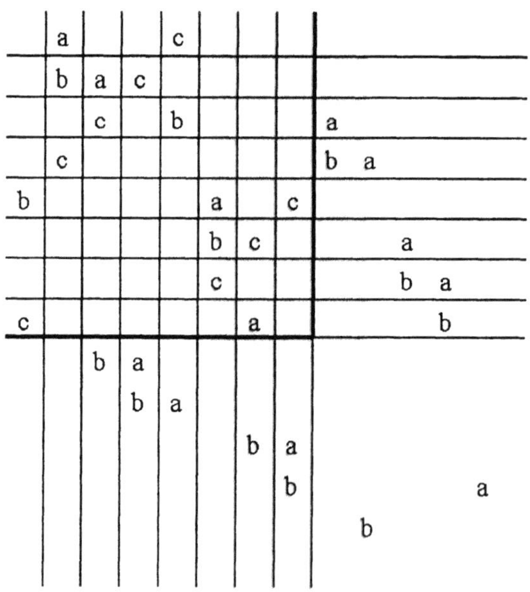

Figure 5: $k = 4$

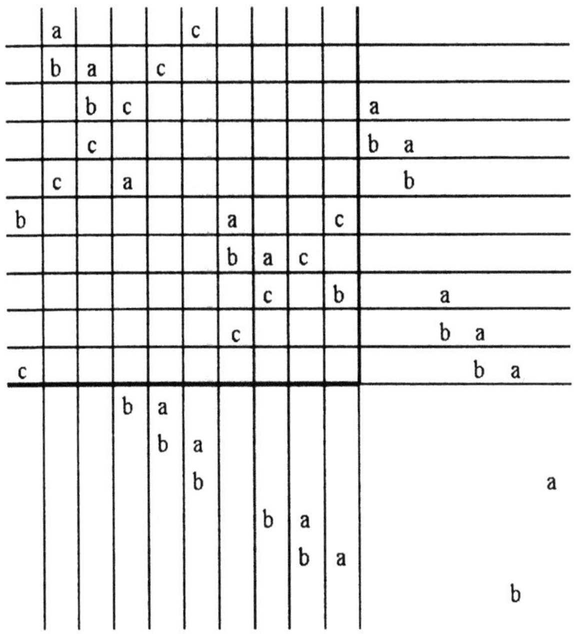

Figure 6: $k = 5$

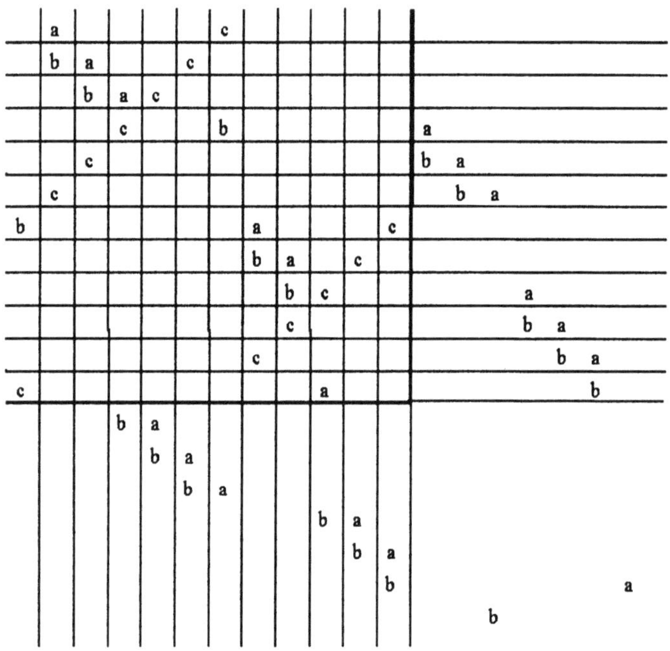

Figure 7: $k = 6$

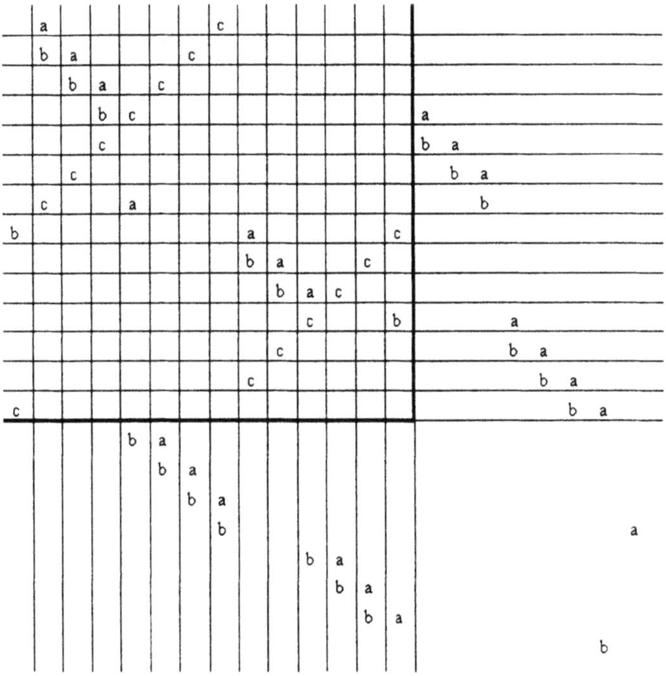

Figure 8: $k = 7$

Consider the size of the Paull's matrix, we have Theorem 5.

Theorem 5 $\varphi_1(n, n, r; s) \geq \min(3 \times \lfloor s/2 \rfloor, r - 1)$

Proof. By the result of Benes [2] and the above construction, the inequality holds, and the equality holds if $3 \times \lfloor s/2 \rfloor \geq r - 1$. ■

Note that $3 \times \lfloor s/2 \rfloor \geq s + \sqrt{2s} + 1$ for all $s \geq 16$, contradicting Bassalygo's claim.

Theorem 6 $\varphi_1(n, n, r; s) \leq \min(2s, r - 1)$

Proof. Assume that (I_1, O_1) is the next request, row 1 contains a but not b, and column 1 contains b but not a. Define $D =$ (first s rows) $\cup$ (first s columns). Then we can find two rerouting paths with null second parts and starting with the above a and b respectively. As argued before, the two paths are disjoint since at any point of the path the tracing back of a path containing only a's and b's are unique. Since there are at most $4s - 2$ a's and b's in D, one of the two paths must get out of D after $2s - 1$ turns. Change the last letter to c and interchange a's and b's in the path (of length $2s$). Thus the request can be routed in $2s$ rearrangements. ■

4 Two-path algorithms

Again, assume row 1 contains no b and column 1 no a, where c is a third middle switch. A *two-path algorithm* using a, b, c consists of a cb-path starting from the c in the first row, a ca-path starting from the c in the first column, and a connection of the (I_1, O_1) request by c. Note that a path can end as soon as it gets out of C.

Sometimes a two-path algorithm can be better than one-path algorithms. For example, we can route the request (I_1, O_1) for the network state in Fig. 6 under a one-path algorithm, which requires $3 \times \lfloor 10/2 \rfloor = 15$ rearrangements if $r \geq 16$. But we can rearrange the connections lying on the two paths as shown in Fig. 9 (interchange the symbols at the two ends of each broken line), and then route the request through c after only 4 rearrangements.

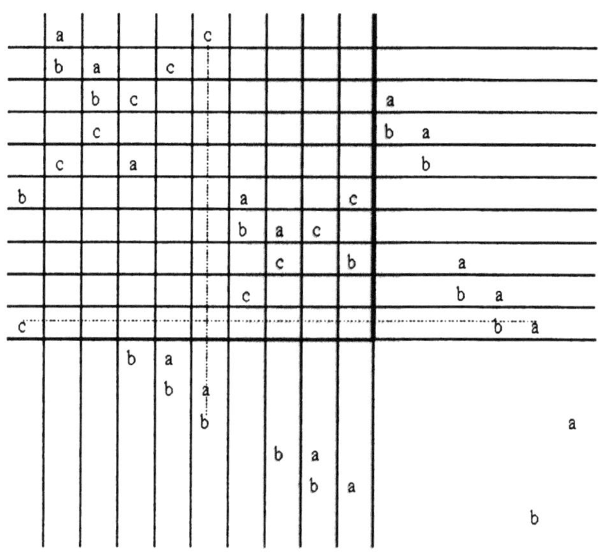

Figure 9: $k = 5$

However, the number of rearrangements under a two-path algorithm is not always fewer than $\varphi_1(n, n, r; s)$. For example, we can route the request (I_1, O_1) for the network state in Fig. 4 by rearranging the connections lying on the two paths (*cbcb* and *caca*) as in Fig. 10, but 8 rearrangements are needed. Clearly, the request can be routed after 6 rearrangements under a one-path algorithm. Moreover, this is a counterexample against $\varphi(n, n, r; s) = s + 1$.

It is hoped that by considering the ab-path, the ba-path, the cb-path and the ca-path together, we can obtain a better upper bound than $2s$.

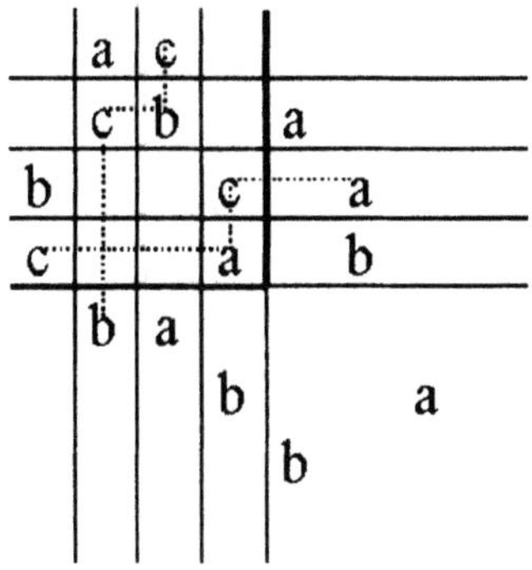

Figure 10: $s = 4$

Acknowledgement 7 *We thank Mr. S.C. Liu for helpful discussions and lament his untimely death.*

References

[1] L.A. Bassalygo, On a number of reswitching in a three-stage connecting network, Inter. Teletraffic Cong. 7, pp. 231/1-231/4, 1973.

[2] V.E. Benes, Mathematical Theory of Connecting Networks and Telephone Traffic, Academic, New York, 1965.

[3] M.C. Paull, Reswitching of connection networks, Bell Syst. Tech. J. 4, pp. 833-855, 1962.

SWITCHING NETWORKS: RECENT ADVANCES
D.-Z. Du and H.Q. Ngo (Eds.) pp. 191-214

Periodic Shifting Networks

Przemka Kanarek
Krzysztof Loryś
Institute of Computer Science
Wrocław University, Poland
E-mail: pka,lorys@tcs.uni.wroc.pl

Contents

1 Introduction

A network can be viewed as a graph, where each node corresponds to a processor and edges represent communication links between processors. Each processor can directly access only its private, local resources (e.g. memory). An exchange of information between processors is via paths of the graph. There is a variety of possible network architectures (graphs) but only some of them can be considered as practical. We usually demand that the degree of the network should be reasonably bounded and the communication traffic should be easy to control. On the other hand we wish that the network is able to realize at least typical communication tasks efficiently.

A basic communication task occurring in parallel computing is *routing*. Generally it can be formulated as delivering packets (messages) from distinguished source sites to destination sites according to given delivery addresses. Various modifications of this task arise when we consider: a type of transformation between source an destination sites, methods of computing destination addresses, sizes of packets, flexibility of communication routes, ways of resolving conflicts between packets on routes, capacities of routes and on-way sites, etc. (e.g. see [15], [4], [11].) We focus our considerations on a typical situation: packets are to be delivered one-to-one between sources and destinations, capacities of sites and routes are sufficient to handle constant (usually one or two) number of packets, and addresses of packets are known ahead. Under these conditions the mathematical formulation of a routing problem is *performing a permutation.*

Switching networks resolve this kind of problems achieving a satisfactory efficiency thanks to their flexibility and simple structure. The well-known example of a switching network was presented by Beneš ([2], [3]) and Waksman ([13]). Their network performs all permutations in time logarithmic in number of permuted packets. (This bound is optimal for constant degree networks, as one packet can reach only d^t different nodes in a network of degree d in time t.) However, in many applications it is not necessary to perform all permutations. Thus it is worth to seek for smaller but still reasonable families of permutations, which preserve wide communication capabilities while allowing to reduce time and size complexity of the networks meaningfully. Such a useful family of permutations are shifts by powers of two. We call them *hypercube shifts*, as they can be used for simulation hypercube networks. They also owe their importance to applications in efficient realization of all shifts ([12]) and gossiping and broadcasting algorithms ([6]).

There are only $\log n$ such shifts between n sites, so the same trivial argument as for all permutations implies that to perform hypercube shifts, i.e. by $\{2^i \mid 0 \leq i < \log n\}$, the switching network needs time $\Omega(\log\log n)$. We have proved in ([5]) that this bound is tight presenting a network of depth (thus also time) $O(\log\log n)$. With aid of this construction it is possible to substitute any n-node hypercube interconnection network by an $O(n\log\log n)$-node, 4-degree network. This allows to simulate a communication step along one dimension of a hypercube in time $O(\log\log n)$. The essential fact is that performing arbitrary sequence of k such communication steps takes time $O(k\log\log n)$, independently of the order in which different dimensions of a hypercube are used. This differs our construction from the shuffle-exchange network of size $O(n)$ ([10]), which simulates hypercube communication with $O(\log\log n)$ delay factor, assuming that the algorithm uses dimensions of a hypercube in some order convenient for the simulation.

One of the goals while constructing networks is their low complexity. In this issue an interesting alternative for "classical" constructions seems to be periodication. Its idea is to substitute a complex network by a shallow one which performs the same task when repeated several times. Periodification has been already successfully used for several problems and resulted in remarkable reduction of size complexity of networks. For instance, sorting of n elements can be performed by a non-periodic comparator network of depth $O(\log n)$ ([1]). A periodic approach to the sorting problem resulted in the network of constant depth and size $O(n)$ (comparing to $O(n\log n)$ size of the AKS network) working in time $O(\log^2 n)$ ([9]). Periodicity was also applied for merging problem. In this case there were constructed optimal constant depth networks that work in logarithmic time ([8]). Another problem for which periodic solution matches the time complexity of the non-periodic construction is realization of all permutations on switching networks. A simple three-layer shuffle-exchange network ([14]) substitutes the well known construction of Beneš and Waksman, while its size is less by factor $\log n$.

In this paper we achieve a full trade-off between network depth and number of repetitions for the problem hypercube shifts. We present two distinct approaches towards this task. The first is an embedding technique converting non-periodic shifting networks into periodic networks of a constant depth and the same time complexity. The method is very general and can be applied to many different classes of shifts. In particular, for hypercube shifts it returns constant depth, linear size network working in time $O(\log\log n)$. For generality of embedding method we pay with nonpractically large constants hidden under big O notation. Therefore we develop another method which is especially dedicated to hypercube shifts. It makes use of their algebraic properties to reduce the problem of realizing hypercube shifts to the problem of realizing shifts by consecutive numbers. We obtain an optimal solution to the last problem by some nontrivial and interesting constructions.

1.1 Outline of the Paper

The preliminary section contains definitions and simple constructions of switching networks. Then we present a method of embedding a sequential shifting network in a periodic schema. This allows to convert existing switching networks for shifts into constant depth periodic networks realizing the same tasks. Afterwards, in the last section, we present a direct construction for hypercube shifts based on the reduction technique.

Acknowledgments

This research was supported by KBN grants 8T 11C 04419 and 8T 11C 03215.

2 Preliminaries

A *network* is an acyclic, directed graph $G = (V, E)$ with distinguished subsets of nodes $I, O \subseteq V$ called respectively *input* and *output nodes.* We restrict our considerations to networks of rather simple and regular (thus practical) structure, namely to *networks of 2-switches.*

Definition 2.1 *A* 2-switching network *is a* $(k+1)$*-partite directed graph* $G = (V_0 \cup V_1 \cup \cdots \cup V_k, E)$, *where*

(i) $E = E_0 \cup \cdots E_{k-1}$ *and* $E_i \subseteq V_i \times V_{i+1}$, *for* $i = 0, \ldots, k-1$; *i.e. arcs are only between consecutive* V_i*s;*

(ii) $I = V_0$ *are* inputs *of the network and for all* $v \in V_0$ *in-degree*$(v) = 0$ *and out-degree*$(v) = 1$*;*

(iii) $O = V_k$ *are* outputs *of the network and for all* $v \in V_k$ *in-degree*$(v) = 1$ *and out-degree*$(v) = 0$*;*

(iv) if $v \in V_i$, *for* $1 \leq i \leq k$, *then* $1 \leq$ *in-degree*$(v) =$ *out-degree*$(v) \leq 2$; *if in-degree*$(v) = 2$ *then we call it a switch (2-switch more accurately), otherwise it is an* idle node.

Intermediate V_i*s are called* layers *of the network. We say that the network is an* n *input network, if* $n = |V_0|$ *(it is easy to see that also* $n = |V_k|$*). The depth of the network is the length of paths from inputs to outputs, thus* k *in this case. In what follows we consider only 2-switching networks and 2-switches, therefore we omit number two in their names. (See Figure 1 for a sample network of 9 inputs and depth 3.)*

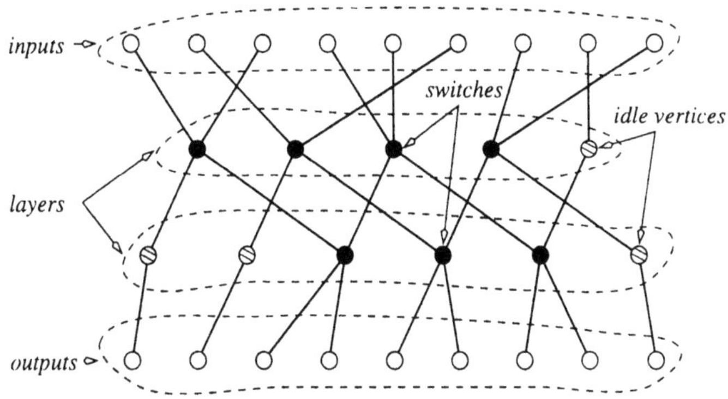

Figure 1: The sample graph of interconnections of a switching network.

Switching networks are used to perform communicational tasks as follows. We assume that initially there is a packet placed in each input node. Each packet has prescribed a delivery address, i.e. an output node that it is heading for. We consider the case when all delivery addresses are different, thus packets are to be "permuted". More formally, assuming that both input and output nodes are numbered with integers $\{0, 1, \ldots, n-1\}$, we say that the permutation π is performed, if the packet from the i-th input is delivered to the $\pi(i)$-th output, for $0 \leq i < n$.

We also require that a packet, before it starts, should have a *route* (path) joining its input with its output node. Once routes are fixed, we assume that all packets start simultaneously, cover a distance of one arc in a time unit and are not allowed to stop at nodes, unless it is an output node. The routes of different packets have to be edge-disjointed. Therefore each 2-switch can direct its two incoming packets in one of two ways. The way is determined by a *state* of the switch (we call one of them *active* and the other *inactive*).

To simplify descriptions of our constructions we prefer a bit different way of presenting switching networks. For a network $G = (V, E)$, where $E = E_0 \cup \ldots \cup E_{k-1}$, let us consider its *dual graph* $H = (E, F)$, where edges of G become nodes of H and there is an arc between $e, f \in E$ if and only if they are incident, i.e. there exist $u, v, w \in V$ such that $e = (u, v)$ and $f = (v, w)$. See that the switch v of G becomes in H a quadruple of arcs $((e_1, f_1), (e_1, f_2), (e_2, f_1), (e_2, f_2))$, where e_1, e_2 are incoming and f_1, f_2 are outgoing arcs of v (in G). Hence, the active state of v corresponds to using arcs (e_1, f_2) and (e_2, f_1) for routing packets, while the inactive state

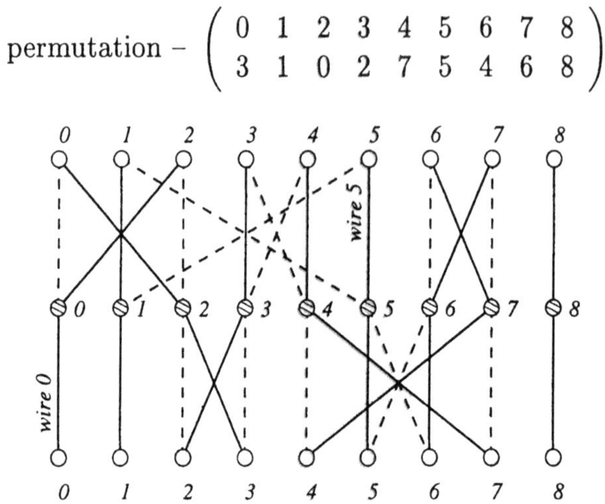

Figure 2: The network with its setting and the performed permutation (wires are drawn as vertical lines, while routes are solid lines).

corresponds to using the remaining ones.

The convention that we have just taken, allows a very clear presentation of switching networks. Notice that $|E_i| = n$, for $i = 0, \ldots, k-1$, and we can enumerate nodes of each E_i with integers $0, 1, \ldots, n-1$ as follows. We arbitrarily number inputs (elements of E_0) and then send packets through the network in which all switches are inactive. The path traversed by the packet from the i-th input is called the i-th *wire*.

Let $Perm(N)$ be the set of all permutations that the network N can realize under different settings of switches. Let us notice that the identity permutation always belongs to $Perm(N)$ (this trivial observation will be extensively used further on). We say that N performs a set of permutations A if $A \subseteq Perm(N)$. We also recall here a simple observation on lower bounds on the depth of switching networks.

Proposition 2.2 *If a network N performs the set of permutations P on input X and $d = \max_{x \in X} |\{\pi(x) : \pi \in P\}|$ (which means that some input should be able to reach d different outputs in N), then $depth(N) \geq \log d$.*

Periodic Switching Networks

We say that a network is used in a *periodic mode* if it is applied several times to realize one communication task. Each application of a network is called *a round.* The round starts with determining the setting of the switches. We assume that at the beginning of each round all switches are inactive. Therefore setting consists in pointing only those switches that are to be set active; all remaining ones are inactive by default. Afterwards packets are delivered from inputs to outputs along routes created. At the end of each round (except the last one) the contents of the ith output is moved in one step to the ith input (for $0 \leq i < n$). Thus the permutation obtained in a periodic mode is a composition of permutations obtainable by a single use of a network; more formally, permutations that the network N can perform in r rounds are $\{P_1 \circ \cdots \circ P_r : P_i \in Perm(N), 1 \leq i \leq r\}$.

2.1 Symmetries and Shifts

Due to reasons explained in the introductory section, we are especially interested in rather simple sets of permutations, namely shifts. So now, let us distinguish and define *symmetries* and *shifts*; classes of permutations which will be useful in defining our target permutations and still can be realized very efficiently.

Definition 2.3 *For integers k, t ($k > 0$) and a sequence of pairwise different integers $\mathcal{S} = (i_0, i_1, \ldots, i_{k-1})$, where $i_j \in \{0, 1, \ldots, n-1\}$, for $0 \leq j < k$, a* symmetry with a center *$t/2$ (further on, informally but conveniently, also called a* symmetry by *t) on $\mathcal{S}$ is a permutation*

$$Sym_{t,\mathcal{S}}(l) = \begin{cases} i_{(t-j) \bmod k} & \text{for } l = i_j \\ l & \text{for } l \notin \mathcal{S} \end{cases} .$$

a shift *by t on $\mathcal{S}$ is a permutation*

$$Shift_{t,\mathcal{S}}(l) = \begin{cases} i_{(t+j) \bmod k} & \text{for } l = i_j \\ l & \text{for } l \notin \mathcal{S} \end{cases}$$

The next proposition states that for each particular permutation there exists a very efficient switching network.

Proposition 2.4

(i) Each symmetry can be realized by a network of depth 1.

(ii) Each shift can be realized by a network of depth 2.

(iii) For every permutation there exists a switching network of depth 2 *performing this permutation.*

Proof:
(i) Notice that a symmetry by t on $\mathcal{S}$, where $\mathcal{S} = \{i_0, \ldots, i_{k-1}\}$, is a permutation consisting of $\lfloor t/2 \rfloor$ disjointed transpositions: $(i_j, i_{(t-j) \bmod k})$, for $0 \le j \le t/2$. Thus we can built a one-layer network containing a switch for each transposition. Certainly, this network performs the symmetry in hand, when all its switches are set active, and the identity permutation when all are inactive.

(ii) For the proof of this part notice that for every integer t and a sequence $\mathcal{S}$, it is $Shift_{t,\mathcal{S}} = Sym_{t+j,\mathcal{S}} \circ Sym_{j,\mathcal{S}}$, for any integer j. Thus two layers, each performing a single symmetry, are sufficient to perform the shift. Moreover, we are free to choose a value of j while realizing a shift.

(iii) Hence, since for every permutation π there are disjointed cycles $C_1, \ldots, C_m$, such that $\pi = C_1 \circ \cdots \circ C_m$, therefore $\pi = Shift_{1,C_1} \circ \cdots \circ Shift_{1,C_m}$ (here, for notational simplicity, we have the same notation for cycles and sequences of their elements). □

The next simple, but useful result of this section is a switching network realizing consecutive shifts, i.e. shifts by numbers $0, 1, \ldots, n-1$ on $\{0, 1, \ldots, N-1\}$ (due to Proposition 2.2, its depth has to be at least $\log n$).

Proposition 2.5 *There is a switching network of depth* $\lceil \log n \rceil + 1$ *performing shifts* $\{Shift_{t,(0,1,\ldots,N-1)} : 0 \le t < n\}$, *for* $0 \le n < N$.

Proof: Let $0 \le t \le n-1$ and $t = t_q t_{q-1} \cdots t_1 t_0$, for $t_i \in \{0, 1\}$, be the binary representation of t. Notice that $Shift_{t,(0,1,\ldots,N-1)}$ can be decomposed into symmetries as follows (we omit the second parameter of all permutations in the formulas below, as it is always $(0, 1, \ldots, N-1)$):

$$\begin{aligned} Shift_t &= \circ_{i=0}^{q} (Shift_{2^i})^{t_i} \\ &= \circ_{i=0}^{q} (Sym_{2^i+1} \circ Sym_{2^i})^{t_i} \\ &= (Sym_{2^q+1})^{t_q} \circ (\circ_{i=1}^{q} (Sym_{2^i})^{t_i + t_{i-1}}) \circ Sym_1^{t_0} \\ &= (Sym_{2^q+1})^{t_q} \circ (\circ_{i=1}^{q} (Sym_{2^i})^{(t_i + t_{i-1}) \bmod 2}) \circ Sym_1^{t_0}, \end{aligned}$$

where Sym^r and $Shift^r$ denote a symmetry or a shift applied r times. Thus a network which can perform symmetries Sym_{2^0}, $Sym_{2^1}, \ldots$, $Sym_{2^{\lceil \log n \rceil}}$, can also perform any of the required shifts. By Proposition 2.4(i) there exists such a network of depth $\lceil \log n \rceil + 1$. □

Finally, let us notice a trivial property:

Proposition 2.6 *Permutations* $\{Shift_{t,X}$*: for* $t \in T\}$*, may be performed by a network of depth* $|T| + 1$.

Proof: It is enough to have a network performing symmetries $\{Sym_{t,X} : t = 0 \text{ or } t \in T\}$. □

Propositions 2.5 and 2.6 have their periodic counterparts. As we shall show there are constant depth networks that perform shifts by n consecutive integers in $O(\log n)$ rounds and arbitrary set T of shifts in $O(|T|)$ rounds.

3 Depth versus Number of Rounds

In this section we show how to convert a non-periodic network N of depth d into a constant depth network N', which can simulate N in $O(d)$ rounds preserving a wide class of shifts performed by N.

We apply the technique, which was previously successfully used for other periodic constructions, e.g. sorting networks [9]. The main idea is very simple. We place all layers of the network N, side by side, in one layer of the network N', then appropriate setting of switches before each round allows us to simulate N. But to make this new construction work we must require sets of performed permutations to have some nice properties, which, saying informally, allow to compose target permutations from permutations on subsets of inputs. Fortunately, shifts possess this property. Furthermore, to maintain our construction effective, we may not allow to exist too many permutations, that send their input for a long distance (they can not be handled by the embedded network). Fortunately, long hypercube shifts (the same holds also other interesting families of shifts, like e.g. Fibonacci shifts) are rare and thus we can handle them separately without significant increase of the depth of the whole construction.

Now we describe in details requirements that shifts must fulfil to allow our embedding technique. Let $\mathcal{A} = \{a_1, a_2, \ldots\}$ be an infinite set of positive integers and let $\mathcal{A}_n = \mathcal{A} \cap \{1, 2, \ldots, n-1\}$. We assume that shifts by $\mathcal{A}_n$, for $n > 0$, can be performed by non-periodic networks of depth $d(n)$, i.e. for each n we have a family of networks $\{Net_{n,m} \mid m \geq n\}$, such that $Net_{n,m}$ performs $\{Shift_{t,(0,1,\ldots,m-1)}$, for $t \in \mathcal{A}_n\}$, and $depth(Net_{n,m}) \leq d(n)$, for any m. We require that the density of sets $\mathcal{A}_n$ and the depth of networks $Net_{n,m}$ do not grow too rapidly:

$$(\forall c > 0) \quad |\mathcal{A}_{cn} \setminus \mathcal{A}_n| \leq c \tag{1}$$

$$(\exists D)\ (\forall k \geq 1) \quad d(k+1) - d(k) \leq D. \tag{2}$$

Under the above assumptions we can show a trade-off between network depth and the number of rounds:

Theorem 3.1 *For each N there exists a network $\mathcal{P}_N$ of constant depth performing shifts by A_N on N inputs in $O(d(N))$ rounds.*

As we have already mentioned the construction of $\mathcal{P}_N$ proceeds separately for short and long shifts. By "short" we mean shifts of length not greater than n, that is defined as follows:

$$n = \max\{k \mid kd(k) \leq N - k\}. \tag{3}$$

This value determines almost maximal network $Net_{n,n}$, that can be embedded in one layer on N wires. For technical reasons we do not embed the maximal one. The next subsection contains description of this embedding. Note that by condition (1) there are at most $O(d(n))$ long shifts, so to complete the construction it suffices to prove a periodic counterpart of Proposition 2.6.

Before the proof of Theorem 3.1 let us notice that hypercube shifts $\mathcal{A} = \{2^i \mid i \geq 0\}$ satisfy the conditions (1) and (2), as there exist (non-periodic) networks of depth $\lceil 6 \log\log n + 3 \rceil$ performing shifts by $\mathcal{A}_n$ on $(0, 1, \ldots, m-1)$, for $n > 0$ and arbitrary $m \geq n$ (see [5]). Thus Theorem 3.1 can be applied in this case. A careful analysis of the construction given in its proof results in the following corollary:

Corollary 3.2 *There exists a switching network of depth 18 performing shifts by $\{2^i \mid 0 \leq i \leq \log N\}$ on the set $(0, 1, \ldots, N-1)$ in $12 \log\log N + 3 \log\log\log N + 9$ rounds.*

Although the constants reveal purely theoretical character of this result, we write them explicitly to emphasis the improvement obtained in the last section.

3.1 Parallel Embedding

In this subsection we show a construction of a network $T_{N,n}$, a part of $\mathcal{P}_N$, that is resposible for performing shifts from $\mathcal{A}_n$.

For a convenience we use the following notation: let $R = Net_{n,n}$ and $R' = Net_{n,m}$, and let $R_1, R_2, \ldots, R_{d(n)}$ and $R'_1, R'_2, \ldots R'_{d(n)}$ denote consecutive layers of R and R', respectively. We also write $\bar{R}[X]$ if a sequence X of inputs (numbers of wires) is the input of the subnetwork $\bar{R}$. If X_j is a sequence of inputs then by $X_{j,i}$ we denote its i-th element.

Construction 3.1 *The parallel embedding.*
The network $T_{N,n}$ consists of five functionally separate subnetworks arranged in consecutive layers: $\mathcal{S}_1$, $\mathcal{S}_2$, $\mathcal{R}$, $\mathcal{R}'$ and $\mathcal{C}$.

$\mathcal{S}_1$ – performs $Shift_{-n,(0,1,\ldots,N-1)}$;

$\mathcal{S}_2$ – performs $Shift_{(2d(n)-1)n \ mod \ N,(0,1,\ldots,N-1)} = Shift_{2N-2m-n,(0,1,\ldots,N-1)}$;

$\mathcal{R}$ – consists of $R_1[X_1], R_2[X_2], \ldots, R_{d(n)}[X_{d(n)}]$, where $X_i = (N-(i-1)n+1, \ldots, N-in)$. Certainly, as X_is are disjointed, the whole $\mathcal{R}$ can be embedded in one layer. (See the Figure 3(a).)

$\mathcal{R}'$ – we take $R'_1[Y_1], R'_2[Y_2], \ldots, R'_{d(n)}[Y_{d(n)}]$, where $Y_i = (N-in-m, \ldots, N-in-1)$ (elements in Y_i should be taken modulo N, if necessary). Unfortunately different R'_i may share inputs, so we can not assume that $\mathcal{R}'$ has depth 1. We will estimate this depth later. (See the Figure 3(b).)

$\mathcal{C}$ – consists of n independent (that is on disjointed sets of inputs) parts, each performing $Shift_{1,(i,i+n,i+2n,\ldots,i+d(n)n)}$, for $i = 0, \ldots, n-1$.

$\square_{\text{Construction 3.1}}$

Lemma 3.3

(i) $depth(\mathcal{T}_{N,n}) \in O(1)$,

(ii) $\mathcal{T}_{N,n}$ *performs* $Shift_{t,(0,1,\ldots,N-1)}$, *for each* $t \in \mathcal{A}_n$, *in* $O(d(n))$ *rounds.*

Proof:
(i) Observe that the definition of n and the assumption (2) imply that $m = N - nd(n) < (n+1) + n(d(n+1) - d(n)) + d(n+1) < (D+3)n$,

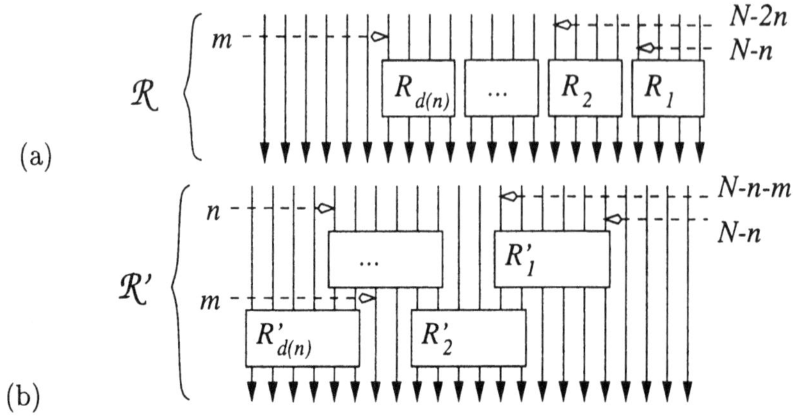

Figure 3: Embedding (a) $\mathcal{R}$ and (b) $\mathcal{R}'$ in $\mathcal{T}_{N,n}$

thus $R'_1, R'_2, \ldots, R'_{d(n)}$ can be embedded in $m/n \in O(1)$ layers. Remaining subnetworks of $\mathcal{T}_{N,n}$ are evidently of depth 1 or 2.

(ii) Suppose $t \in \mathcal{A}_n$. We say that R_i (resp. R'_i) is *active* if it is set according to the setting which enables the network R (resp. R') to perform shift by t on n (resp. m) inputs. Recall that parts of the network that are not explicitly mentioned to be active before executing a round remain inactive. To perform $Shift_{t,(0,1,\ldots,N-1)}$ we shall use $\mathcal{T}_{N,n}$ according to the following algorithm.

Algorithm 3.1

```
for i from 1 to d(n) do
    set active S_1;
    set active R_j, for 1 ≤ j ≤ i;
    set active R'_i;
    execute a round of T_{N,n}
endfor
for i from 1 to d(n) − 1 do
    set active S_1;
    set active R_j, for i ≤ j ≤ d(n);
    execute a round of T_{N,n}
endfor
set active S_2;
```

`set active` $C_i \in \mathcal{C}$, for $0 \leq i < t$;
`execute a round of` $\mathcal{T}_{N,n}$.

For a proof of correctness let us consider a set of n adjacent inputs $B_j = \{jn, jn+1, \ldots, (j+1)n-1\}$, for some $0 \leq j < d(n)$. The last group of m inputs $B_{d(n)} = \{N-m, N-m+1, \ldots, N-1\}$ we analyze later.

Notice that the group B_j is shifted by $\mathcal{S}_1$, so that at the $(j+i)$-th round, for $0 \leq i < d(n)$, it is the input of the active R_i. During remaining rounds, the group is only shifted by $\mathcal{S}_1$, because appropriate R_is and R_i's, which may influence it, are inactive. Hence during the loops `for` the group is permuted consecutively by R_1, R_2,..., $R_{d(n)}$ and $2d(n)-1$ times shifted by $-n$. With the shift performed by $\mathcal{S}_2$ it results in performing Shift_{t,B_j}, for $0 \leq j < d(n)$. For $B_{d(n)}$ a similar analysis shows that $\mathit{Shift}_{t,B_{d(n)}}$ is performed.

Finally, notice that C_i influences $(B_{1,i}, B_{2,i}, \ldots, B_{d(n),i})$ and we activate C_i only for $i < t$. Thus $\mathcal{C}$ affects only the "wrapped" (by $\bmod\ n$) part of B_j moving it to B_{j+1} (or to B_0 in case of $B_{d(n)}$). Therefore the shift by t of the whole interval $(0, 1, \ldots, N-1)$ is completed. □

3.2 Periodic Linear Network

In this section we prove a periodic analogue of the Proposition 2.6, which is not that obvious.

Theorem 3.4 *Let $N > 0$ and Q be a set of positive integers. There exists a network $\mathcal{U}_Q$ of constant depth performing $\mathit{Shift}_{q,(0,1,\ldots,N-1)}$, for $q \in Q$, in $O(|Q|)$ rounds.*

Proof: Let $Q = \{q_1, q_2 \ldots, q_r\}$ and $0 < q_1 < q_2 < \cdots < q_r$. We can assume that $q_{i+1} - q_i \geq r$, for $1 \leq i < r$, since in case of "closer" neighbours: q and q', we can realize $\mathit{Shift}_{q',(0,1,\ldots,N-1)}$ as $\mathit{Shift}^{(q'-q)}_{1,(0,1,\ldots,N-1)} \circ \mathit{Shift}_{q,(0,1,\ldots,N-1)}$. This may require at most r rounds and 2 layers for $\mathit{Shift}_{1,(0,1,\ldots,N-1)}$ extra, thus we will be still within required bounds. Next, for $q = q_t \in Q$ we denote $p = \lfloor N/r \rfloor$, $N' = rp$, $r' = N \bmod r$ and $q_0 = 0$. As in the previous proof, we split inputs into appropriate groups: $D_i = \{ir, \ldots, (i+1)r - 1\}$, for $0 \leq i < p$, and $D_p = \{N', \ldots, N-1\}$, and observe that these groups can be permuted correctly.

To realize $\mathit{Shift}_{q,(0,1,\ldots,N-1)}$ we first permute all groups, but the last one, so that $\mathit{Shift}_{q,(0,1,\ldots,N'-1)}$ is performed. It is done in two phases. The first one – *approximation phase* performs $\mathit{Shift}_{r\lfloor q/r \rfloor,(0,1,\ldots,N'-1)}$. Then the *small shift phase* performs $\mathit{Shift}_{q \bmod r,(0,1,\ldots,N'-1)}$. Finally, *the final interval phase*, places D_p in proper position, and we are done. Below we describe the construction and more formally argue that it is able to perform the required permutation.

Construction 3.2
The network $\mathcal{U}_Q$ consists of subnetworks $\mathcal{L}$, $\mathcal{C}$, $\mathcal{S}_1$, $\mathcal{F}$, arranged in consecutive layers.

$\mathcal{L}$ - consists of r independent subnetworks L_i, for $1 \leq i \leq r$. L_i is a network performing $Shift_{\lfloor q_i/r \rfloor,(D_{0,i},D_{1,i},\ldots,D_{p-1,i})}$, where $D_{j,i}$ is the i-th element of the j-th group.

$\mathcal{C}$ - consists of subnetworks C_i, for $0 \leq i < p$, where C_i performs $Shift_{1,D_i}$.

$\mathcal{S}_1$ - is a network performing $Shift_{1,(0,1,\ldots,N'-1)}$.

$\mathcal{F}$ - consists of subnetworks F_i, where F_i performs $Shift_{-r',(q_{i-1}-r',\ldots,q_i-1)} = Shift_{q_i-q_{i-1},(q_{i-1}-r',\ldots,q_i-1)}$, for $0 < i \leq r$. Notice that different F_i's may have common inputs, so we cannot place them in 2 layers as previous subnetworks. But as adjacent F_is overlap exactly at r' positions and $q_i - q_{i-1} \geq r > r'$, thus one element may be an input of at most two F_is. So we can embed $\mathcal{F}$ in 4 layers (e.g. F_1, F_3, F_5,... in the first layers, and remaining F_is in next two layers.).

□ Construction 3.2

Directly by the construction, we have that $\mathcal{U}_Q$ has a constant depth. So it remains to present an algorithm realizing the required shifts.

Algorithm 3.2 *Performing* $Shift_{q,(0,1,\ldots,N-1)}$, *for* $q = q_t \in Q$.

```
for i from 1 to r do {approximation phase}
    set active L_t;
    set active C;
    execute a round of U_Q
endfor
for i from 1 to q mod r do {small shift phase}
    set active S_1;
    execute a round of U_Q
endfor
for i from 1 to t do {final interval phase}
```

```
    set active F_i ∈ 𝓕;
    execute a round of 𝒰_Q
endfor
```

The permutation ϕ_A performed during the approximation phase is equal to

$$\begin{aligned}\phi_A &= \mathit{Shift}_{1,D_{\lfloor q/r\rfloor+j}}^{(i-t) \bmod r} \circ \mathit{Shift}_{\lfloor q/r\rfloor,(D_{0,t},\ldots,D_{p-1,t})} \circ \mathit{Shift}_{1,D_j}^{(t-i) \bmod r}\\ &= \mathit{Shift}_{r\lfloor q/r\rfloor,(0,1,\ldots,N'-1)}\end{aligned}$$

During the small shift phase the permutation $\phi_S(x) = (x + q \bmod r) \bmod N'$ on $(0, 1, \ldots, N' - 1)$ is performed. Thus $\phi_S(\phi_A(x)) = x + q \bmod N'$, for $0 \le x < N'$. Finally, ϕ_F – the permutation performed by the final phase, is such that

$$\phi_F(i) = \begin{cases} (i - r') \bmod N & \text{for } 0 \le i < q, \\ i & \text{for } q \le i < N', \\ i + \sum_{j=t}^{1}(q_j - q_{j-1} \bmod N) = i + q \bmod N & \text{for } i \in D_p. \end{cases}$$

It is easy to check, that composed with shift $\phi_S \circ \phi_A$, it yields the required permutation. The bound on the number of rounds follows directly from the algorithm.

$\square_{\text{Proof of Theorem 3.4}}$

4 Direct Construction for Hypercube Shifts

In this section we improve the result of Corollary 3.2. As before, we present a constant depth periodic network realizing shifts by powers of 2. Certainly, in comparison with the previous result, only constant factors in bounds of the complexity of the network may be improved and in this section we reduce them to more "practical" values.

The presented construction depends tightly on the kind of shifts performed and cannot be easily generalized for shifts other than by powers of 2 (this explains our interest in embedding technique of the previous section, in spite of worse constants in time and size bounds for hypercube shifts). Besides the main result, this section contains several interesting network constructions, e.g. a constant depth network working in logarithmic time for consecutive shifts (Lemma 4.1) – a result mentioned earlier as a periodic counterpart of Proposition 2.5. This construction shows that in case of consecutive shifts, there is also a trade-off between the depth of the network and the number of rounds (although this case can not be handled by parallel embedding due to bad features of consecutive shifts).

In the next subsection we present main steps of the construction We show, that efficient realization of hypercube shifts on $(0, 1, \ldots, N-1)$, for odd N, can be reduced to efficient realization of shifts by consecutive numbers. Hence, in the next subsection we show the construction for consecutive shifts. Finally, we generalize the construction, so that it works also for even N.

4.1 Outline of the Construction

Let N be an odd positive integer and let e_2 be the integer $0 < e_2 < N$, such that $2 \cdot e_2 = 1 \bmod N$ (we can consider e_2 as $2^{-1} \bmod N$). We split the set $\{0, 1, \ldots, N-1\}$ into disjointed subsets G_0, $G_1, \ldots$, G_r, for $r \geq 1$, such that:

(i) if $g \in G_i$ then $2g \bmod N \in G_i$, for $1 \leq i \leq r$; that is G_is are closed under multiplication by 2;

(ii) none of the groups $G_0, \ldots$, G_r can be split into smaller subsets having the property *(i)*.

In particular we take $G_0 = \{0\}$ and $G_1 = \{2^i \bmod N \mid i \geq 0\}$. Notice that the groups G_i are also closed under multiplication by e_2.

In each group G_i we fix an arbitrary element g_i, which we call a *generator* of G_i. Since all elements of G_i can be written as a product of the generator and an appropriate power of 2, let us define, for $0 \leq i \leq r$, functions:

$$ord_i : G_i \to \{0, \ldots, |G_i| - 1\}, \quad \text{such that} \quad x = 2^{ord_i(x)} g_i \bmod N,$$
$$rev_i : G_i \to G_i, \quad \text{such that} \quad rev_i(x) = g_i \cdot 2^{-ord_i(x)} \bmod N.$$

From now on, whenever we treat G_i as an ordered sequence of its elements, we consider the ordering $G_i = (h_1, h_2, \ldots)$, where $ord_i(h_j) < ord_i(h_k)$ for $j < k$. In particular, by performing shift by t on G_i, we mean $Shift_{t,(h_1,h_2,\ldots)}$

The above definitions allow us to formulate a natural correlation between addition of orders in defined groups and multiplication by powers of 2 in the interval $(0, 1, \ldots, N-1)$. Notice that, for $x \in G_i$, we have

$$Shift_{t,G_i}(x) \quad = \quad x \cdot 2^t \bmod N. \tag{4}$$

To perform $\{Shift_{2^t,(0,1,\ldots,N-1)} : 0 \leq t < \log N\}$ we apply the following algorithm using a "procedure" for $\{Shift_{t,G_i} : 0 \leq t \leq \min\{\log N, |G_i| - 1\}\}$, for each $0 \leq i \leq r$.

Algorithm 4.3 *Performing* $Shift_{2^t,(0,1,\ldots,N-1)}$ *for an odd* N.

```
    procedure hypercube_shift_on_odd(2^t,(0,1,...,N-1))
(1)     perform rev_i, for 0 ≤ i ≤ r;
(2)     perform Shift_{t,G_i}, for 0 ≤ i ≤ r;
(3)     perform rev_i, for 0 ≤ i ≤ r;
(4)     perform Shift_{1,(0,1,...,N-1)};
(5)     perform Shift_{t,G_i}, for 0 ≤ i ≤ r.
```

To see that Algorithm 4.3 performs $Shift_{2^t,(0,1,\ldots,N-1)}$, let $x \in G_i$ and $k = ord_i(x)$, hence $x = g_i \cdot 2^k \bmod N$. By property (4) we have, that Algorithm 4.3 performs the following permutation π:

$$
\begin{aligned}
\pi(x) &= Shift_{t,G_j}(Shift_{1,(0,1,\ldots,N-1)}(rev_i(Shift_{t,G_i}(rev_i(x))))) \\
&= Shift_{t,G_j}(Shift_{1,(0,1,\ldots,N-1)}(rev_i(Shift_{t,G_i}(g_i \cdot e_2^k)))) \\
&= Shift_{t,G_j}(Shift_{1,(0,1,\ldots,N-1)}(rev_i((g_i \cdot e_2^k \cdot 2^t) \bmod N))) \\
&= Shift_{t,G_j}(Shift_{1,(0,1,\ldots,N-1)}((g_i \cdot e_2^{-k} \cdot 2^{-t}) \bmod N)) \\
&= Shift_{t,G_j}(Shift_{1,(0,1,\ldots,N-1)}((x \cdot e_2^t) \bmod N)) \\
&= Shift_{t,G_j}((x \cdot e_2^t + 1) \bmod N)) \\
&= ((x \cdot e_2^t + 1) \bmod N) \cdot 2^t \bmod N \\
&= (x + 2^t) \bmod N.
\end{aligned}
$$

□

As $\circ_{i=0}^{r} rev_i$ is a set of disjointed transpositions, we need a constant number of layers to perform steps (1), (3) and (4) of the algorithm. Thus we are left with solving the problem of efficient realization of $Shift_{t,G_i}$, for $t \in \{0, 1, \ldots, \lfloor \log N \rfloor\}$.

4.2 Periodic Realization of Consecutive Shifts

To simplify the notation, let us assume that the problem is to perform $Shift_{t,(0,1,\ldots,M-1)}$, for $t \in \{0, 1, \ldots, m-1\}$, in $O(m)$ rounds. Substituting $M = |G_i|$ and $m = \min\{\lfloor \log N \rfloor, |G_i| - 1\}$ and taking into account the specific orders of G_is, we will be done with the problem of the previous section.

We deal with the problem in two stages. First, we describe a construction dependent on factorization of a number of inputs M. In particular, this construction is very efficient for odd Ms. Later, we apply it to an even number of inputs in a small additional cost of depth.

Lemma 4.1 *$Shift_{t,(0,1,\ldots,M-1)}$, for $t \in \{0,1,\ldots,m-1\}$, may be performed by a switching network in periodic mode*

(i) in depth $3+\lceil \log d \rceil$ and $2\log_d(m-1)+1$ rounds, where $d = \min\{k > 1 : k$ does not divide $M\}$;

(ii) in depth 4 and $2\log m + 1$ rounds, if M is odd;

(iii) in depth 5 and $4\log m$ rounds, if M is even.

Proof:
(i) Since d and M are relatively prime, there exists an integer e $(0 < e < M)$, such that $de = 1(\bmod M)$. So let us define permutations $\pi_d(i) = id \bmod M$ and $\pi_e(i) = ie \bmod N$, for $i \in \{0,1,\ldots,M-1\}$. Also let $t \in \{0,1,\ldots,m-1\}$ and $(t_k t_{k-1} \ldots t_1 t_0)_{(d)}$ be a d-ary representation of t $(t_k > 0)$. We will show that Algorithm 4.4 performs $Shift_{t,(0,1,\ldots,M-1)}$.

Algorithm 4.4 *Performing $Shift_{t,(0,1,\ldots,M-1)}$ for an odd M.*

```
procedure sequential_shift_on_odd(t,(0,1,...,M-1))
   for i from 1 to k do perform π_e;
   for i from k downto 1 do
      perform Shift_{t_i,(0,1,...,M-1)}
      perform π_d
   endfor
   perform Shift_{t_0,(0,1,...,M-1)}
```

Observe that in the above algorithm the first loop performs permutation $\rho_1(x) = x \cdot e^k \bmod M$. Hence, the second loop performs $\rho_2(x) = (\cdots((x + t_k)d + t_{k-1})d + \cdots + t_1)d \bmod M = x \cdot d^k + \sum_{i=1}^{k} t_i d^i \bmod M = x \cdot d^k + t \bmod M$. So it is easy to see, that $Shift_{t_0,(0,1,\ldots,M-1)}(\rho_2(\rho_1(x))) = (x \cdot e^k \cdot d^k + t - t_0) \bmod M$, so we have obtained $Shift_{t,(0,1,\ldots,M-1)}$.

Each step of Algorithm 4.4 is a permutation, thus we can build a network $\mathcal{N}_d(M)$ of depth $3+\lceil \log(d-1) \rceil$ (2 layers for π_d and π_e as $\pi_e = \pi_d^{-1}$ and $\lceil \log(d-1) \rceil + 1$ layers for $Shift_{s,(0,1,\ldots,M-1)}$, where $1 \le s < d$, as in

Proposition 2.5) realizing the algorithm. Also appropriately arranging the order of layers we can see that $2k+1 \leq 2\log_d(m-1)+1$ rounds suffice to perform required shifts.
(ii) This point evidently follows from *(i)*. For odd M we have $d=2$, thus $\mathcal{N}_2(M)$ has depth 4 and works in $2\log(m-1)+1$ rounds.
(iii) Let us start with the following simple observation: for an arbitrary $t \in \{0,1,\ldots,N-1\}$

$$\mathit{Shift}_{t,(0,1,\ldots,M-1)} = \mathit{Shift}_{-1,(0,1,\ldots,t-1)} \circ (0, M-1) \circ \mathit{Shift}_{t,(0,1,\ldots,M-2)},$$

where $(0, M-1)$ denotes a single transposition. To perform $\mathit{Shift}_{t,(0,1,\ldots,M-2)}$ we can use $\mathcal{N}_2(M-1)$ (from (ii)). So it remains to realize $\mathit{Shift}_{-1,(0,1,\ldots,t-1)}$, for any $0 \leq t < m$. For this purpose we define the following permutations. Let

$$\pi_2(i) = 2i \bmod (M-1)$$

and $\text{œ} = \{(0,1),(2,3)\ldots\}$ be a set of odd-even transpositions in $(0,1,\ldots,M-1)$. We define $\text{œ}_{[0..i]}$ as a permutation obtained by realizing only those transpositions, which concern elements of the interval $[0..i]$, i.e.:

$$\text{œ}_{[0..i]}(j) = \begin{cases} j+1 & \text{for } j < i \text{ and } j \text{ even,} \\ j & \text{for } j = i \text{ and } j \text{ even,} \\ j-1 & \text{for } j \leq i \text{ and } j \text{ odd,} \\ j & \text{for } j > i. \end{cases}$$

We shall show that permutations π_2, π_2^{-1} and $\text{œ}_{[0..i]}$ are sufficient to perform $\mathit{Shift}_{-1,(0,1,\ldots,t-1)}$. To this end we apply Algorithm 4.5.

Algorithm 4.5 *Performing* $\mathit{Shift}_{-1}(0,1,\ldots,t-1)$, *for* $1 \leq t < m$.

```
procedure back_shift(−1,(0,1,...,t − 1))
    {Let k be a length of a binary representation of t − 1.}
    for i from 1 to k do perform π2^(−1);
    for i from k downto 1 do perform œ[0..⌊(t−1)/2^i⌋];
                                perform π2;
    perform œ[0..t−1].
```

By induction on t we can easily check, that Algorithm 4.5 indeed performs $\mathit{Shift}_{-1,(0,1,\ldots,t-1)}$, for $1 \leq t < m$. For $t=1$ it evidently holds. For $t>1$, notice that we can regroup instructions of the algorithm to emphasis its recursive structure and see that all but first executions of the first loop, and all but last executions of the second loop, should (by inductive assumption) perform shift by -1 on $(0, 1, \ldots, \lfloor (t-1)/2 \rfloor)$. Hence the permutation performed by the whole algorithm is:

$$\text{œ}_{[0..t-1]} \circ \pi_2 \circ \mathit{Shift}_{-1,(0,1,\ldots,\lfloor (t-1)/2 \rfloor)} \circ \pi_2^{-1},$$

To see that it equals $\mathit{Shift}_{-1,(0,1,\ldots,t-1)}$ we consider the following cases:

(i) $x > t-1$.
Then $\pi_2^{-1}(x) \notin \{0, 1, \ldots, \lfloor (t-1)/2 \rfloor\}$ and thus $\pi_2 \circ \mathit{Shift}_{-1,(0,1,\ldots,\lfloor (t-1)/2 \rfloor)} \circ \pi_2^{-1}(x) = x$ and evidently $\text{œ}_{[0..t-1]}(x) = x$.

(ii) $x \leq t-1$ and x is odd.
Then, as previously, $\pi_2^{-1}(x) \notin \{0, 1, \ldots, \lfloor (t-1)/2 \rfloor\}$, so $\pi_2 \circ \mathit{Shift}_{-1,(0,1,\ldots,\lfloor (t-1)/2 \rfloor)} \circ \pi_2^{-1}(x) = x$. But then $\text{œ}_{[0..t-1]}(x) = x - 1$.

(iii) $0 < x \leq t-1$ and x is even.
Then $\pi_2^{-1}(x) = x/2$ and $0 < x/2 \leq \lfloor (t-1)/2 \rfloor\}$. Hence $\mathit{Shift}_{-1,(0,1,\ldots,\lfloor (t-1)/2 \rfloor)}(x/2) = x/2 - 1$ and $\pi_2(x/2 - 1) = x - 2$. Evidently $0 \leq x - 2 < t - 1$ and $x - 2$ is even, so $\text{œ}_{[0..t-1]}(x-2) = (x-2) + 1 = x - 1$.

(iv) $x = 0$.
Then $\pi_2^{-1}(0) = 0$ and $\mathit{Shift}_{-1,(0,1,\ldots,\lfloor (t-1)/2 \rfloor)}(0) = \lfloor (t-1)/2 \rfloor$. Hence $\pi_2 \lfloor (t-1)/2 \rfloor) = \lfloor (t-1)/2 \rfloor$. An easy inspection based on definition of oe shows that $\text{œ}_{[0..t-1]}(2 \cdot \lfloor (t-1)/2 \rfloor) = t - 1$.

Summarizing, to realize consecutive shifts, for the even M, we require a network performing both Algorithm 4.4 and Algorithm 4.5. See that for this purpose we need permutations $\pi_2(0, 1, \ldots, M-2)$, $\pi_2^{-1}(0, 1, \ldots, M-2)$, $\mathit{Shift}_{1,(0,1,\ldots,M-2)}$, $\text{œ}_{[0..i]}(0, 1, \ldots, M-2)$, for $1 \leq i < m - 1$, and the transposition $(0, M-1)$. It is easy to see, that 5 layers are sufficient to realize them and $4 \log m$ rounds will do. Thus we have proved Lemma 4.1.

$\square_{\text{Lemma 4.1}}$

4.3 Proof of Theorem 4.2

In the previous sections we have shown almost all details of the construction. It remains to deal with the even number of inputs and to estimate parameters of our construction in details. We start with the first problem.

Let N be an even integer. We choose an odd integer N_1, as close to $N/2$ as possible, so that either $N_1 = N/2$ (when $N/2$ is an odd integer) or $N_1 = N/2 - 1$ (when $N/2$ is even). Hence we build networks performing $\mathit{Shift}_{2^k,(0,1,\ldots,N_1-1)}$ and $\mathit{Shift}_{2^k,(N_1,\ldots,N-1)}$ as described in the previous sections. Then we complement the network with a layer of transpositions: $(0, N_1), (1, N_1+1), \ldots, (N_1-1, 2N_1)$. This allows us to perform each shift by $2^k \leq N_1$ as follows:

$$\mathit{Shift}_{2^k,(0,1,\ldots,N-1)} = \circ_{i=0}^{2^k-1}(i, N_1+i) \circ \mathit{Shift}_{2^k,(0,1,\ldots,N_1-1)} \circ \mathit{Shift}_{2^k,(N_1,\ldots,N-1)}.$$

So we are left with the last, biggest shift by $2^{\lfloor \log N \rfloor}$ on $(0, 1, \ldots, N-1)$. We add separate layers performing it and in this way we are done with the case of even number of inputs.

So we can finally write a complete algorithm performing hypercube shifts on odd number of inputs, and hence regardless of parity of number of inputs:

Algorithm 4.6 *Performing* $\mathit{Shift}_{2^t,(0,1,\ldots,N-1)}$, *for* $t = 0, \ldots, \lfloor \log N \rfloor$.

```
procedure hypercube_shift_on_odd(2^t,(0,1,...,N-1));
   for each G_i preform rev_i(G_i);
   if even |G_i| then H_i ← G_i − {g_i 2^(|G_i|−1)} else H_i ← G_i;
   for each H_i perform sequential_shift_on_odd(t,H_i);
   if even |G_i| then
      transpose(g_i,g_i 2^(|G_i|−1));
      back_shift(−1,(0,1,...,t−1));
   for each G_i preform rev_i(G_i);
   perform Shift_1,(0,1,...,N−1);
   if even |G_i| then
      transpose(g_i,g_i 2^(|G_i|−1));
      back_shift(−1,(0,1,...,t−1));

procedure hypercube_shift(2^t,(0,1,...,N−1));
   if t = ⌊log N⌋ then perform Shift_2^t,(0,1,...,N−1); exit
   if odd N then hypercube_shift_on_odd(2^t,(0,1,...,N−1)); exit
   if odd N/2 then N_1 ← N/2 else N_1 ← N/2 − 1;
   hypercube_shift_on_odd(2^t,(0,1,...,N_1−1));
   hypercube_shift_on_odd(2^t,(N_1,...,N−1));
   for 0 ≤ i < 2^t transpose(i,i+N_1);
```

We can realize the above algorithm on the network consisting of 9 layers $L_1, \ldots, L_9$ as follows (groups G_i and H_i are defined either for N (for odd N) or separately for N_1 and $N - N_1$ (for even N)):

(a) $rev_i(G_i)$, for $0 \le i \le r$ performed by L_1,

(b) $Shift_{1,(0,1,\ldots,N-1)}$ or $Shift_{1,(0,\ldots,N_1-1)}$ and $Shift_{1,(N_1,\ldots,N-1)}$, in case of even N, performed by L_2 and L_3 (in case of even N we choose L_2 so that symmetries of $(0,\ldots,N_1-1)$, $(N_1,\ldots,N-1)$ are also a symmetry of $(0,1,\ldots,N-1)$; it is easy to see, that this can be done;

(c) π_2 on H_i, for $0 \le i \le r$ performed by L_4 and L_5;

(d) π_2^{-1} on H_i, for $0 \le i \le r$ performed by L_5 and L_4;

(e) $Shift_{1,H_i}$ performed by L_1 and L_6 (see, that rev_i is a symmetry on H_i);

(f) in case of even $|G_i|$ – $\text{œ}_{[0..j]}(H_i)$, for $0 \le i \le r$ and $1 \le j < |H_i|$ performed by L_7;

(g) in case of even $|G_i|$ – the transposition $(g_i, g_i 2^{|G_i|-1})$, for $0 \le i \le r$; this switch can be embedded e.g. in L_1;

(h) in case of even N – transpositions (i, N_1+i), for $0 \le i < N_1$ performed by L_8;

(i) in case of even N – $Shift_{\lfloor \log N \rfloor,(0,1,\ldots,N-1)}$ performed by L_2 and L_9.

Finally, we have the following theorem:

Theorem 4.2 *For a positive integer N, there exists a constant depth switching network performing $Shift_{t,(0,1,\ldots,N-1)}$, for $t \in \{2^i : 0 \le i \log N\}$ in periodic mode. The depth of the network is 9 and hence the size is at most $4.5N$. The number of rounds is $8 \log\log N + 4$, which implies that time is at most $64 \log\log N + 32$. In case of odd number of inputs N the depth 7 of the network is sufficient.*

References

[1] M. Ajtai, J. Komlós, E. Szemeredi, Sorting in $c \cdot \log n$ Parallel Steps, *Combinatorica* 3 (1983) pp. 1–19.

[2] V. Beneš, Permutation Groups, Complexes, and Rearrangeable Multistage Connecting Networks, *Bell System Tech. Journal* 43 (1964) pp. 1619–1640.

[3] V. Beneš, *Mathematical Theory of Connecting Networks and Telephone Traffic* (Academic Press, New York, 1965).

[4] R. Cole, B. Maggs, R. Sitaraman, Routing on Butterfly Networks with Random Faults, *Proc. of the 36th Annual Symposium on Foundations of Computer Science*, (IEEE, 1995) pp. 558–570.

[5] J. Hromkovič, K. Loryś, P. Kanarek, R. Klasing, W. Unger, H. Wagener, On the Sizes of Permutation Networks and Consequences for Efficient Simulation of Hypercube Algorithms on Bounded-Degree Networks (extended abstract), *Proc. of the 12th Symposium on Theoretical Aspects of Computer Science* (Springer-Verlag LNCS 900, 1995) pp.255–266.

[6] J. Hromkovič, R. Klasing, W. Unger, H. Wagener, Optimal Algorithms for Broadcast and Gossip in the Edge-Disjoint Path Modes, *Proc. of the 4th Scandinavian Workshop on Algorithm Theory* (Springer-Verlag LNCS 824, 1994) pp. 219–230.

[7] P. Kanarek, K. Loryś, Shallow Switching Networks for Special Families of Shifts, *Tech. Rep. tr-ri-95-162* (University of Paderborn, June 1995).

[8] M. Kutyłowski, K. Loryś, B. Oesterdiekhoff, Periodic Merging Networks, *Proc. of the 7th ISAAC*, (Springer-Verlag LNCS 1178, 1996), pp. 336-345.

[9] M. Kutyłowski, K. Loryś, B. Oesterdiekhoff, R. Wanka, Fast and Feasible Periodic Sorting Networks of Constant Depth, *Proc. of the 35th Annual Symposium on Foundations of Computer Science* (IEEE 1994) pp. 369–380.

[10] F.T. Leighton, *Introduction to Parallel Algorithms and Architectures: Arrays, Trees, Hypercubes* (Morgan Kaufmann Publisher, 1992).

[11] N. Pippenger, Self-routing Superconcentrators, *Proc. of the 25th Annual ACM Symposium on the Theory of Computing* (ACM Press, New York, 1993) pp. 355–361.

[12] N. Pippenger, L. Valiant, Shifting Graphs and their Applications, *J. of the ACM* 23(3) (1976) pp. 423–432.

[13] A. Waksman, A Permutation Network, *J. of the ACM* Vol.15 No.1 (1968) pp. 159-163.

[14] R. Wanka, Paralleles Sortieren auf mehrdimensionalen Gittern, *Ph.D. Dissertation* (Universität Paderborn, 1994).

[15] E. Upfal, On the Theory of Interconnection Networks for Parallel Computers, *Proc. of the 21st Annual International Colloquium on Automata, Languages and Programming* (Springer-Verlag LNCS 320, 1994) pp. 473–486 .

SWITCHING NETWORKS: RECENT ADVANCES
D.-Z. Du and H.Q. Ngo (Eds.) pp. 215-256

Analysis of Shuffle-Exchange Networks under Permutation Trafic

Hasan Çam
Computer Science and Computer Engineering Department
University of Arkansas, Fayetteville, AR 72701
E-mail: cam@engr.uark.edu

Contents

1 Introduction

Multistage interconnection networks (MINs) such as shuffle-exchange, baseline, and banyan networks have been widely investigated for use in parallel processing systems and high speed communication switches. Among the main advantages of a shuffle-exchange type network are self-routing, low hardware complexity compared to crossbar switch, distributed control, suitable for VLSI implementation, modular and scalable. The main disadvantage of an n-stage shuffle-exchange type network with N inputs and N outputs is to have a limited number of connection patterns bacause of *internal conflicts* in switching elements, where $N = 2^n$.

In an n-stage shuffle-exchange type network with N inputs/outputs, there is a unique path between any input and output, and only a limited number of permutations can be implemented. The permutation capability of an n-stage MIN can be improved by adding extra stages. Many researchers have been interested in the question of how many more stages are needed to make an n-stage shuffle-exchange network *rearrangeable* [1, 2, 3, 4, 5, 6, 7, 8, 9, 10, 11, 12, 13]. (A network is called rearrangeable (or rearrangeably non-blocking) if the network can implement each and every permutation of its input ports to output ports in a single pass through the network). Indeed, the question of whether a $(2n-1)$-stage shuffle-exchange is rearrangeable or not has remained an open question until very recently [14]. The basic idea behind the proof for the rearrangeability of the $(2n-1)$-stage shuffle-exchange networks is presented in this paper.

Asynchronous Transfer Mode (ATM) is the switching and multiplexing technique chosen for Broadband Integrated Services Digital Network (B-ISDN). In ATM, user information is transferred through the network in small fixed size packets of 53 bytes called *cells*, and line speeds equal to 155 Mbps, 622 Mbps and above. An ATM switch routes cells arriving at its input ports to their desired output ports. A multistage ATM switch fabric is usually constructed by interconnecting switching boxes (SBs) of size (2×2) each in a specific topology using links. When more than one cell compete for the same internal link of ATM switch, an *internal conflict* (or blocking) occurs. When more than one cell compete for the same output port, an *external* (or output) conflict occurs.

Shuffle-exchange type networks are very attractive to ATM switch designers due to their desirable features. In particular, they have constant latency for any input-output pair, are self-routing and modular, support synchronous and asynchronous modes of operation, preserve cell-sequence due to path-uniqueness, are very suitable for VLSI implementation, and have a complexity of $O(N\, log_2 N)$ compared to $O(N^2)$ of the crossbar switch. The major drawback of shuffle-exchange type networks is to have a very limited number of connection patterns, which causes them to suffer from internal conflicts, in addition to external conflicts. This paper discusses an input-buffered ATM switch based on an n-stage shuffle-exchange type network. Our objective is to alleviate this drawback to the extent that the throughput of an input-buffered ATM switch based on an n-stage shuffle-exchange type network with content addressable FIFO (CAFIFO) input buffers can reach the throughput of a nonblocking network. Our algorithm uses parallel windowing for preventing external blocking and alleviating head-of-line blocking, and employs the notion of frames and balanced matrices to prevent internal blocking in a shuffle-exchange type network. The fact that we replace all stages of a nonblocking switching network by an n-stage shuffle-exchange network yields a significant economy in implementation.

If two or more packets in an ATM switch may arrive simultaneously on different inputs destined for the same output and, hence, external blocking is unavoidable in a packet switch. In order to smooth the statistical fluctuations in packet arrivals to the switch, three main approaches were proposed: *input buffering*, *output buffering*, and *completely shared buffering* [15]. To be able to implement output buffering, either the switch must operate as fast as the sum of the speeds of its input ports, or the buffers must have as many write ports as there are input ports to the switch, to be able to handle simultaneous packet arrivals [16]. Because implementing buffers with multiple write ports increases their size and performance, the Knockout Switch [17] implements output buffering by using a concentrator at each output port to reduce the number of buffers needed to receive simultaneously arriving packets. Therefore, although output buffering provides the best performance in terms of switch throughput and cell delay under uniform traffic patterns, the implementation of input buffering is appealing because only one packet at a time arrives at the input port so that the buffer needs only a single write port.

Input-buffered ATM switches resolve contention for external conflict prior to switching. Some of the external conflict resolution algorithms are three-phase algorithm [18], scheduling CAM [19], and ring reservation [20]. When input buffers are FIFO (first in first out) queues, a serious problem called head-of-line (HOL) blocking occurs: a single packet at the head of the queue whose destination output port is busy can block all other packets in that queue from being transmitted even if their destination output ports are idle. This limits the maximum throughput of a nonblocking switch to 0.56 under uniform traffic, regardless of the selection policy used among the contending HOL cells [21]. Contention resolution algorithms including ring reservation algorithm suffer from HOL blocking. Three methods, namely, output expansion [22], channel grouping [23] and windowing [15] are the most known techniques to alleviate HOL blocking. In *output expansion*, extra internal switch links and output buffers are added in order to be able to forward more than one cell to each output port during the same cell interval. In *channel grouping*, the switch output ports are partitioned into a number of groups and cells are allowed to contend for a group, rather than a specific port. In *windowing* (also called bypass queue [24, 25, 26]), cells behind the HOL cell are also allowed to contend for output port access.

Two types of windowing, namely, sequential windowing (also called serial bypass) and parallel windowing (also called parallel bypass) are possible. In *sequential windowing*, the first cells in a switch's input buffers first contend for outputs, then the losing input buffers that contain a second cell are allowed to contend a second time, those that lose in the second time and have a third cell are allowed to contend a third time, and so on until w times are contended, where w is the depth of window. In *parallel windowing*, all cells in the windows of all buffers contend in a single round with the winners proceeding to the outputs, with the condition that only one cell can be selected at most from a buffer. To preserve the cell sequence, among those cells destined to the same output port in the window of a buffer the cell closest to the head of the buffer is selected if one of them needs to be chosen.

The rest of the paper is organized as follows. The basic terminology, some preliminary results and definitions used throughout the paper are introduced in Section 2. Section 3 presents the basic idea behind the proof for the rearrangeability of the $(2n-1)$-stage shuffle-exchange networks. In Section 4, a modified shuffle-exchange network (MSN), along with a fast destination-tag routing scheme, is proposed. Section 5 discusses a parallel windowing algorithm for alleviating head-of-line blocking and preventing both internal and external conflicts on reverse baseline network with CAFIFO input buffers. Concluding remarks are made in Section 6.

2 Frames and Balanced Permutation Matrices

In this section, we present the definitions of balanced permutations, balanced matrices, frames, some related terms and preliminary results. Somewhat different versions of some of these definitions and a motivational example appeared in [27, 28, 29]. Some preliminary results that will be used throughout the paper are also presented in this section.

Definition 2.1 *(balanced permutation matrix, balanced permutation)*: Let the N rows of the first k columns of a permutation matrix representing a permutation on $\{0, 1, \cdots, N-1\}$ be partitioned into 2^{n-k} so-called *blocks*, each having 2^k consecutive rows, for $1 \leq k \leq n$ and $N = 2^n$. A permutation matrix is called a *balanced permutation matrix* if, for each k, every block of the first k columns of the permutation matrix contains the binary representation of a permutation on $\{0, 1, \ldots, 2^k - 1\}$. Any permutation whose binary representation is a balanced permutation matrix is called a *balanced permutation* [28].

The construction of a balanced permutation on $\{0, 1, \ldots, 15\}$ is illustrated in Figure 1. A block with k columns and 2^k rows contains the *binary representation* of a permutation on $\{0, 1, \ldots, 2^k - 1\}$, for $1 \leq k \leq 4$. Figure 1(a) shows 8 blocks each having a column vector of 2 entries that represents a permutation on $\{0, 1\}$. Figure 1(b) illustrates 4 blocks each having a matrix of order 4×2 that represents a permutation on $\{0, 1, 2, 3\}$. Each of two blocks shown in Figure 1(c) has a matrix of order 8×3 that represents a permutation on $\{0, 1, 2, 3, 4, 5, 6, 7\}$. As an example, Figure 1(e) shows a balanced permutation matrix representing a balanced permutation $\pi = (0\ 14\ 15\ 5\ 13\ 2\ 7\ 4\ 1\ 3\ 8)(6\ 10)(9)(11\ 12)$ in cyclic notation.

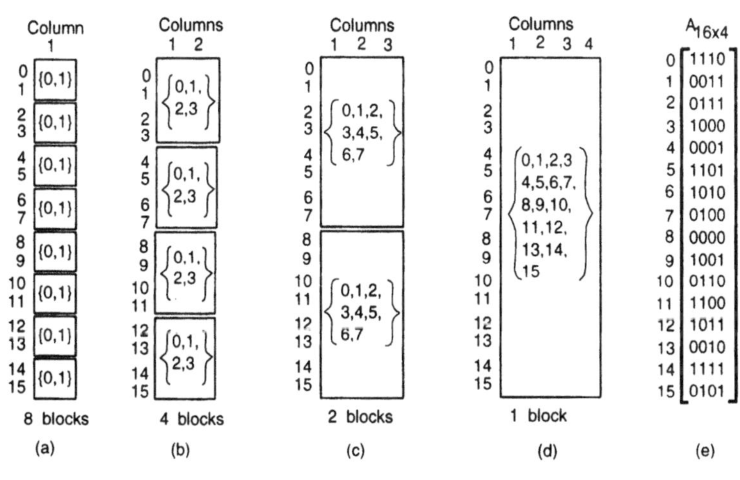

Figure 1: The construction of a balanced permutation on $\{0, 1, \ldots 15\}$.

The concept of balanced matrices was introduced by Linial and Tarsi [2]. Next, we rephrase their definition using permutation matrices.

Definition 2.2 *(balanced matrix)*: Let $N = 2^n$ and $n \geq 1$. A matrix $A_{N\times k}$ containing is *balanced* if either one of the following conditions is satisfied:

1. For $1 \leq k \leq n$, $A_{N\times k}$ is a submatrix of a permutation matrix of order $N \times n$.

2. For $k > n$, every n consecutive columns of $A_{N\times k}$ form a permutation matrix.

EXAMPLE. Two balanced matrices D and E are shown below. But, notice that the matrix $[D\ E]$ is not balanced because not every consecutive three columns of $[D\ E]$ form a permutation matrix of order 8×3.

$$D = [d_1\ d_2\ d_3] = \begin{array}{c} 0\\1\\2\\3\\4\\5\\6\\7 \end{array}\begin{bmatrix} 1 & 1 & 1\\ 1 & 0 & 0\\ 0 & 0 & 0\\ 0 & 1 & 0\\ 1 & 1 & 0\\ 0 & 0 & 1\\ 1 & 0 & 1\\ 0 & 1 & 1 \end{bmatrix} \qquad E = [e_1\ e_2] = \begin{array}{c} 0\\1\\2\\3\\4\\5\\6\\7 \end{array}\begin{bmatrix} 1 & 0\\ 1 & 0\\ 1 & 1\\ 1 & 1\\ 0 & 1\\ 0 & 1\\ 0 & 0\\ 0 & 0 \end{bmatrix}$$

A subset M of edges in a graph G is called a *matching* if no two edges of M have a vertex in common. A matching M is said to be a *perfect matching* if it covers all vertices of G [31].

Definition 2.3 *(perfect matching graph of a balanced matrix of order $N \times (n-1)$)*: The *perfect matching graph* PG_A of a balanced matrix $A_{N\times(n-1)}$ on N vertices is a graph whose vertices have degree one and vertices V_i and V_j are joined by an edge if the ith and jth rows of $A_{N\times(n-1)}$ are identical.

EXAMPLE. The perfect matching graph corresponding to the submatrix $[d_1\ d_2]$ of the above balanced matrix D has four edges, and each of the following four sets contains the two vertices of an edge: $\{V_0,\ V_4\}$, $\{V_1,\ V_6\}$, $\{V_2,\ V_5\}$, and $\{V_3,\ V_7\}$.

Definition 2.4 *(2-labeling)*: The *2-labeling* (or 2-coloring) of a graph refers to the assignment of integers 0 and 1 to its vertices such that the assigned integers, called 2-labels, of those two vertices incident with an edge constitute the set $\{0, 1\}$.

The notation $A_{x:y}$ refers to the submatrix that contains columns x, $x+1$, ..., y of matrix A, where $1 \leq x \leq y$. Unless specifically stated, $A_{x:y}$

is assumed to have N rows. We show in Theorem 2.1 that for any given permutation matrix $A_{N\times n}$, there are $2^{Nn/2}$ permutation matrices $B_{N\times n}$'s such that every n consecutive columns of any matrix $[A_{N\times n}\ B_{N\times n}]$ form a permutation matrix.

Theorem 2.1 *Given a permutation matrix $A_{N\times n}$, there are $2^{Nn/2}$ permutation matrices $B_{N\times n} = [b_1\ b_2 \ldots b_n]$ such that $[A_{N\times n}\ B_{N\times n}]$ is a balanced matrix for each $B_{N\times n}$.*

Proof. In order for $[A_{1:n}\ B_{1:n}]$ to be a balanced matrix, its every n consecutive columns must form a permutation matrix. Since $A_{1:n}$ is already a permutation matrix, the number of balanced matrices $[A_{2:n}\ b_1]$ will be determined first. The unique perfect matching graph $PG_{A_{2:n}}$ of $A_{2:n}$ has $N/2$ edges. There are $2^{N/2}$ distinct possible 2-labelings of this graph because each pair of adjacent vertices can be labeled in two different ways (0-1 or 1-0). Each 2-labeling of the graph corresponds to a distinct b_1 of length N. Thus, the total number of those b_1's that are balanced with $A_{2:n}$ in $[A_{2:n}\ b_1]$ is $2^{N/2}$. This reasoning is applied to the construction of other columns of $B_{1:n}$ as follows.

The unique perfect matching graph of any submatrix $[A_{3:n}\ b_1]$ has $N/2$ edges. There are $2^{N/2}$ distinct possible 2-labelings of this graph and, therefore, there are $2^{N/2}$ distinct b_2's such that $[A_{3:n}\ b_1\ b_2]$ is balanced for each b_2. Similarly, from the perfect matching graph of any submatrix $[A_{4:n}\ b_1\ b_2]$, $2^{N/2}$ distinct b_3's can be constructed. This process of construction is continued until all distinct b_n's are determined. Thus, for a given $A_{1:n}$, there are $2^{Nn/2}$ $B_{1:n}$'s such that $[A_{1:n}\ B_{1:n}]$ is a balanced matrix for each $B_{1:n}$. □

Another important concept is *frame* that looks like a grid with horizontal links of different lengths. Frame is used to identify balanced permutation matrices.

Definition 2.5 *(frame, block of a frame)*: For $k \geq 1$, a *frame* $F_{N\times k;\gamma}$ consists of k columns labeled f_j, $j = 1, 2, \ldots, k$, and N rows labeled according to a permutation γ such that the label of the ith row from top equals $\gamma(i)$, for $0 \leq i \leq N-1$. For $1 \leq j \leq n$, column f_j consists of 2^{n-j} blocks of 2^j rows (entries) each. If γ is the identity permutation, then the frame is denoted by $F_{N\times k}$. In the graphical representation of a frame, any polygon with four sides and four right angles is called a *block* (or *rectangle*) of a frame. (See Fig. 2(a) for a frame).

Definition 2.6 *(fit)*: Let $k \geq 1$ and $1 \leq j \leq k$. Consider a matrix $A_{N\times k}$ and a frame $F_{N\times k}$. When row i of $A_{N\times k}$ is inserted into the row i of $F_{N\times k}$, the matrix $A_{N\times k}$ is said to *fit* the frame $F_{N\times k}$ if and only if for each j, each and every j column rectangle whose entries are located in columns f_1, $f_2, \cdots, f_j$ of $F_{N\times k}$ contains a balanced matrix.

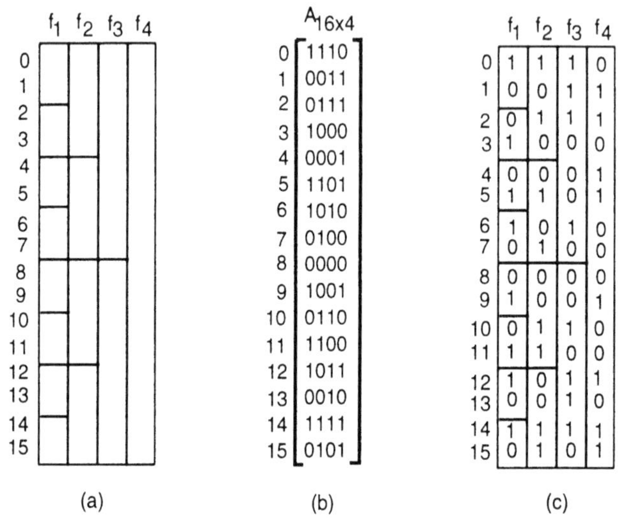

Figure 2: (a) Frame $F_{16\times 4}$ (or denoted $F_{1:4}$ with $N = 16$). (b) and (c) A balanced permutation matrix $A_{16\times 4}$ that fits $F_{16\times 4}$.

EXAMPLE. Consider the balanced matrix $A_{16\times 4}$ corresponding to the binary representation of a balanced permutation
$\pi = (0\ 14\ 15\ 5\ 13\ 2\ 7\ 4\ 1\ 3\ 8)(6\ 10)(9)(11\ 12)$, shown in Figure 2(b). Let us insert the ith row of $A_{16\times 4}$ into the ith row of the frame $F_{16\times 4}$ as shown in Fig. 2(c). The first k columns, $1 \leq k \leq 3$, of the frame consists of 2^{3-k} rectangles, each having 2^k rows and k columns. Note that any submatrix of order $2^k \times k$ that is located in any rectangle of the frame $F_{16\times 4}$ is balanced and represents a permutation on $\{0, 1, \cdots,\ 2^k - 1\}$. Therefore, $A_{16\times 4}$ fits the frame.

Figure 3.(g) also shows some frames. Note that the reverse permutation matrix $R_{16\times 4}$ shown in Figure 3.(g) fits $F_{16\times 4}$ shown in Figure 3.(a). But, $R_{16\times 4}$ does not fit $F_{16\times 4;\gamma_1}$ shown in Figure 3.(c) due to that many rectangles of $F_{16\times 4;\gamma_1}$ do not contain balanced matrices when row i of $F_{16\times 4;\gamma_1}$ is filled in by the row i of $R_{16\times 4}$. For instance, in $F_{16\times 4;\gamma_1}$ the topmost rectangle of

f_1 does not contain a balanced matrix.

An interconnection network (IN) with N inputs/outputs and k stages is denoted by both $IN_{N\times k}$ and $IN_{1:k}$, for $k \geq 1$. The subnetwork that consists of the stages x through y of $IN_{1:k}$ is denoted by $IN_{x:y}$, where $1 \leq x < y \leq k$. If $x > y$, then $IN_{x:y}$ refers to a nil network. The notation used for networks is different from that used for matrices because matrices are always denoted by single letters, as opposed to double letters used for INs. The symbols RB and SE in this paper refer to the networks reverse baseline and shuffle-exchange, respectively.

In the remainder of this section, some preliminary results are presented. The following lemma computes the number of matrices that fit $F_{N\times k}$.

Lemma 2.2 *The number of those matrices of order $N \times k$ each of which fits $F_{N\times k}$ equals $2^{Nk/2}$, for $k \geq 1$.*

Proof. The first column of $F_{N\times k}$ consists of $N/2$ blocks (or rectangles) each having 2 rows only. Because each block can constitute the set $\{0, 1\}$ in two different ways independently of the other blocks, there exist $2^{N/2}$ column vectors that fit the first column of $F_{N\times k}$. Each of these column vectors, say x, generates $2^{N/2}$ column vectors, say y's, in the sense that each of $[x\ y]$ fits the first two columns of $F_{N\times k}$. Similarly, each $[x\ y]$ generates $2^{N/2}$ column vectors, say z's such that any $[x\ y\ z]$ fits the first three columns of $F_{N\times k}$. This generation of column vectors is continued until all matrices that fit $F_{N\times k}$ are generated. Thus, the lemma holds. □

Note that any permutation matrix $A_{N\times n}$ that fits $F_{N\times n}$ is indeed a *balanced permutation matrix.* Therefore, the number of permutation matrices that fit $F_{N\times n}$ is equal to the number of balanced permutation matrices of order $N \times n$, as stated in the following theorem, whose proof is given in [28].

Theorem 2.3 *The number of balanced permutation matrices of order $N\times n$ equals $2^{Nn/2}$.*

The following theorems establish relations between frames and permutations that pass shuffle-exchange (SE) and reverse baseline (RB) networks. The basic ideas behind the proofs of these theorems appear in [27], and the complete proofs are provided in [14].

Theorem 2.4 *A matrix $D_{1:k} = [d_1\ d_2\ \ldots\ d_k]$ fits $F_{1:k}$ if and only if $D_{1:k}$ passes $RB_{1:k}$, $1 \leq k \leq n$. Moreover, $RB_{1:k}$ sends its* i*th input to its* j*th output, where j is equal to the sum of $(\lfloor i/2^k \rfloor \times 2^k)$ and the value of $D_{1:k}(i)$.*

Theorem 2.4 implies that for $k = n$, a matrix $D_{1:k}$ fits $F_{1:k}$ if and only if $D_{1:k}$ is *realized* by $RB_{1:k}$. It is shown in [2] how balanced matrices can

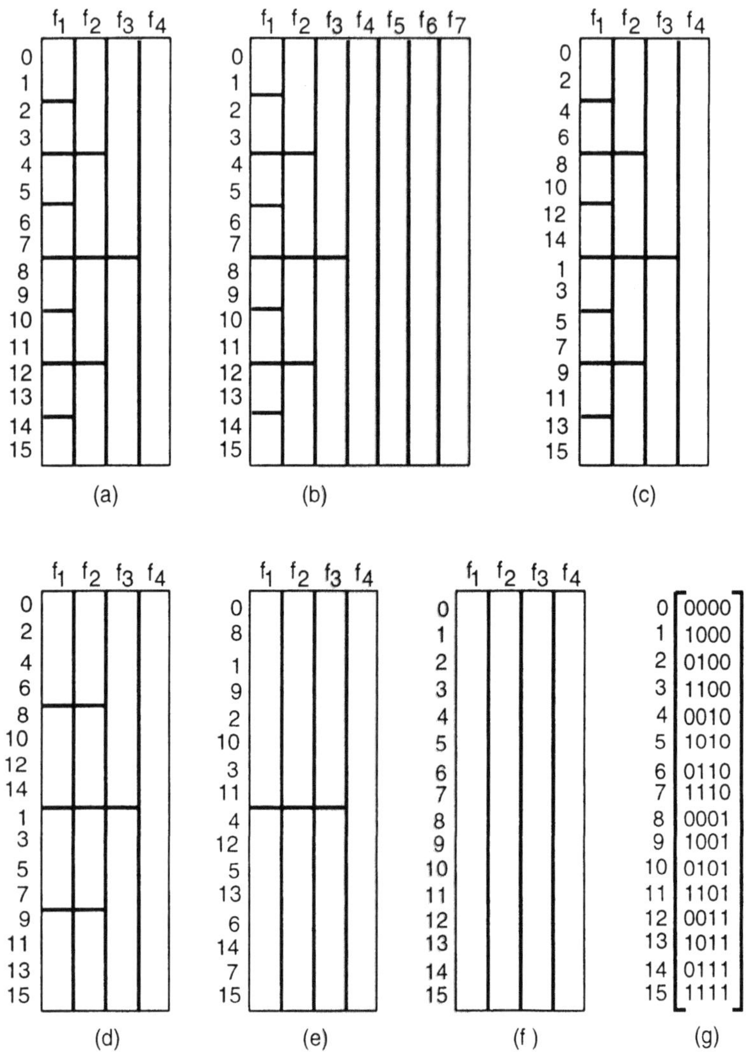

Figure 3: Let $N = 16$. (a) $F_{16\times 4} = F_{16\times 4;\gamma_0} = F_{1:4} = F^0_{1:4} = F^0_{1:4;\gamma_0}$. (b) $F_{16\times 7} = F_{16\times 7;\gamma_0} = F_{1:7} = F^0_{1:7;\gamma_0}$. (c) $F_{16\times 4;\gamma_1} = F_{1:4;\gamma_1} = F^0_{1:4;\gamma_1}$. (d) $F^1_{16\times 4;\gamma_1} = F^1_{1:4;\gamma_1}$. (e) $F^2_{16\times 4;\gamma_3} = F^2_{1:4;\gamma_3}$. (f) The *universal frame*, $F^3_{16\times 4;\gamma_0} = F^3_{1:4;\gamma_0}$. (g) The reverse permutation matrix $R_{16\times 4}$.

be used to determine the number of SE stages (or the number of passes through a single SE stage) necessary to realize a given permutation. Lemma 2.5 below restates their result using the notation of this paper.

Lemma 2.5 [2] *Let $M_{N\times m}$ and $C_{N\times k}$ be balanced matrices such that $M_{N\times m} = [I_{N\times n}\ C_{N\times k}]$, $k \geq 1$ and $n+k = m$. The k-stage shuffle-exchange network (i.e., $SE_{1:k}$) realizes the permutation represented by $M_{(m+1-n):m}$.*

To illustrate Lemma 2.5, consider the identity permutation matrix $I_{8\times 3}$ $= [i_1\ i_2\ i_3]$ and some balanced matrices $M_{8\times 4} = [I_{8\times 3}\ i_1]$, $M_{8\times 5} = [I_{8\times 3}\ i_1\ i_2]$ and $M_{8\times 6} = [I_{8\times 3}\ I_{8\times 3}]$. Because $M_{8\times 4}$, $M_{8\times 5}$ and $M_{8\times 6}$ are balanced, the permutations represented in binary by $[i_2\ i_3\ i_1]$, $[i_3\ i_1\ i_2]$ and $[i_1\ i_2\ i_3]$ are realized by the single-stage SE, 2-stage SE and 3-stage SE with $N = 8$ inputs/outputs, respectively. The following theorem shows that $F_{1:(n+m-1)}$ can be used to characterize the permutations realized by $RB_{1:(n-1)}SE_{1:m}$, for $m \geq 1$.

Theorem 2.6 *A balanced matrix $D_{1:(n+m-1)}$, $m \geq 1$, fits the frame $F_{1:(n+m-1)}$ if and only if $D_{1:(n+m-1)}$ passes the network $RB_{1:(n-1)}SE_{1:m}$. Moreover, $RB_{1:(n-1)}SE_{1:m}$ realizes the permutation represented by the last n columns of $D_{1:(n+m-1)}$, denoted by $D_{m:(n+m-1)}$.*

Now, we introduce Figure 4 and explain below the basic idea behind Theorem 2.6. Let $(x_1^i x_2^i \ldots x_n^i \ldots x_{n+m-1}^i)$ denote the row i of a balanced matrix $D_{1:(n+m-1)}$ fitting $F_{1:(n+m-1)}$, where $0 \leq i \leq N-1$. As stated above, Theorem 2.4 implies that for $k = n$, a matrix $D_{1:k}$ fits $F_{1:k}$ if and only if $D_{1:k}$ is *realized* by $RB_{1:k}$. Thus, any permutation whose binary representation fits $F_{1:n}$ is realized by $RB_{1:n}$, and vice versa. This implies that when $(x_1^i x_2^i \ldots x_n^i)$ is used as the routing tag of input i of $RB_{1:n}$, this routing tag (along with input i) reaches the output with label $(x_1^i x_2^i \ldots x_n^i)$ of $RB_{1:n}$. (Note that the bit x_j^i, $1 \leq j \leq n$, is used as the control bit for setting SB at stage j of $RB_{1:n}$). This also means that the matrix formed by all the N routing tags $(x_1^i x_2^i \ldots x_n^i)$ at the outputs of $RB_{1:n}$ is the identity permutation matrix. Recall that the last stage of $RB_{1:n}$ is an SE stage and, hence, $RB_{1:n}$ is identical to $RB_{1:(n-1)}SE_1$. Therefore, it follows that the permutation whose binary representation is the same as the submatrix $D_{1:n}$ of $D_{1:(n+m-1)}$ is realized by $RB_{1:(n-1)}SE_1$, and all the N routing tags $(x_1^i x_2^i \ldots x_n^i)$ form the identity permutation matrix at the outputs of $RB_{1:(n-1)}SE_1$. When a SE stage is appended to the right of $RB_{1:(n-1)}SE_1$ in order to form $RB_{1:(n-1)}SE_{1:2}$, the routing tag $(x_1^i x_2^i \ldots x_n^i)$ is first shuffled through the shuffle pattern to the output with label $(x_2^i x_3^i \dots x_n^i x_1^i)$ of the shuffle pattern and is, then, switched by the exchange stage to the output with label $(x_2^i x_3^i \dots x_n^i x_{n+1}^i)$ of $RB_{1:(n-1)}SE_{1:2}$, where x_{n+1}^i is used as the control bit for setting the SB of the exchange stage. Because no conflict occurs in the SBs, the matrix formed by all the N routing tags $(x_2^i x_3^i \dots x_n^i x_{n+1}^i)$ at the outputs of $RB_{1:(n-1)}SE_{1:2}$ is now the identity permutation matrix. This implies that the permutation whose binary representation is the same as the submatrix $D_{2:(n+1)}$ is realized by $RB_{1:(n-1)}SE_{1:2}$. In general, when a SE stage is appended to the right of $RB_{1:(n-1)}SE_{1:(m-1)}$ in order to form $RB_{1:(n-1)}SE_{1:m}$, the routing tag $(x_{m-1}^i x_m^i \cdots x_{n+m-2}^i)$ is first shuffled through the shuffle pattern to the output with label $(x_m^i x_{m+1}^i \cdots x_{n+m-2}^i x_{m-1}^i)$ of the shuffle pattern and is, then, switched by the exchange stage to the output with label $(x_m^i x_{m+1}^i \cdots x_{n+m-2}^i x_{n+m-1}^i)$ of $RB_{1:(n-1)}SE_{1:m}$. Because the matrix formed by all the N routing tags $(x_m^i x_{m+1}^i \cdots x_{n+m-2}^i x_{n+m-1}^i)$ at the outputs of $RB_{1:(n-1)}SE_{1:m}$ is the identity permutation matrix, the permutation whose binary representation is the same as the submatrix $D_{m:(n+m-1)}$ is realized by $RB_{1:(n-1)}SE_{1:m}$. In other words, the permutation whose binary representation is the same as the last n columns of $D_{1:(n+m-1)}$ is realized by $RB_{1:(n-1)}SE_{1:m}$.

For $N = 8$ and $m = 3$, Figure 4 shows how the rows of a $D_{1:(n+m-1)}$ are used as the routing tags of the inputs of $RB_{1:(n-1)}SE_{1:m}$. This figure also illustrates how the permutation corresponding to the last n columns of $D_{1:(n+m-1)}$ is realized by $RB_{1:(n-1)}SE_{1:m}$. The matrices formed by the routing tags at the third, fourth, and fifth stages of $RB_{1:2}SE_{1:3}$ are shown such that each identity permutation matrix formed by the routing tags is encircled by a dotted line box. The dotted line box at the outputs of $RB_{1:2}SE_{1:3}$ contains the last three columns of the matrix formed by the routing tags. This implies that the permutation whose binary representation is $D_{3:5}$ is realized by $RB_{1:2}SE_{1:3}$.

3 Rearrangeability of $(2n-1)$-Stage Shuffle-Exchange Networks

An interconnection network (IN) with $N = 2^n$ inputs/outputs is called a rearrangeable network if it realizes each and every of $N!$ permutations in a single pass. It is known that the lower bound for the number of stages of a multistage IN with (2×2) switching boxes (SBs) to be rearrangeable is $2n - 1$. The question of whether this lower bound holds for $(2n - 1)$-

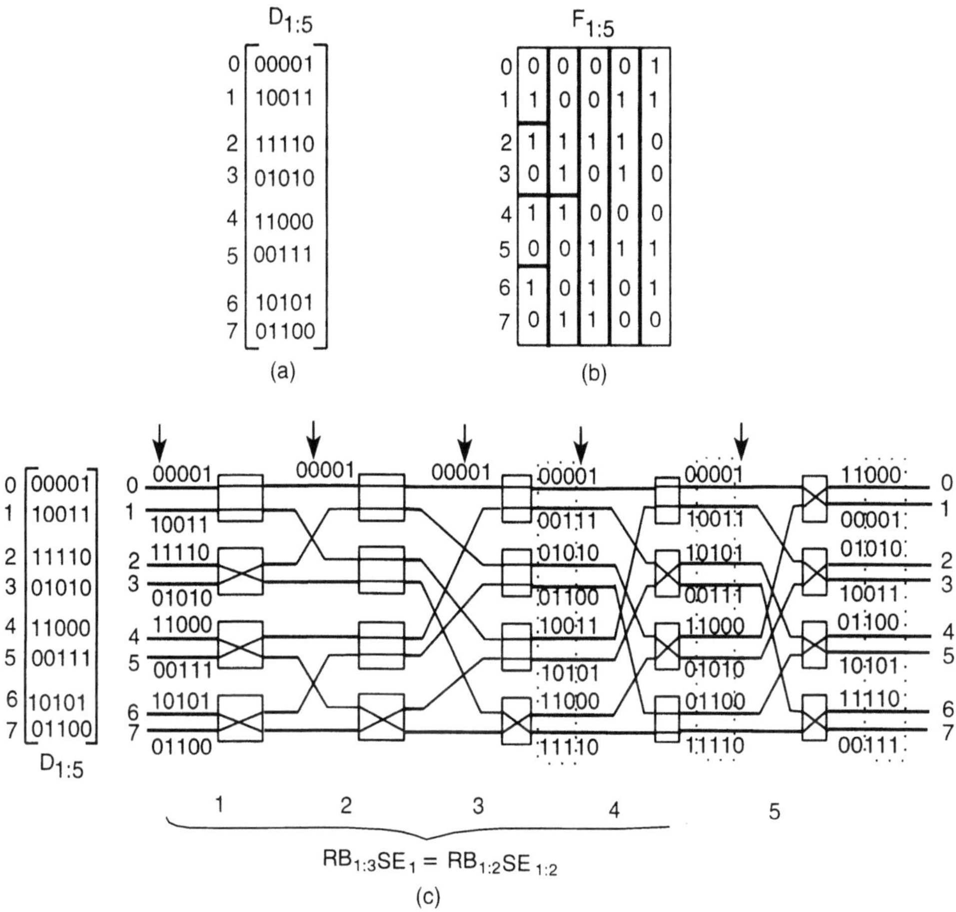

Figure 4: In this figure, $N = 8$, $n = 3$, and $m = 3$. (a) A balanced matrix $D_{1:5}$ fitting $F_{1:5}$. (b) $F_{1:5}$ along with the matrix $D_{1:5}$. (c) The network $RB_{1:2}SE_{1:3}$ realizes the permutation corresponding to the balanced submatrix $D_{3:5}$ (i.e., the last n columns of $D_{1:5}$), where the row i of $D_{1:5}$ is used as the routing tag of the input i. Arrows point to the control bits used to set the switching boxes. The matrices formed by the routing tags at the stages 4, 5 and the outputs of $RB_{1:2}SE_{1:3}$ are shown, and each identity permutation matrix is encircled by a dotted line box.

stage shuffle-exchange networks remained an open question for few decades [1, 2, 3, 4, 5, 6, 7, 8, 9, 10, 11, 12, 13] until it has been proven recently in [14] that the $(2n-1)$-stage shuffle-exchange network is rearrangeable. In this section, we present the basic idea behind the proof for the rearrangeability of $(2n-1)$-stage shuffle-exchange networks.

Basic idea behind the proof of the rearrangeability of $(2n-1)$-stage SE networks is as follows. We consider all permutation matrices of order $N \times n$ that correspond to the binary representation of all $N!$ permutations. The main proof is based on a routing process of $n-1$ steps. In the first step, a column vector r of $N/2$ zeros followed by $N/2$ 1s is added to the left of every one of $N!$ balanced matrices that are the binary representations of $N!$ permutations. Since prepending the column vector r makes some resulting matrices unbalanced, the total number of distinct balanced matrices of $n+1$ columns become less than $N!$. Then, each of those balanced matrices with $n+1$ columns is routed through an inverse shuffle-exchange (ISE) stage of $N/2$ switches in $2^{N/2}$ ways from its outputs to inputs. This (inverse) routing results in a set α of balanced matrices with $n+1$ columns such that any balanced matrix of order $N \times n$ is contained in the last n columns of at least one matrix of the set α. The remaining $n-2$ steps are similar to step 1. Next, we present an example to show all three steps for $N = 16$.

EXAMPLE. Let $N = 16$ and $n = 4$. Consider a three-stage inverse shuffle-exchange network with 16 inputs/outputs whose stages are labeled from left to right by ISE_1, ISE_2, and ISE_3. All those 16! balanced matrices corresponding to $N!$ permutations are generated in three steps. For $1 \leq k \leq 3$, step k uses stage ISE_{4-k}.

Initially, the set α_0 contains all 16! balanced matrices of order 16×4. In *step 1*, column r_3 shown in Figure 5.(a) is placed to the left of every member of α_0. If the resulting matrix obtained from the addition of r_3 to the left of a member of α_0 is balanced, then it remains in α_0; otherwise, it is removed from α_0. Because any matrix that fits the frame $F^2_{4:7}$ is balanced with r_3, it remains in α_0. ($F^2_{4:7}$ consists of a pile of two universal frames each having $N/2$ rows and $n-1$ columns, followed by a single block of N rows). Then, each member of the latest α_0 is reverse-routed through ISE_3 in all $2^{16/2}$ different ways since each switching box of ISE_3 can be set in two different ways (i.e., either cross or straight). (By reverse-routing through ISE_3, it is meant that balanced matrix is routed from outputs to inputs). The matrices that are obtained at the inputs side of ISE_3 constitute the set α_1. Each member of α_1 fits the frame shown in Figure 5.(b). Any submatrix that is formed by the last n columns of any member of α_1 is inserted into another set called β_1. It is shown in [14] that β_1 contains all $N!$ balanced matrices of order $N \times n$. Note that each of the following eight sets contains the input/output labels of a switch of ISE_3: $\{0,4\}$, $\{8,12\}$, $\{1,5\}$, $\{9,13\}$, $\{2,6\}$, $\{10,14\}$, $\{3,7\}$ and $\{11,15\}$.

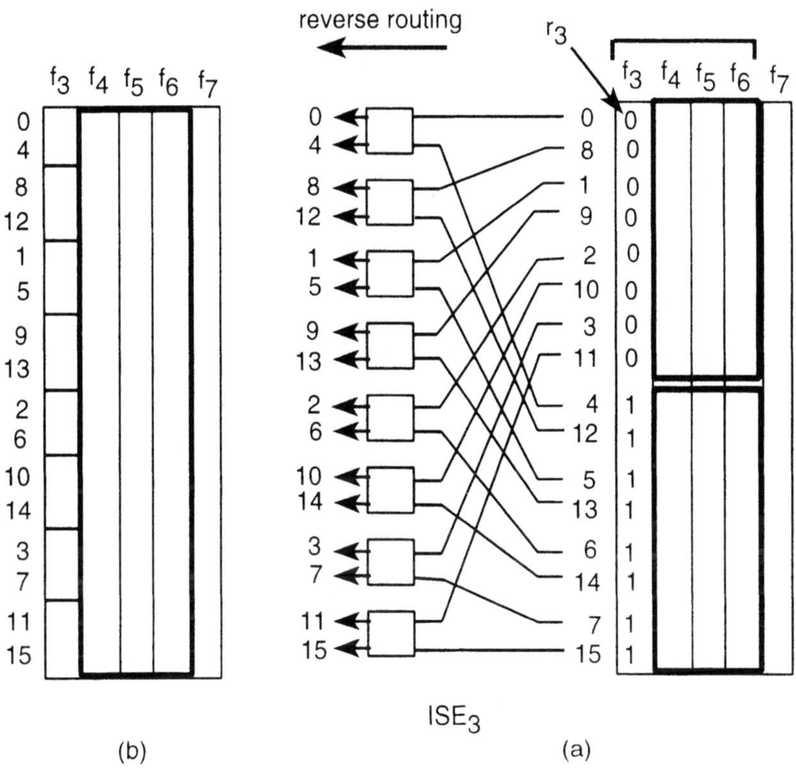

Figure 5: The implementation of the first step for $N = 16$.

In *step 2*, the frames and matrices of α_1 that are obtained at the end of step 1 are now located at the outputs side of ISE_2, as shown in Figure 6.(a). Column r_2 is placed to the left of every member of α_1. If the resulting matrix is balanced, then it remains in α_1; otherwise, it is removed from α_1. Then, each member of the latest α_1 is reverse-routed through ISE_2 in all $2^{16/2}$ different ways, and the matrices that are obtained at the inputs side of ISE_2 constitute the set α_2. Each member of α_2 fits the frame shown in

Figure 6.(b). Any submatrix that is formed by the last n columns of any member of α_2 is inserted into another set β_2. It is shown in [14] that β_2 also contains any balanced matrix of order $N \times n$. Note that each of the following eight sets contains the input/output labels of a switch of ISE_2: $\{0, 2\}$, $\{4, 6\}$, $\{8, 10\}$, $\{12, 14\}$, $\{1, 3\}$, $\{5, 7\}$, $\{9, 11\}$ and $\{13, 15\}$.

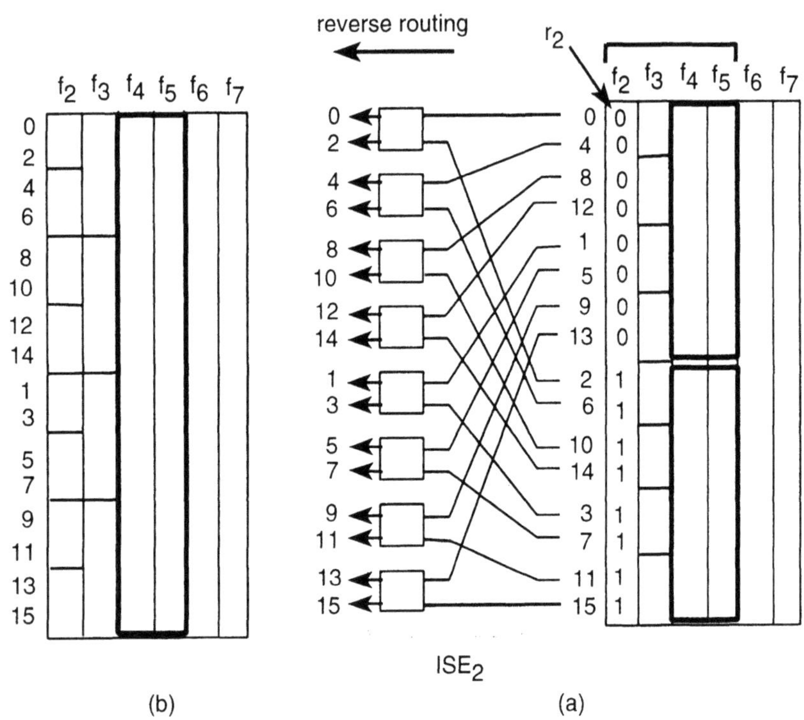

Figure 6: The implementation of the second step for $N = 16$.

In *step 3*, column r_1 is placed to the left of each member of α_2, as shown in Figure 7.(a). If the resulting matrix is balanced, then it remains in α_2; otherwise, it is removed from α_2. Then, each member of the latest α_2 is reverse-routed through ISE_1 in all $2^{16/2}$ different ways, and the matrices that are obtained at the inputs side of ISE_1 constitute the set α_3. Each member of α_3 fits the frame $F^0_{1:7}$ shown in Figure 7.(b). Any submatrix that is formed by the last n columns of any member of α_3 is inserted into β_3.

β_3 also contains any balanced matrix of order 16×4, which implies that 7-stage shuffle-exchange network is rearrangeable. In general, β_{n-1} contains any balanced matrix of order $N \times n$, which implies that $(2n-1)$-stage shuffle-exchange network is rearrangeable. End of Example.

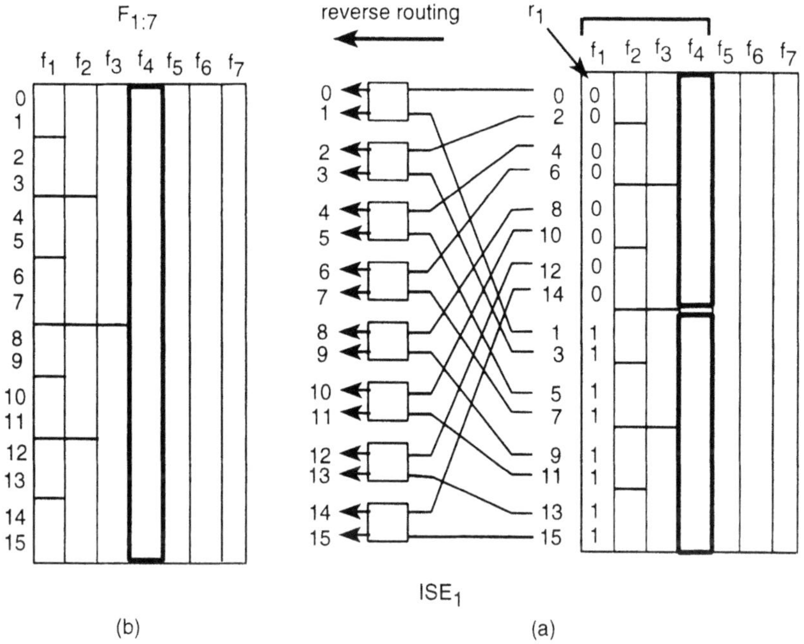

Figure 7: The implementation of the third step for $N = 16$.

4 Modified Shuffle-Exchange Network (MSN) as an ATM Switch

Asynchronous Transfer Mode (ATM) is the switching and multiplexing technique chosen for Broadband Integrated Services Digital Network (B-ISDN). In ATM, user information is transferred through the network in small fixed size packets of 53 bytes called *cells*, and line speeds equal to 155 Mbps, 622 Mbps and above. An ATM switch routes cells arriving at its input ports to their desired output ports. A multistage ATM switch is usually constructed by interconnecting switching elements (SEs) of size (2×2) each in a specific topology using links. When more than one cell compete for the same internal link of ATM switch, an *internal conflict* (or blocking) occurs. When more than one cell compete for the same output port, an *external* (or output) conflict occurs. In this section, we present a modified shuffle-exchange network (MSN) [30].

4.1 Architecture of MSN

MSN is designed with the following goals in mind:

1. reduce the number of internal conflicts that occur in the SEs of a shuffle-exchange network.

2. whenever it is not possible to avoid having internal conflicts among four packets arriving at two SEs of a stage, route correctly those packets that are closest to their destinations.

3. whenever a packet is misrouted (or deflected) to a wrong outlet of an SE due to conflict, first compute its new routing tag based on the minimum distance between the packet and its destination and, then, start immediately re-routing the packet.

To achieve the first part of goal 3, we developed the fast destination tag routing scheme (FDR) to compute a new routing tag based on the minimum distance. To start rerouting immediately and easily, the stages of a network should be identical and, therefore, we have chosen shuffle-exchange network to realize the second part of goal 3. To accomplish goals 1 and 2, we introduce an α-exchange stage that is inserted just after every other shuffle-exchange stage. These three goals together aim to improve the network performance by preventing especially those packets that are very close to their destinations from being misrouted in case of a conflict, thereby increasing the number of successful packets and reducing the possibility of occurring conflicts in SEs of the successive stages.

Let α denote a connection pattern such that, for $0 \leq i \leq N-1$, input i of the pattern is connected to output i or $i + (N/2)\ mod\ N$ based on that i is even or odd, respectively. $\alpha - exchange$ is defined as being a stage consisting of the connection pattern α followed by a column of $N/2$ bit-controlled SEs of size (2×2) each. As shown in Figure 8, MSN is constructed by inserting an $\alpha - exchange$ stage just after every other shuffle-exchange stage, where each stage has $N/2$ unbuffered binary SEs. MSN stages are labeled from left to right starting with 1 up to K. The odd-numbered and even-numbered stages of MSN are identical to the shuffle-exchange and α-exchange stages, respectively. Any SE of an even-numbered (respectively,

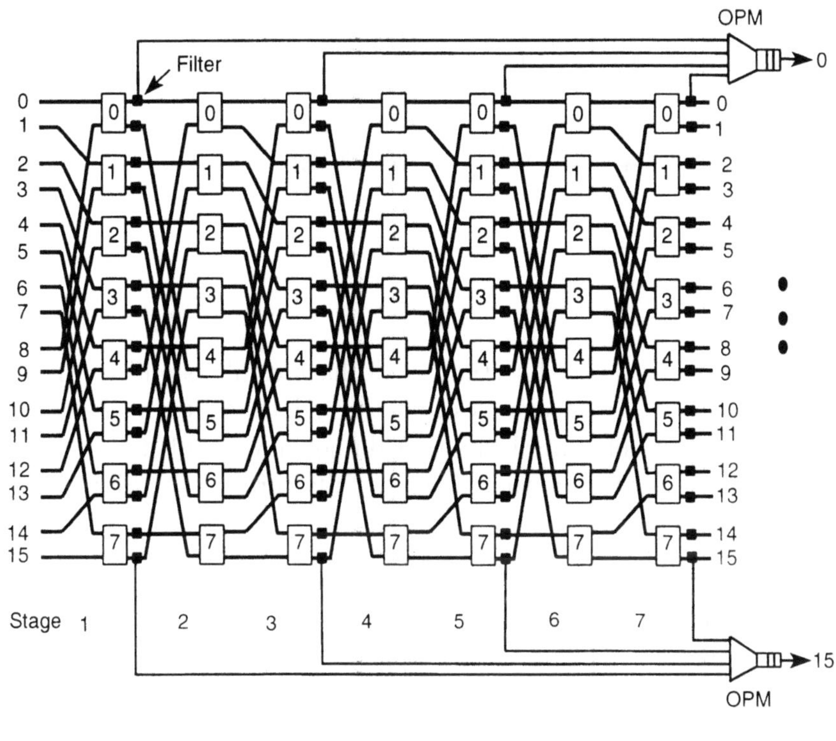

Figure 8: A 7-stage MSN with 16 inputs/outputs. The odd-numbered and even-numbered stages of MSN are identical to the shuffle-exchange and α-exchange stages, respectively.

odd-numbered) stage is called *even-SE* (respectively, *odd-SE*). The outlets of every odd-SE are connected to the inputs of the same number output port modules (OPMs) through filters.

The SEs of each stage in MSN are labeled from top to bottom, starting with 0 until $2^{n-1}-1$. For $0 \leq j \leq 2^{n-2}-1$, the jth and $(j+2^{n-2})$th SEs of each stage form a pair j. Thus, each stage has 2^{n-2} pairs of SEs. Note that the four outlets of pair j of odd-SEs are connected to the inlets of pair j of even-SEs through the connection pattern α. These two pairs of odd-SEs and even-SEs are said to form a core-network with 4 inputs/outputs, as shown in Figure 9. The odd-SEs of a core-network act somewhat like sorting cells

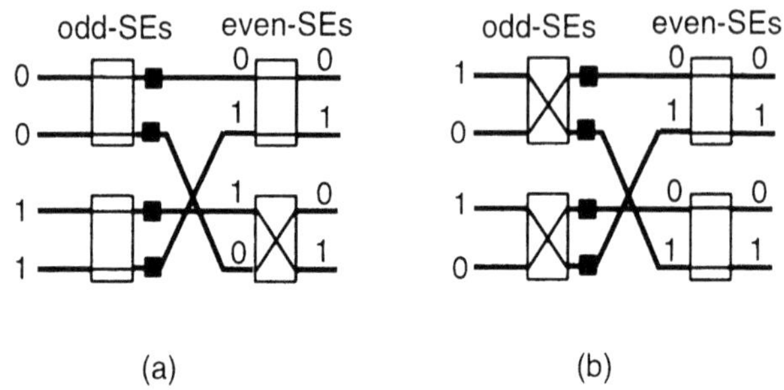

Figure 9: (a) Those control bits that would have conflicts on two SEs of a banyan network can be routed successfully through the SEs of a core network in MSN. (b) Any set of control bits that do not have conflicts on two SEs of a banyan network are also routed successfully through the SEs of a core network in MSN.

introduced in [40] in the sense that their main function is to prevent even-SEs from having some conflicts. Therefore, the number of internal conflicts in MSN depends on how much a core-network is successful in preventing internal conflicts on its even-SEs. Let a and b (respectively, c and d) be the control bits of an odd-SE (respectively, the other odd-SE) of a core-network, where control bit refers to a destination address bit used in setting SE. If $\{a, b\}=\{0,1\}$ and $\{c, d\}=\{0,1\}$, no conflict occurs in the even-SEs of the core-network similar to SEs of any banyan network, as shown in Figure 9.(b). But, if $\{a, b\}=\{0,0\}$ and $\{c, d\}=\{1,1\}$, no conflict occurs in the even-SEs either, as shown in Figure 9.(a). Also note that some active packets arriving at inputs of a core-network may not reach any outlet of an even-SE of the network because filters at the outlets of odd-SEs may route them to OPMs due to that their counters are zero, which helps reduce the possibility of having internal conflict in even-SEs.

Algorithm SET-SE, shown in Figure 10, describes how to set both an even-SE and odd-SE, and shows when a filter extracts successful packets to their destinations. Each packet header in MSN comprises an activity bit a,

Algorithm **SET-SE**

Inputs: activity bits (a^u, a^l), counter bits (C^u, C^l), conflict bits (w^u, w^l), destination address bits ($d^u_{n-C^u+1}$, $d^l_{n-C^l+1}$).

Output: An SE of MSN is set either straight or cross.

begin

1   **if** ($a^u = 0$ and $a^l = 0$) **then**
        leave SE in its present state;

2   **if** ($a^u = 1$ and $a^l = 0$) **then**
        set SE according to $d^u_{n-C^u+1}$; $w^u \leftarrow 0$;

3   **if** ($a^u = 0$ and $a^l = 1$) **then**
        set SE according to $d^l_{n-C^l+1}$; $w^l \leftarrow 0$;

4   **if** ($a^u = 1$ and $a^l = 1$) **then**

4.1   **if** ($d^u_{n-C^u+1} \neq d^l_{n-C^l+1}$) **then**
            set SE according to $d^u_{n-C^u+1}$;
            $w^u \leftarrow 0$; $w^l \leftarrow 0$; **goto** 5;

4.2   **else if** ($C^u > C^l$) **then**
            $winner \leftarrow l$; $loser \leftarrow u$;

4.3   **else if** ($C^u < C^l$) **then**
            $winner \leftarrow u$; $loser \leftarrow l$;

4.4   **else** $winner \leftarrow random(u, l)$;
            $loser \leftarrow$ the other packet;

4.5   set SE according to $d^{winner}_{n-C^{winner}+1}$;

4.6 $w^{loser} \leftarrow 1$; $w^{winner} \leftarrow 0$;

5    $k \leftarrow u$;

6   **for** $i \leftarrow 1$ **to** 2 **do**

6.1   **if** (SE is an even-SE and $a^k = 1$) **then**

6.1.1   **if** ($w^k = 0$) **then**
                $C^k \leftarrow C^k - 1$;

6.1.2   **else**
                initialize C^k; $w^k \leftarrow 0$;

6.2   **if** (SE is an odd-SE and $a^k = 1$) **then**

6.2.1   **if** ($w^k = 0$ and $C^k \leq 1$) **then**
                $C^k \leftarrow 0$; filter routes packet to OPM;

6.3 $k \leftarrow l$;

     endfor

end

Figure 10: Algorithm SET-SE used for setting each switching element of MSN.

its binary destination address $d_1d_2\cdots d_n$, a counter field C of $\lceil log_2 n\rceil$ bits, and a conflict bit w. The activity bit a indicates whether an input receives a packet ($a = 1$) or does not receive ($a = 0$). Whether an SE will be set to cross or straight depends upon the activity bits (a^u, a^l), the counters (C^u, C^l), and the control bits $d^u_{n-C^u+1}$ and $d^l_{n-C^l+1}$ of the packets arriving at the inputs of the SE, where superscripts u and l stand for upper and lower, respectively, to distinguish the packets of the upper and lower inputs of an SE. In case of an internal conflict, the packet with smaller counter value is routed correctly because it is closer to the destination.

If no packet arrives at the inputs of an SE of MSN, then no action is taken and the SE is left in its present state (line 1 of SET-SE). If one of the inputs of an SE does not receive a packet, the SE is set according to the control bit of the available packet (lines 2 to 3). If both inputs receive a packet and the control bits form the set {0,1}, then the SE does not have a conflict and is set accordingly (line 4.1). If SE has a conflict due to that the control bits of the packets are the same, then the packet with smaller counter value is routed correctly (lines 4.2 and 4.3); if the counter values are the same, then the packet to be routed correctly is selected randomly (line 4.4). Initially, all conflict (i.e., w) bits are reset to zero. Whenever a packet loses contention, its conflict bit is set to 1 (line 4.6). If the counter of a packet that is routed correctly in an odd-SE is less than or equal to 1, then the packet is distinguished and routed to its destination by a filter (line 6.2.1). If a packet reaches an outlet of an even-SE with a conflict bit of 0 (respectively, 1), then its counter is decremented (respectively, initialized by FDR), as shown in lines 6.1.1 and 6.1.2 of SET-SE.

FDR

The destination tag routing scheme works in omega network as follows [?]. Assume that the stages of omega network are labeled in ascending order from left to right, starting with 1, and the outputs of each stage are labeled in ascending order from 0 to N-1. Let $i^m = i_1^m i_2^m \cdots i_n^m$ and $d^r = d_1^r d_2^r \cdots d_n^r$ denote the input m and the output r, respectively, of omega network, where $0 \le m, r \le N-1$. Now, let us suppose that a packet, say A, arriving at input port i^m is destined to the output port d^r. The destination tag routing scheme initializes the routing tag of packet A to d^r. If packet A does not encounter any internal conflict, stage k switches packet A from its input $i_k^m i_{k+1}^m \cdots i_n^m d_1^r \cdots d_{k-1}^r$ into the output $i_{k+1}^m i_{k+2}^m \cdots i_n^m d_1^r \cdots d_k^r$, for k, $1 \le k \le n$. Thus, stage n finally switches packet A into the output port d^r.

The basic idea behind FDR is to eliminate the need for routing packet A through the first k stages if the following already holds: $i^m_{n-k+1} i^m_{n-k+2} \cdots i^m_n = d^r_1 d^r_2 \cdots d^r_k$ and $i^m_{n-k} i^m_{n-k+1} \cdots i^m_n \neq d^r_1 d^r_2 \cdots d^r_k d^r_{k+1}$. After determining k, FDR initializes the routing tag of packet A to $d^r_{k+1} d^r_{k+2} \cdots d^r_n$ and sets the counter field C of packet A to $n - k$. As an example for $N = 16$, if a packet with destination address 1100 arrives at input port with address 0110, the value of k for the packet will be 3 because the last three bits $i_2 i_3 i_4 (= 110)$ of the input port address are the same as the first three bits $d_1 d_2 d_3 (= 110)$ of the destination address. So, the packet needs to be routed correctly through one stage only according to $d_4 = 0$ and, therefore, the counter field C of the packet is initialized to 1.

4.2 Simulation Results

To capture the correlation in the time domain among packets originating from the same source, packet generation is usually assumed to be a succession of active and idle periods. Packet generation occurs only during active periods, and the packets generated during the same active period form a *burst*. The most popular traffic source model exhibiting this behavior is called the ON/OFF model which is simple and analytically tractable. The terms ON and OFF correspond to the active and idle periods, respectively. In BT, packets arrive at an input port in the form of bursts of random length such that all packets of a burst are destined to the same output port. BT is characterized by the distribution of burst length (i.e., the length of active period), the gap between consecutive bursts (i.e., the length of idle period), and the process describing the output port requests for bursts. We assume that the output port requested by all packets of a burst is chosen uniformly among all output ports independently from all other bursts.

The peak cell arrival rate, p, refers to the cell arrival rate when the source is in the ON state and is equal to $1/T$, where T is the time between two consecutive cell arrivals during the ON period. In accordance with the recommendation by ITU-T, we assume that the active and idle periods are exponentially distributed with parameters a and b whose mean values are $1/a$ and $1/b$, respectively. We also assume that the following three parameters are known for each source: 1) the average packet rate, denoted (m), that is equal to $(p \times a^{-1})/(a^{-1} + b^{-1})$, 2) the mean burst length in packets, (B), being equal to a^{-1}/T, and 3) the burstiness factor (β) that equals p/m. Each of active and idle periods can be taken either as exponential or a geometric random variable, based on the choice of time axis as either continuous or slotted, respectively. The parameters a^{-1}, b^{-1}, and T (packet interarrival time) can be calculated when the parameters m, B, and β of a source are known [35]. Once a^{-1} and b^{-1} are calculated, the exponentially distributed durations for active and idle periods in a given timeslot can be computed using the exponential distribution functions.

Various traffic sources to be serviced by ATM networks belong to three main classes: 1) constant bit rate (CBR) sources (e.g., 64 kbit/sec voice and fixed-rate video networks) which do not have any idle period in packet generation, 2) variable bit rate (VBR) sources such as file transfer, electronic mail, and terminal emulation in services of connectionless and connection-oriented data, and 3) VBR-video sources. We compared the packet loss rates of the proposed MSN with MS4 [36], and TBSF [37] under an ATM traffic in which the sources generate 20% video, 40% voice, and 40% data traffic. This ATM traffic is generated by using trace-driven simulation (i.e., the traffic of sources is first generated and, then, applied to each of three networks). Figure 11 shows the packet loss rate against the number of stages for MSN, MS4, and TBSF under the ATM traffic for $N = 64$. As seen from the figures, MSN exhibited lower packet loss rates than MS4 and TBSF. As N increases, the difference between packet loss rates of MSN and the switches MS4 and TBSF is expected to grow in favour of MSN.

5 Shuffle-Exchange Type Networks for Input-Queued ATM Switches

It is common to use a graph model to represent a multistage network. Since shuffle-exchange, banyan, baseline, reverse-baseline, and many other $log_2 N$-stage networks are topologically equivalent, without loss of generality we choose the reverse baseline network for examining the graph properties and connection patterns in this section.

In a packet switch, two or more packets may arrive simultaneously on different inputs destined for the same output and, hence, external blocking is unavoidable in a packet switch. In order to smooth the statistical fluctuations in packet arrivals to the switch, three main approaches were proposed: *input buffering*, *output buffering*, and *completely shared buffering* [15]. To be able to implement output buffering, either the switch must operate as fast as the sum of the speeds of its input ports, or the buffers must have as many write ports as there are input ports to the switch, to be able to handle simultaneous packet arrivals [16]. That is, the internal data transfer rate of

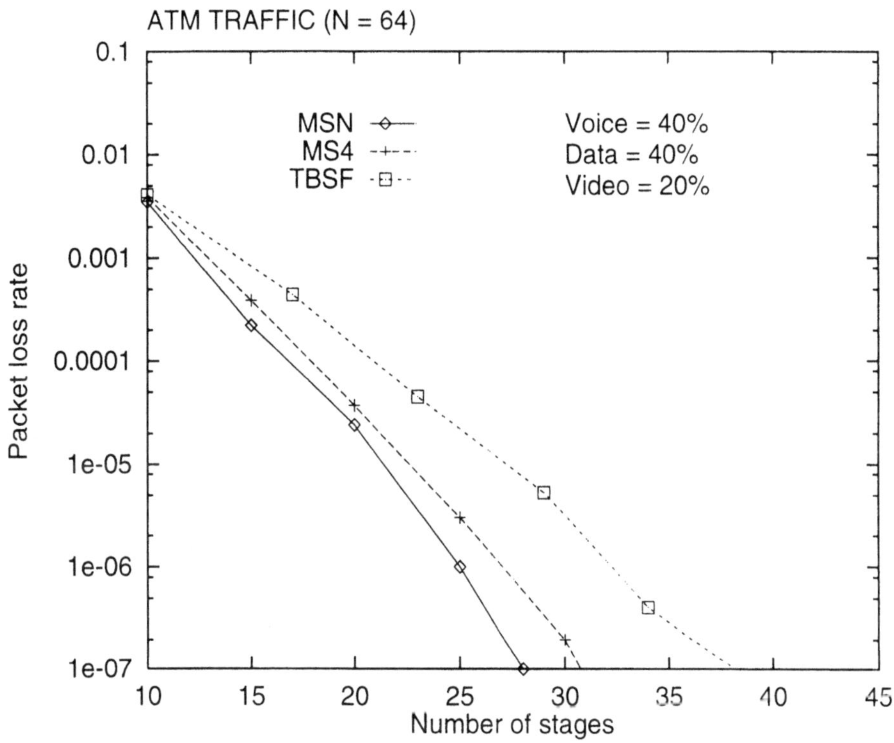

Figure 11: Comparison of the packet loss rates of MSN, MS4, and TBSF under an ATM traffic and full load for $N = 64$.

switch fabric with N inputs and N outputs must be N times higher than the external link speed. In addition, since all N cells of input ports may be destined to the same output port, a memory with access rate of N times the external link speed of or a complex parallel memory architecture is required. As the data rates on external links such as fiber optic links increase very rapidly, implementing output buffered switches with high speed links becomes very difficult. Since implementing buffers with multiple write ports increases their size and performance, the Knockout Switch [17] implements output buffering by using a concentrator at each output port to reduce the number of buffers needed to receive simultaneously arriving packets. Therefore, although output buffering provides the best performance in terms of switch throughput and cell delay under uniform traffic patterns, the implementation of input buffering is appealing because only one packet at a time arrives at the input port so that the buffer needs only a single write port.

Input buffering switches resolve contention for external conflict prior to switching. Some of the external conflict resolution algorithms are three-phase algorithm [18], scheduling CAM [19], and ring reservation [20]. When input buffers are FIFO (first in first out) queues, a serious problem called head-of-line (HOL) blocking occurs: a single packet at the head of the queue whose destination output port is busy can block all other packets in that queue from being transmitted even if their destination output ports are idle. This limits the maximum throughput of a nonblocking switch to 0.56 under uniform traffic, regardless of the selection policy used among the contending HOL cells [21]. Contention resolution algorithms including ring reservation algorithm suffer from HOL blocking. Three methods, namely, output expansion [22], channel grouping [23] and windowing [15] are the most known techniques to alleviate HOL blocking. In *output expansion*, extra internal switch links and output buffers are added in order to be able to forward more than one cell to each output port during the same cell interval. In *channel grouping*, the switch output ports are partitioned into a number of groups and cells are allowed to contend for a group, rather than a specific port. In *windowing* (also called bypass queue [24, 25, 26]), cells behind the HOL cell are also allowed to contend for output port access.

Two types of windowing, namely, sequential windowing (also called serial bypass) and parallel windowing (also called parallel bypass) are possible. In *sequential windowing*, the first cells in a switch's input buffers first contend for outputs, then the losing input buffers that contain a second cell are allowed to contend a second time, those that lose in the second time and have a third cell are allowed to contend a third time, and so on until w times are contended, where w is the depth of window. In *parallel windowing*, all cells in the windows of all buffers contend in a single round with the winners proceeding to the outputs, with the condition that only one cell can be selected at most from a buffer. To preserve the cell sequence, among those cells destined to the same output port in the window of a buffer the cell closest to the head of the buffer is selected if one of them needs to be chosen. Although output expansion and channel grouping require fundamental changes in the switch, they were preferred over windowing because extra contention cycles were needed in windowing to query cells behind the HOL cell. But, Shultz and Gulak [42] have recently introduced a cost-efficient hardware for implementing parallel windowing such that w cells can be queried in every input buffer during a single contention (or arbitration) cycle, without any delay overhead. In this paper, we shall explain how to modify their hardware to implement parallel windowing on reverse baseline network.

Shuffle-exchange type networks are very attractive to ATM switch designers due to their desirable features. In particular, they have constant latency for any input-output pair, are self-routing and modular, support synchronous and asynchronous modes of operation, preserve cell-sequence due to path-uniqueness, are very suitable for VLSI implementation, and have a complexity of $O(N\ log_2 N)$ compared to $O(N^2)$ of the crossbar switch. The major drawback of baseline-type networks is to have a very limited number of connection patterns, which causes them to suffer from internal conflicts, in addition to external conflicts. Our objective is to alleviate this drawback to the extent that the throughput of a reverse baseline network with content addressable input buffers can reach the throughput of a nonblocking network. Our algorithm uses parallel windowing for preventing both external blocking and internal blocking in a reverse baseline network. The fact that we replace all stages of a nonblocking switching network by $n = log_2 N$ stages of a baseline-type network yields a significant economy in implementation.

Scheduling algorithm selects those cells that must be transferred to output ports within a cell time, with the constraints that only one cell is selected from each input port and each selected cell is destined to a distinct output port. We will propose a scheduling algorithm for multicast traffic that can transfer a multicast cell to more than one output port. A scheduling algorithm must be able to support QoS parameters, in particular giving high priorities to cells belonging to time-sensitive applications.

Figure 12 shows the model of an input buffering reverse baseline network that we consider in this section. The switching elements (SEs) of the network do not have any buffer. Because reverse baseline network is a self-routing

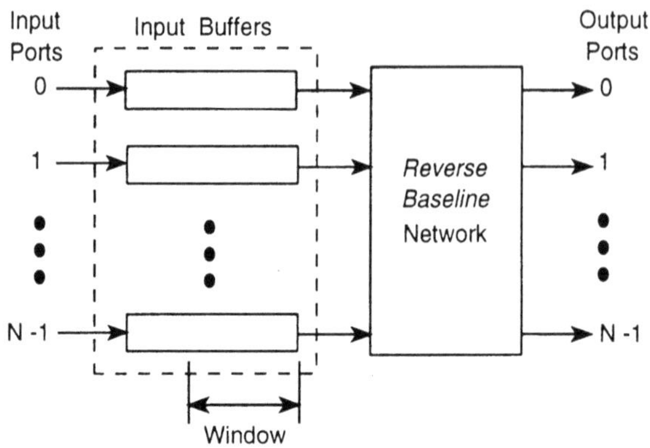

Figure 12: An input buffering reverse baseline network.

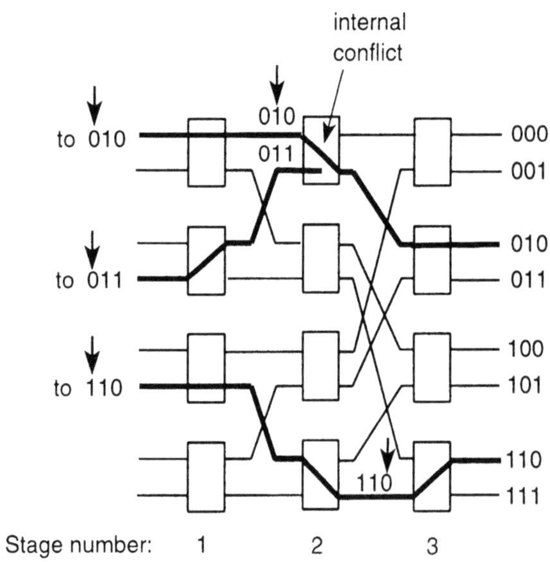

Figure 13: Internal blocking and self-routing in a reverse baseline network.

network, routing cells through the network is very simple: the destination address bits of cells are put into their headers and are used successively for routing as follows. To route a cell destined to an output port with binary representation $(d_1 d_2 \cdots d_n)$ in a self-routing ATM switch with $n = log_2 N$ stages and N inputs/outputs, an SE at stage k, $1 \leq k \leq n$ is set according to the bit d_k of the destination address of the cell arriving at an input of the SE: the cell is routed to the upper output of the SE if $d_k = 0$ and to the lower output if $d_k = 1$, regardless of whether the cell has arrived at lower or upper input of the SE. Internal conflict occurs when two cells arriving at a SE of stage k request the same output port of the SE because the d_k bits of their destination addresses are the same. Figure 13 illustrates the self-routing mechanism and internal conflict in a reverse baseline network.

In this section, we discuss how to grant admissions to the cells of input buffers of a reverse baseline network so that no internal (and external) conflict occurs. To avoid conflicts, the destination addresses of cells must meet the requirements identified by incomplete frames defined next [27].

Definition 5.1 *(balanced incomplete matrix)*: Let $M < N$ and $k \geq 1$. A matrix $A_{N \times k}$ is a *balanced incomplete matrix* if $A_{N \times k}$ consists of any M rows of a balanced matrix $A_{N \times k}$ plus $N - M$ blank rows that do not have any entry.

Figure 14.(d) shows a balanced incomplete matrix $B_{8 \times 3}$ whose rows 3 and 6 are blank and the other six rows are identical to those $A_{8 \times 3}$'s rows with the same row labels, shown in Figure 14.(b).

As an example, consider the reverse baseline network with 8 inputs/outputs shown in Figure 14.(a), the balanced matrix $A_{8 \times 3}$ corresponding to the binary representation of the permutation $\pi = (0\ 6)(1\ 2)(3\ 5\ 4)(7)$, and the frame $F_{8 \times 3}$ of Figure 14.(c) corresponding to the reverse baseline network. When row i, $0 \leq i \leq 7$, of $A_{8 \times 3}$ is used as the routing tag for the ith input of the reverse baseline network, no conflict occurs in the SEs of the network and a path is established between input i and output $A_{8 \times 3}(i)$ (i.e., the output whose label equals the contents of row i of $A_{8 \times 3}$). Therefore, the reverse baseline network realizes the permutation π. Now, let us place the ith row of $A_{8 \times 3}$ into the ith row of the frame as shown in Fig. 14.(c). The first k columns, $1 \leq k \leq 3$, of the frame consists of 2^{3-k} rectangles, each having 2^k rows and k columns. Because $A_{8 \times 3}$ fits the frame and the balanced incomplete matrix $B_{8 \times 3}$ is obtained from $A_{8 \times 3}$ by removing all entries of rows 3 and 6, $B_{6 \times 3}$ also fits the frame. Note that the matrix enclosed by any rectangle of the frames is balanced and represents a complete or partial permutation on $\{0, 1, \cdots, 2^k\}$.

We relate balanced matrix, balanced incomplete matrix, and frame to the problem of cell scheduling in input buffers of a reverse baseline network as follows. The objective in cell scheduling is to form a balanced matrix from the selected cells' destination addresses such that the balanced matrix fits the frame. If it is not possible to form a balanced matrix with N rows, then those cells whose destination addresses violate the conditions of fit are ignored in the selection process so that a balanced incomplete matrix is formed.

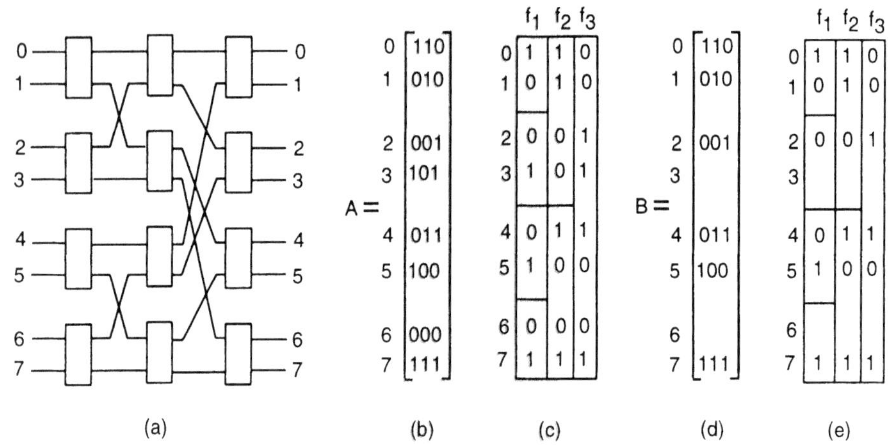

Figure 14: (a) A reverse baseline network with 8 inputs/outputs. (b) A balanced matrix $A_{8\times3}$ which is the binary representation of permutation $\pi = (0\ 6)(1\ 2)(3\ 5\ 4)(7)$. (c) The frame corresponding to the reverse baseline network. Row i of the frame contains row i of $A_{8\times3}$. $A_{8\times3}$ fits the frame. (d) A balanced incomplete matrix, $B_{8\times3}$, whose rows 3 and 6 do not have any entry. (e) Row i of the frame contains row i of $B_{8\times3}$. $B_{8\times3}$ fits the frame.

5.1 Parallel Windowing for Alleviating HOL Blocking

Shultz and Gulak [42] introduced hardware that allows parallel windowing in a single arbitration cycle for nonblocking switches with CAFIFO input buffers. Their hardware basically implements ring reservation algorithm [20] which resolves external conflicts on a nonblocking switch with input buffers. Ring reservation mechanism has a distributed architecture in which a string of 1-bit tokens rotates around a ring of input buffers in pipelined fashion (one bit per clock tick), with each bit representing the status of an output port (0= FREE, 1= TAKEN). Each token is associated with an output port. At each buffer, a counter keeps track of the output port associated with the token bit. That is, if 1-bit tokens are, for instance, rotating in the direction of buffer j to buffer $(j+1)$ *mod* N, then the counter of buffer j is initially set to j and is reduced cyclically by 1 at every bit clock cycle. If the output port corresponding to the present 1-bit token is the destination of the first cell of buffer cell and the 1-bit token equals 1, then the cell is blocked. If the output port is the destination of the first cell of buffer and the corresponding 1-bit token is 0, then the cell is granted admission and the 1-bit token is set to 1. This technique can be pipelined with the switching function.

In the case of CAFIFO-based input buffers, all cells in the windows are searched in an arbitration cycle to determine if their destination addresses match the output port associated with the present token. If a cell's destination address in a buffer matches the output port represented by the current 1-bit token, the cell is marked and the 1-bit token is set to 1 so that no other cell from any other buffer can be selected for the same destination in an arbitration cycle. All token bits pass through all input CAFIFOs during one *arbitration cycle.* The cells selected in an arbitration cycle enter the switch fabric during the next arbitration cycle. The length of an arbitration cycle is the same as that of a *time slot.* An arbitration cycle takes N bit clock cycles for N input buffers simply because each 1-bit token visits every buffer. The maximum value of N is determined by the ratio of the frequency of the bit clock to the frequency of cell interval clock [42]. Figure 15 shows the cells selected in an arbitration cycle from CAFIFO buffers of a nonblocking switch and the proposed blocking switch.

We propose algorithm RB-CAFIFO, shown in Figure 16, for selecting cells in an arbitration cycle from the CAFIFO input buffers of a baseline network. In RB-CAFIFO, an arbitration cycle is implemented in n passes, and pass k requires 2^k bit clock cycles, for $1 \leq k \leq n$. (See three passes in Figure 15.(b) for $N = 8$). Thus, the total number of bit clock cycles required in an arbitration cycle is $2N-2$. Pass k guarantees that when the k leftmost bit(s) of the destination address of the selected cell from buffer i are placed into row i of the frame corresponding to reverse baseline network, the resultant matrix of order $N \times k$ fits the first k columns of the frame.

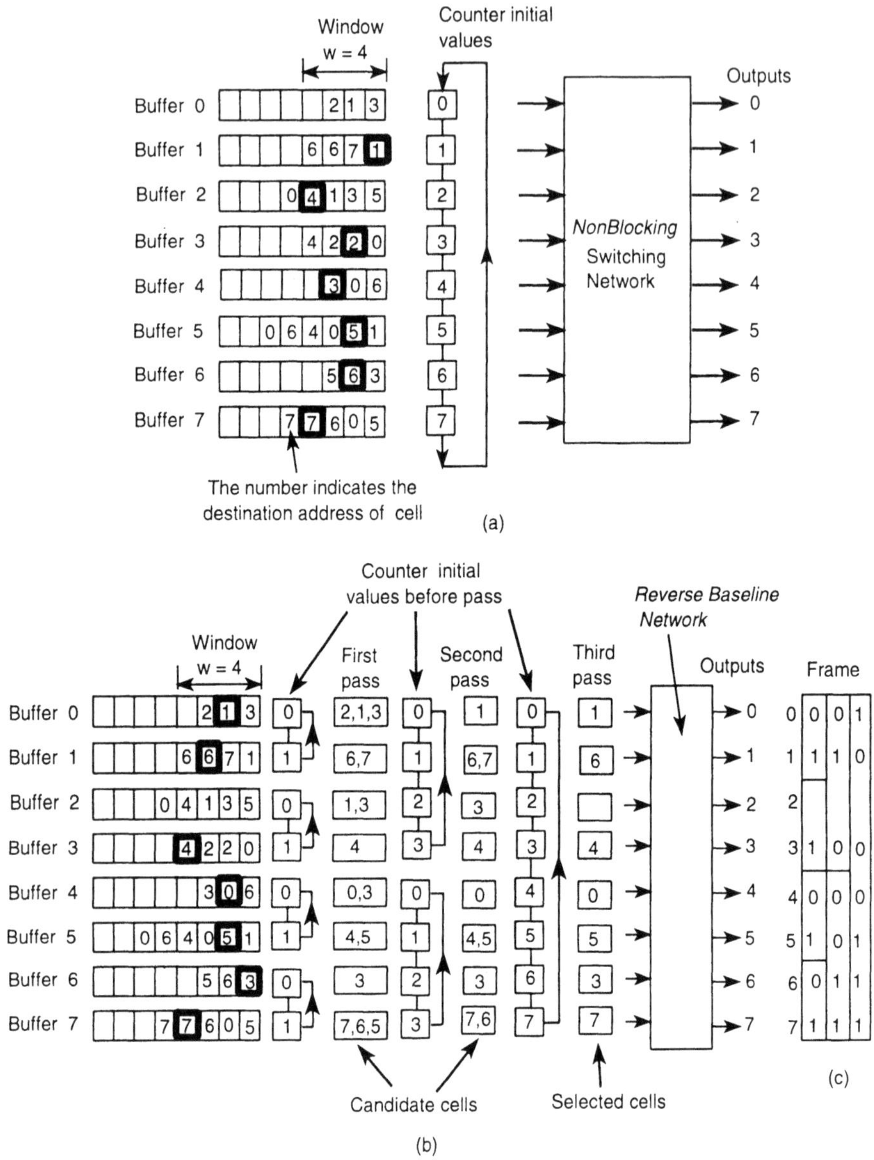

Figure 15: (a) Thick solid squares denote the cells selected in arbitration cycle of the CAFIFO algorithm of Shultz and Gulak for *nonblocking* switch. (b) Thick solid squares denote the cells selected in arbitration cycle. (c) The balanced incomplete matrix formed by the destination addresses of selected cells.

Algorithm RB-CAFIFO

```
begin
1.  Initially, all the valid bit entry fields of all cells in each buffer are zero;
    all 1-bit tokens are zero; output ports are lebeled from 0 to N - 1;
2.  w ← window_size;
3.  for each buffer i, 0 ≤ i ≤ N - 1, in each arbitration cycle do in parallel with other
    buffers
        comment: k denotes the label of a pass.
4.      for k ← 1 to n do
5.          buffer[i] ← 0;
6.          buffer_counter ← i mod 2^k;
7.          set the current 1-bit token to 0;
8.          for j ← 1 to 2^k do
9.              if (buffer[i] = 0) and (the current 1-bit token is 0) then
10.                 consider those cells that are located within the window_size w and their
                      all valid bit entry fields are set to 1 in each one of the previous passes;
11.                 if (the k leftmost bits of these cells' destination addresses are equal to
                      the binary representation of buffer_counter) then
12.                     buffer[i] ← 1;
13.                     set the current 1-bit token to 1;
14.                     for each matching cell of the buffer, set its kth pass valid bit entry
                        field to 1;
15.             endif
16.             send the current 1-bit token to the next buffer in the cycle of 2^k elements;
17.             if (buffer_counter > 0) then
18.                 buffer_counter ← (buffer_counter - 1) mod 2^k;
19.             else buffer_counter ← 2^k - 1;
20.         endfor
21.     endfor
22.     if (buffer[i] = 1) then
23.         select the cell whose all valid bit entry fields
                are 1 and it is the closest to the front of buffer;
24. endfor
```

Figure 16: The proposed algorithm RB-CAFIFO for reverse baseline network with CAFIFO input buffers.

5.2 Simulation Results

The performance of an interconnection network is measured by its throughput and delay. In evaluating the performance of a self-routing ATM switch, cell loss rate, being the fraction of cells lost within the switch, is usually more important than delay. In the proposed input buffering switch, no cell is lost within the switch fabric because only those cells that will not have any internal and external conflict enter the switch fabric. In the proposed switch, a cell is lost only if it arrives in a full input buffer. In the simulations, we assumed that an input buffer can store up to 20 cells. Throughput (also called unnormalized throughput) is defined as the average number of cells which are successfully delivered to the output ports by the switch per time slot per input line [41]. The normalized throughput is the fraction of incoming cells that are successfully delivered to the output ports per time slot per input line. Simulation results are presented in this section by plotting unnormalized throughput against cell arrival rate. Because those cells that are not selected in an arbitration cycle are kept in the input buffers, almost all the inputs of an ATM switch have at least one cell in their input buffers after some transient period, regardless of the value of cell arrival rate. Cell arrival rates somewhat affect the number of cells in buffers. By controlling cell arrival rates, cell lost probability in input buffers can be improved [44].

We ran simulations under uniform and Poisson traffic patterns. The message length is expressed in terms of the number of cells. If the input traffic is Poisson traffic, messages arrive at each input according to Poisson process in the sense that the message length is exponentially distributed. Thus, cell arrival rate in Poisson traffic is equal to the product of message arrival rate and the mean length of a message in units of cells. In our simulations, *all the cells of a message* are assigned *the same destination address.* In the simulations, the nonblocking switch such as the crossbar switch uses the ring reservation scheme with CAFIFO implementation proposed by Shultz and Gulak [42], while the reverse baseline network employs algorithm RB-CAFIFO for cell selection. Each simulation result is an average of 8,000 time slots after a transient period of 2,000 time slots.

Figure 17 illustrates the throughputs of the nonblocking and the reverse baseline networks against cell arrival rate under uniform traffic for $N = 16$. In contrast with the reverse baseline network, the throughput of the nonblocking switch shows little improvement at the high cell arrival rates as the window size increases.

Figure 18 shows the throughput versus cell arrival rate characteristics of

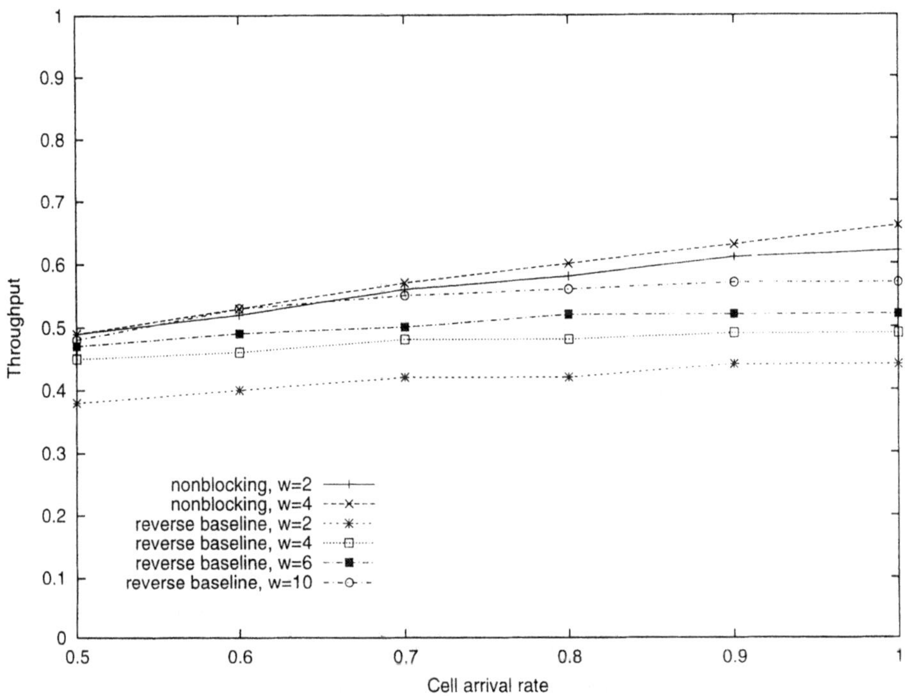

Figure 17: Throughput versus cell arrival rate characteristics for the baseline and nonblocking switches with input buffers under the uniform input/output traffic patterns and $N = 16$.

the nonblocking and blocking switches with respect to different window sizes under the poisson input traffic and the uniformly distributed output traffic. At high cell arrival rates, the throughput of the reverse baseline with $w = 20$ gets very close to that of the nonblocking switch with $w = 2$. The throughput of the reverse baseline network increases significantly as the window size, w, gets larger. (Note that some throughput values of nonblocking switch for cell arrival rates 0.5 and 0.6 are more than 0.5 and 0.6, respectively, due to that most of the input buffers have cells in their windows after the initial transient period of 2,000 time slots).

6 Conclusion

Shuffle-exchange networks provide an efficient interconnection scheme for parallel processing systems and high-speed communication switches. This paper has presented how the permutations realized by shuffle-exchange networks can be identified using the notion of frames and balanced matrices. The basic idea behind the proof for the rearrangeability of the $(2n-1)$-stage shuffle-exchange networks has been presented. For the complete proof, the reader is referred to [14].

Shuffle-exchange type networks have some desirable features needed in the design of ATM switches. In particular, they have constant latency for any input-output pair, are self-routing and modular, preserve cell-sequence due to path-uniqueness, are very suitable for VLSI implementation, and have an area complexity of $O(N \, log_2 N)$ compared to $O(N^2)$ of the crossbar switch with N inputs. But, path-uniqueness also leads baseline-type networks to be internally blocking. Therefore, we have presented two techniques in this paper: *i*) an efficient deflection routing scheme (FDR) for the modified shuffle-exchange network (MSN) and *ii*) a parallel window-based contention resolution algorithm. MSN improves the performance of banyan networks by reducing their internal conflicts. FDR minimizes the number of stages that a packet has to be routed through to reach its destination.

The window-based contention resolution algorithm presented in this paper not only prevents external blocking and alleviates HOL blocking but also prevents internal blocking with the help of frames and balanced matrices. The simulation results indicate that throughput of the n-stage reverse baseline network can be as good as the throughput of a nonblocking switch if the window size is made sufficiently large. Therefore, the reverse baseline network with CAFIFO input buffers can also be considered as a desirable ATM

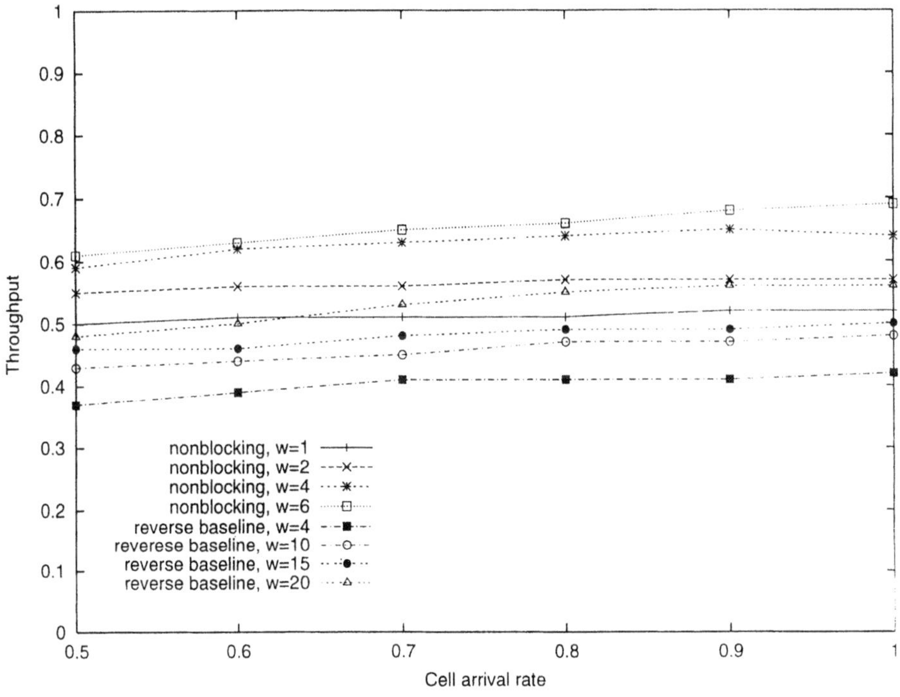

Figure 18: Throughput versus cell arrival rate characteristics with respect to different window sizes for the nonblocking and blocking switches under the poisson input traffic and the uniformly distributed output traffic.

switch fabric. Due to the fact that input buffers are content addressable FIFOs, increasing the window size does not increase the hardware. But, when the window size is increased very much, some cells should be prevented from having excessive latencies by assigning higher priorities to the cells close to the head of buffers.

References

[1] C.S. Raghavendra and A. Varma, "Rearrangeability of the five stage Shuffle-Exchange network for N=8," *IEEE Trans. on Commun.*, Vol. com-35, pp. 808-812, Aug. 1987.

[2] N. Linial and M. Tarsi, "Interpolation between bases and the Shuffle-Exchange network," *Europ. J. Combinatorics*, Vol. 10, pp. 29-39, 1989.

[3] H.S. Stone, "Parallel processing with the perfect shuffle," *IEEE Trans. on Comput.* , Vol. c-20, pp. 153-161, Feb. 1971.

[4] D.S. Parker, "Notes on shuffle/exchange-type switching networks," *IEEE Trans. Comput.*, Vol. C-29, pp. 213-222, March 1980.

[5] H.J. Siegel, "The universality of various types of SIMD machine interconnection networks," *Proceedings of the 4th Annual Symposium on Computer Architecture*, pp. 70-79, March 1977.

[6] C. Wu and T.Y. Feng, "On a class of multistage interconenction networks," *IEEE Trans. on Comput.*, Vol. C-29, pp. 696-702, 1980.

[7] C.S. Babu and C.S. Raghavendra, "On the invariants of shuffle/exchange networks," *Proceedings of the First IEEE Symp. on Parallel and Distributed Processing*, pp. 249-256, May 1989.

[8] C. Wu and T.-Y. Feng, "The universality of the shuffle-exchange network," *IEEE Trans. on Comput.*, Vol. C-30, pp. 324-331, May 1981.

[9] S.T. Huang and S.K. Tripathi, "Finite state model and compatibility theory: new analysis tools for permutation networks," *IEEE Trans. on Comput.*, Vol. C-35, pp. 591-601, July 1986.

[10] C.S. Raghavendra, "On the rearrangeability conjecture of $(2log_2N - 1)$-stage shuffle/exchange network (Position Paper)," *Computer Architecture Technical Committee Newsletter*, pp. 10-12, Winter 1994-1995.

[11] M.K. Kim, H. Yoon and S.R. Maeng, "On the correctness of inside-out routing algorithm," *IEEE Trans. on Comput.*, Vol 46, no 7, pp. 820-823, July 1997.

[12] V.E. Benes, "Proving the rearrangeability of connecting networks by group calculations," *Bell System Tech. J.*, 45, pp. 421-434. 1975.

[13] F.K. Hwang, "A mathematical abstraction of the rearrangeability conjecture for shuffle-exchange networks," *Operations Research Letters*, Vol. 8, no. 2, pp. 85-87, 1989.

[14] Hasan Çam, "Rearrangeability of $(2n-1)$-Stage Shuffle-Exchange Networks," To appear in *SIAM Journal on Computing*.

[15] M.G. Hluchyj and M.J. Karol, "Queueing in high-performance packet switching," *IEEE J. Select. Areas Commun.*, Vol. 6, No. 9, pp. 1587-1597, Dec. 1988.

[16] Y. Tamir and G.L. Frazier, "High-performance multi-queue buffers for VLSI communication switches," in *Proc. of the 15th Annual Int. Symp. on Computer Architecture*, pp. 343-354, 1988.

[17] Y.S. Yeh, M.G. Hluchyj and A.S. Acampora, "The knockout switch: a simple, modular architecture for high-performance packet switching," *IEEE J. Select. Areas Commun.*, Vol. SAC-5. No. 8, pp. 1274-1283, Oct. 1987.

[18] J. Hui and E. Arthurs, "A broadband packet switch for integrated transport," *IEEE J. Select. Areas Commun.*, Vol. 5, No. 8. pp. 1264-1273, Oct. 1987.

[19] M. Akata, S. Karube, T. Sakamoto, T. Saito, S. Wakasugi, S. Yoshida, H. Matsuno and H. Shibata, "A scheduling content-addressable memory for ATM space-division switch control," in *ISSCC Dig. Tech. Papers*, pp. 244-245, 1991.

[20] B. Bingham and H. Bussey, "Reservation-based contention resolution mechanism for batcher-banyan packet switches," *Electron. Lett.*, Vol. 24, No. 13, pp. 772-773, June 1988.

[21] M.J. Karol, M.G. Hluchyj and S.P. Morgan, "Input versus output queueing on a space-division packet switch," *IEEE Trans. Commun.*, Vol. com-35, No. 12, pp. 1347-1356, Dec. 1987.

[22] T.T. Lee, "A modular architecture for very large packet switches," in *Proc. GLOBECOM,* pp. 1801-1809, 1989.

[23] A. Pattavina, "Multichannel bandwidth allocation in a broadband packet switch," *IEEE J. Select. Areas Commun.*, Vol. 6, No. 9, pp. 1489-1499, Dec. 1988.

[24] K.W. Sarkies , "The bypass queue in fast packet switching," *IEEE Trans. Commun.*, Vol. 39, No. 5, pp. 766-774, May 1991.

[25] J.S. Turner, "Queueing analysis of buffered switching networks," *IEEE Trans. Commun.*, Vol. 41, No. 2, pp. 412-420, Feb. 1993.

[26] J.H. Huang and Y.S. Su, "Analysis and design of banyan and crossbar switches with bypass queues," *Computer Communications,* Vol. 20, pp. 115-124, 1997.

[27] H. Çam and J.A.B. Fortes, "Frames: a simple characterization of permutations realized by frequently used networks," *IEEE Trans. on Comput.*, Vol. 44, No. 5, pp. 695-697, May 1995.

[28] Hasan Çam, "Balanced Permutations and Multistage Interconnection Networks," *International Journal of Computer Mathematics*, vol. 73, no. 1, pp. 125-137, Nov. 1999.

[29] Hasan Çam, "Preventing Internal and External Conflicts in an Input Buffering Reverse Baseline ATM Switch," *Int. Journal of Communication Systems* vol. 13, no. 4, pp. 317-334, June 2000.

[30] Hasan Çam, "A High-Performance ATM Switch Based on Modified Shuffle-Exchange Network," *Computer Communications*, vol. 22, no. 2, pp. 110-118, 25 Jan. 1999.

[31] J.A. Bondy and U.S.R. Murty, *Graph Theory with Applications,* North-Holland, New York, 1979.

[32] S.F. Oktuğ and M.U. Çağlayan, "Design and performance evaluation of a banyan network based interconnection structure for ATM switches," *IEEE Journal on Selected Areas in Communications,* Vol. 15, No. 5, pp. 807-816, June 1997.

[33] D. Basak, A.K. Choudhury and E.L. Hahne, "Sharing memory in banyan-based ATM switches," *IEEE Journal on Selected Areas in Communications,* Vol. 15, No. 5, pp. 881-891, June 1997.

[34] M.D. Marco and A. Pattavina, "Distributed routing protocols for ATM extended banyan networks," *IEEE Journal on Selected Areas in Communications,* Vol. 15, No. 5, pp. 925-937, June 1997.

[35] G.D. Stamoulis, M.E. Anagnostou and A.D. Georgantas, "Traffic source models for ATM networks: a survey," *Computer Communications* , Vol. 17, pp. 428-438, 1994.

[36] R.Y. Awdeh and H.T. Mouftah, "MS4 - A high performance output buffering ATM switch," *Computer Commun.,* Vol. 18, No. 9, pp. 631-643, Sept. 1995.

[37] F.A. Tobagi, T. Kwok and F.M. Chiussi, "Architecture, performance, and implementation of the tandem banyan fast packet switch," *IEEE J. Select. Areas Commun.,* Vol. 9, No. 8, pp. 1173-1193. Oct. 1991.

[38] S. Urushidani, "Rerouting network: a high-performance self-routing switch for B-ISDN," *IEEE J. Select. Areas Commun.,* Vol. 9, No. 8, pp. 1194-1204, Oct. 1991.

[39] F. Ayadi, J.F. Hayes and M. Kavehrad, "Performance analysis of the bilayered ShuffleNet and SR_Net under deflection routing," *Computer Communications* , Vol. 20, pp. 125-137, 1997.

[40] T.H. Lee, "Design and analysis of a new self-routing network," *IEEE Trans. on Commun.* , Vol. 40, pp. 171-177, Jan. 1992.

[41] R.Y. Awdeh and H.T. Mouftah, "Survey of ATM switch architectures," *Computer Networks and ISDN Systems* , Vol. 27, pp. 1567-1613, 1995.

[42] K.J. Schultz and P.G. Gulak, "Multicast contention resolution with single-cycle windowing using content addressable FIFO's," *IEEE/ACM Trans. Networking* , Vol. 4, No. 5, pp. 731-742, Oct. 1996.

[43] Y.K. Park and G. Lee, "NN based ATM cell scheduling with queue length-based priority scheme," *IEEE J. Select. Areas Commun.,* Vol. 15, No. 2, pp. 261-270, Feb. 1997.

[44] A. Pattavina, "Nonblocking architectures for ATM switching," *IEEE Commun. Mag.*, No. 2, pp. 38-48, Feb. 1993.

[45] A. Mekkittikul and N. McKeown, "A practical scheduling algorithm to achieve 100% throughput in input-queued switches," in *Proc. of the INFOCOM'92*, pp. 792-799, 1998.

SWITCHING NETWORKS: RECENT ADVANCES
D.-Z. Du and H.Q. Ngo (Eds.) pp. 257-258

Remarks on Beneš Conjecture

Hung Quang Ngo Ding-Zhu Du
Department of Computer Science and Engineering
University of Minnesota, Minneapolis, MN 55455

The shuffle-exchange network was initially proposed by Stone in 1971 [12]. Beneš conjectured in 1975 [1] that $2n-1$-stages are necessary and sufficient for shuffle-exchange networks to route all $N!$ ($N = 2^n$) perfect matchings from the N inputs to the N outputs, i.e., $m(n) = 2n - 1$, where $m(n)$ is the minimum number of stages for a shuffle-exchange network to be able to rooute all permutations. Parker in 1980 [9] showed that $n + 1 \leq m(n) \leq 3n$. Wu and Feng in 1981 [14] gave an explicit algorithm to route all matchings, proving $m(n) \leq 3n-1$. Huang and Tripathi in 186 [4] improved the bound to $m(n) \leq 3n - 3$. Raghavendra and Varma in 1987 [11] verified the conjecture for $n = 3$ and then showed $m(n) \leq 3n - 4$ [13]. They also showed $2n - 1 \leq m(n)$. Linial and Tarsi in 1989 [7] proposed a different approach to prove the same results of Raghavendra and Varma. Hwang [5] provided an alternative formulation of the conjecture. Feng and Seo in 1994 [3] gave a proof of the conjecture, which was incomplete as pointed out by Kim, Yoon, and Maeng in 1997 [6]. Ngo and Du in 2000 [8] verified the conjecture for $n = 4$, which implies $m(n) \leq 3n - 5$. Two year ago, Çam [2] claimed another proof of the conjecture. So far, it is still an open problem whether Çam's new proof is complete or not.

References

[1] V.E. Beneš, Proving the rearrangeability of connecting networks by group calculations, *Bell System Tech. J.*, 45 (1975) 421-434.

[2] Hasan Çam, Rearrangeability of $(2n-1)$-Stage Shuffle-Exchange Networks, submitted to *SIAM Journal on Computing*.

[3] T.-y. Feng and S.-W. Seo, A new routing algorithm for a class of rearrangeable networks, *IEEE Trans. Comput.* 43 (1994) 1270-1280.

[4] S.T. Huang and S.K. Tripathi, Finite state model and compatibility theory: new analysis tools for permutation networks, *IEEE Trans. on Comput.*, C-35 (1986) 591-601.

[5] F.K. Hwang, A mathematical abstraction of the rearrangeability conjecture for shuffle-exchange networks, *Operations Research Letters*, 8 (1989) 85-87.

[6] M.K. Kim, H. Yoon and S.R. Maeng, On the correctness of inside-out routing algorithm, *IEEE Trans. on Comput.*, 46 (1997) 820-823.

[7] N. Linial and M. Tarsi, Interpolation between bases and the Shuffle-Exchange network, *Europ. J. Combinatorics*, 10 (1989) 29-39.

[8] H.Q. Ngo and D.-Z. Du, On the rearrangeability of shuffle-exchange networks, TR-045, Department of Computer Science and Engineering, University of Minnesota, 2000.

[9] D.S. Parker, Notes on shuffle/exchange-type switching networks, *IEEE Trans. Comput.*, C-29 (1980) 213-222.

[10] C.S. Raghavendra, "On the rearrangeability conjecture of $(2log_2N-1)$-stage shuffle/exchange network (Position Paper)," *Computer Architecture Technical Committee Newsletter*, pp. 10-12, Winter 1994-1995.

[11] C.S. Raghavendra and A. Varma, Rearrangeability of the five stage Shuffle-Exchange network for N=8, *IEEE Trans. on Commun.*, com-35 (1987) 808-812.

[12] H. Stone, Parallel processing with perfect shuffle, *IEEE Trans. Comput.* 20 (1971) 153-161.

[13] A. Varma and C. Raghavendra, Rearrangeability of multistage shuffle/exchange networks, *IEEE Trans. Comm.* COM-36 (1988) 1138-1147.

[14] C. Wu and T.-Y. Feng, The universality of the shuffle-exchange network, *IEEE Trans. on Comput.*, C-30 (1981) 324-331.

SWITCHING NETWORKS: RECENT ADVANCES
D.-Z. Du and H.Q. Ngo (Eds.) pp. 259-276

Layout complexity of bit-permuting exchanges in multi-stage interconnection networks

Shuo-Yen Robert Li
Department of Information Engineering
The Chinese University of Hong Kong, Shatin, N.T., Hong Kong
E-mail: bobli@ie.cuhk.edu.hk

Hui Li
Department of Information Engineering
The Chinese University of Hong Kong, Shatin, N.T., Hong Kong
E-mail: huili_huge@yahoo.com

Abstract

The layout area of a multi-stage interconnection network of 2×2 elements is normally dominated by the inter-stage connection wires rather than the elements. We treat the VLSI layout of an inter-stage exchange pattern as a *channel routing problem* under the popular two-layer Manhattan model. The *layout complexity* of an exchange pattern is defined in terms of the *global density* of the channel routing problem so as to reflect the layout area. We then determine the layout complexity for all bit-permuting exchanges. The result leads to the generalization of the layout optimality of divide-and-conquer networks.

Contents

1 Introduction

Multi-stage interconnection networks are utilized as kernel structure in many applications including packet switching([1], [7], [14], [22], [26]), concentration([13], [15], [16], [18]), sorting([2], [9], [12]), convolution [4], interconnection of parallel processors ([5], [20], [24], [25]), discrete Fourier transform, etc. The building components of multi-stage routing networks are the 2×2 elements connected in stages. In every particular application, these elements are all identical 2×2 switches. Thus the network is specified by the pattern of inter-stage connections. Commonly seen examples of multi-stage interconnection networks include the *shuffle exchange network*(Ω *network*) [10], the *baseline network* [28], the *banyan network* [6] and their mirror images([3], [8], [21]).

In order to connect every input to every output, a $2^n \times 2^n$ multi-stage network consisting of 2×2 elements must include at least n stages. We are particularly interested in the class of $2^n \times 2^n$ n stage networks of 2×2 elements such that the interconnection pattern between two consecutive stages corresponds to a permutation in the hardware address bits. On one hand, the restriction to bit-permuting exchanges has empirically caused little hindrance to practical applications. On the other hand, algebraic tools on permutation groups are very useful in the analysis of bit-permuting networks. All kinds of network reformation and composition become possible or comprehendible. Networks in this class will be called *banyan-type* networks.

The remainder of this paper is organized as follows. Section 2 defines the complexity of an inter-stage exchange as the *global density* of a *channel routing problem* [23] under the popular *two-layer Manhattan model*. Section 3 gives a brief review of the concept of bit permuting exchanges [11] and calculates their *layout complexity.* In Section 4, we apply the result of Section 3 to generalize the theorem of layout optimality of *divide-and-conquer networks.*

2 Layout complexity by the two-layer Manhattan model

Every *banyan-type network* is specified by a sequence of n-1 inter-stage exchanges. All $2^n \times 2^n$ banyan-type networks are equivalent in some sense through intra-stage rearrangement of 2×2 elements([11], [27]). Many of them are also equivalent in stronger senses, i.e., equivalent through intra-stage rearrangement under some constraints. A natural question arises: which among them is the most desirable choice in application? The pattern of line crossings is more complex in some exchanges than others. In chip implementation, the wiring of large exchanges has often been the limiting constraint on the size of a multi-stage interconnection network. We shall formulate the complexity measure by the layout area in VLSI detailed routing. All terminology on VLSI physical designs shall follow [23].

One naive measure for the complexity of a $2^n \times 2^n$ exchange has been the number of line intersections among straight interconnection lines. This measure is of some academic interest due to the simple definition. However, we would prefer a complexity measure to reflect more closely the critical parameter in real applications. Consider the traditional *two-layer Manhattan model with reserved layers*, which has been the most popular model for CMOS technologies. Figure 2.1 illustrates thc typical layout style of a banyan-type network. Every cell is identical and occupies a roughly rectangular area. Cells at a stage are aligned into a vertical file. They form a natural *module* in the *floor planning* in the language of VLSI layout. These modules are horizontally aligned stage by stage. Between two consecutive modules is a *channel* for routing the inter-stage exchange.

The layout of a connection wire by the *two-layer Manhattan model* with reserved layers consists of alternately horizontal and vertical line segments, the vertical line segments along tracks on one layer and the horizontal ones on the other layer. The width of a track is the width of a metal wire plus the spacing between parallel metal wires. Thus the track is typically four times

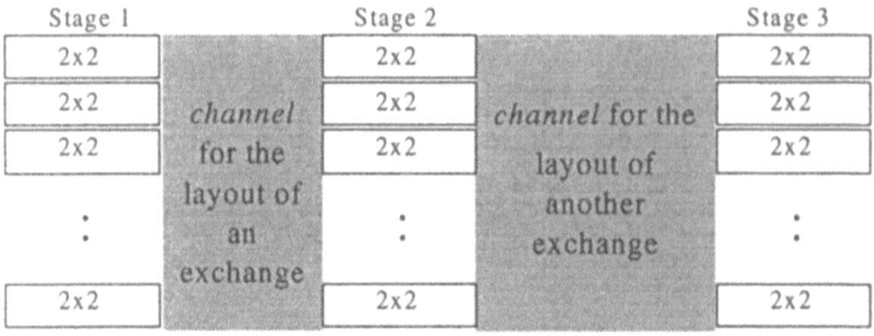

Figure 1: The two-layer Manhattan model of detailed VLSI layout of a multi-stage network.

as wide as the metal wire. The feasibility of the *channel routing problem* for an $N \times N$ exchange is subjected to the following requirements on dimensions of the *channel*:

(1) There are sufficient vertical tracks in the channel for accommodating vertical segments in the connection wires in the exchange.

(2) There are sufficient horizontal tracks in the channel for accommodating horizontal segments.

(3) The height of the channel is at least NH/2, where H is the height of a cell.

Notation An $N \times N$ *exchange* may be viewed as a permutation X on the numbers from 0 to N-1. For $0 \leq b \leq N-1$, there is an interconnection wire in the exchange that leads from the output b at one stage to the input *X(b)* at the next stage. Define the interval

$$IntX(b) = \begin{cases} (b, X(b)], & if\ b < X(b) \\ (X(b), b], & otherwise. \end{cases}$$

If $j \in IntX(b)$, then the interconnection wire from the output b at one stage to the input X(b) at the next stage must intersect the horizontal level j. Let $\vartheta_j^{(X)}$ count those numbers b such that $j \in IntX(b)$. When there is no ambiguity, the notation $\vartheta_j^{(X)}$ will be abbreviated as ϑ_j. In the language of

VLSI layout, $\vartheta_j^{(X)}$ is the *local density* of the aforementioned channel-routing problem for the exchange. It is clearly a lower bound to the required number of vertical tracks in the above condition (1). The *global density* of the channel-routing problem is defined as $max_j\{\vartheta_j^{(X)}\}$. It is also a lower bound, which we want to show to be exact.

Lemma 2.1 *Let $a_1, a_2, \ldots, a_N, z_1, z_2, \ldots, z_N, A_1, A_2, \ldots, A_M$, be positive integers. For every integer j, let ϑ_j count those indices p such that $j \in (a_p, z_p]$ and ξ_j count those index q such that that $j \in (0, A_q]$. If $\vartheta_j \leq \xi_j$ for all j, then the intervals $(a_1, z_1], (a_2, z_2], \ldots, (a_N, z_N]$ can be packed into the intervals $(0, A_1], (0, A_2], \ldots, (0, A_M]$ in the non-overlapping fashion.*

Proof. Constructive proof. Assume without loss of generality that $z_p \leq z_N$ for all p. Choose an index k such that $z_N \leq A_k$. Let ϑ'_j count those among intervals $(a_1, z_1], (a_2, z_2], \ldots, (a_{N-1}, z_{N-1}]$ that contain the point j. On the other hand, define $A'_j = A_j$ for all j except that $A'_k = a_N$. By the maximum of z_N we easily find that, for every positive integer j, at least ϑ'_j among the numbers $A'_1, A'_2, \ldots, A'_M$ are greater than or equal to j. The induction hypothesis on N provides a way to pack the intervals $(a_1, z_1], (a_2, z_2], \ldots, (a_{N-1}, z_{N-1}]$ into the intervals $(0, A'_1], (0, A'_2], \ldots, (0, A'_M]$ in the non-overlapping fashion. Meanwhile, $(0, A_k]$ contains the disjoint union of $(0, A'_k]$ with $(a_N, z_N]$, while $(0, A_j] = (0, A'_j]$ for all $j \neq k$. □

The constructive proof outlines an algorithm for the channel routing of an exchange X over $max_j\{\vartheta_j^{(X)}\}$ vertical tracks. Under this algorithm,

(4) every interconnection wire includes at most one vertical segment. That is, every interconnection wire is either simply a horizontal line or in the horizontal-vertical-horizontal shape.

Thus, the necessary number $max_j\{\vartheta_j^{(X)}\}$ of vertical tracks under the two-layer Manhattan model is sufficient even under the extra restriction (4). The algorithm in the proof can then be translated into channel routing. Figure 2 illustrates the channel routing of the 8×8 shuffle exchange in this way.

The height of the channel is lower bounded by NH/2 and also subjected to the requirement (2). In order to minimize H, the cell is typically laid out short and wide as suggested in Figure 1. Even so, the cell height must at least exceed the transistor height and hence is at least a few times the track

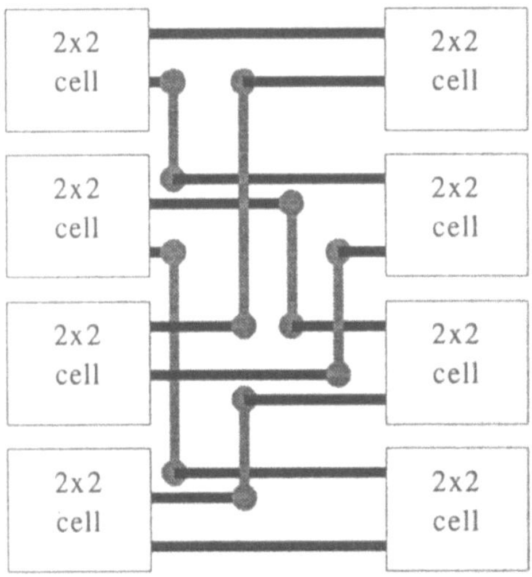

Figure 2: A sample layout of the 8 × 8 shuffle exchange.The two colors of wires indicated two layer of layout.

width. When a layout algorithm abides the restriction (4), a channel height of NH/2 definitely suffices for the requirement (2). In this way, the channel needs only to be at the minimum width of $max_j\{\vartheta_j^{(X)}\}$ and the minimum height of NH/2. Had the definition of the interval IntX(b) been a closed interval rather than the mathematically convenient semi-open interval, some vertical jitter at ends of the interconnection wire would be necessary. However, such jitter could be easily absorbed by the cell height.

In conclusion, the required channel width is exactly $max_j\{\vartheta_j^{(X)}\}$ tracks and the required channel height is exactly N/2 times the cell height. The latter is independent of the particular choice in the $N \times N$ exchange X. Thus the channel routing area for the exchange X is proportional to $max_j\{\vartheta_j^{(X)}\}$. This justifies the following definition.

Definition The *layout complexity* of an exchange X is $max_j\{\vartheta_j^{(X)}\}$.

3 Layout complexity of bit-permuting exchanges

Definition A permutation σ on integers from 1 to n *induces* a $2^n \times 2^n$ exchange $X_\sigma$via

$$X_\sigma : b_{\sigma(1)} b_{\sigma(2)} \dots b_{\sigma(n)} \mapsto b_1 b_2 \dots b_n$$

An exchange so induced is called a *bit-permuting exchange* [11].

The mnemonic interpretation of X_σ is as follows.

• The value of the jth bit before the exchange X_σ gives the value of the $\sigma(j)^{th}$ bit afterwards.

In the channel routing problem for a $2^n \times 2^n$ bit-permuting exchange X, the particular local density $\vartheta_{2^{n-1}}$ may be regarded as the *girth*. The next lemma calculates this girth. It is of course a lower bound to the layout complexity. This lower bound will turn out to be exact.

Definition The *rank* [17] of a non-identity permutation means the smallest number m such that $\sigma(m) \neq m$.

Lemma 3.1 *Let X be a $2^n \times 2^n$ exchange induced by a permutation σ of rank s. Then, $\vartheta_{2^{n-1}} = 2^{n-s}$.*

Proof. By induction on s, we may assume that s=1, i.e., $\sigma(1) \neq 1$. Let b be an n-bit number. Then, IntX(b) contains the point 2^{n-1} if and only if the leading bit of b differs from the leading bit of X(b), which is the $[\sigma^{-1}(1)]^{st}$ bit of b. In the ensemble of all 2^n n-bit numbers, those with the leading bit not equal to the $[\sigma^{-1}(1)]^{st}$ bit account for exactly half of the population. Therefore, $\vartheta_{2^{n-1}} = 2^{n-1}$. □

Theorem 3.2 *A permutation on integers 1 to n with the rank s induces a $2^n \times 2^n$ exchange with the layout complexity 2^{n-s}.*

Proof. Write X = X_σ, where σ is a permutation on integers 1 to n of rank s. The proof is by induction on s. If $s > 1$, then X_σ is the stacking of two disjoint copies of a $2^{n-1} \times 2^{n-1}$ exchange X' induced by a permutation σ' on integers from 1 to n-1 of rank s-1. Thus X shares the same layout complexity as X', which by the induction hypothesis has the layout complexity

$2^{(n-1)-(s-1)}$.

We may therefore proceed by assuming that s=1. Lemma 3.3 below shows that $\vartheta_j \leq 2^{n-1}$ for all j. From Lemma 3.1, we find $max_j\{\vartheta_j\} = 2^{n-1}$ $= \vartheta_{2^{n-1}}$. □

Lemma 3.3 *Let s be a permutation on integers 1 to n with* $\sigma(1) \neq 1$. *Then, for every n-bit number j,*
(a) there can be at most 2^{n-2} *n-bit numbers* x *such that* $x < j - 1/2 < X_\sigma(x)$, *and*
(b) there can be at most 2^{n-2} *n-bit numbers* x *such that* $X_\sigma(x) < j - 1/2 < x$.

Proof. Let $j = binary(j_1 j_2 \dots j_n)$. Write $\sigma(1) = m$ and $\sigma^{-1}(1) = k$. By the symmetry between σ and σ^{-1}, we may assume that $m \leq k$. Moreover, we need only to prove the statement (a), since it becomes the statement (b) after taking the bit-by-bit complement.

Write a generic n-bit number x as binary$(x_1 x_2 \dots x_n)$. When $j_1 = 0$, it suffices to establish the following two statements.

(1) There can be at most binary$(j_2 \dots j_{k-1} j_{k+1} \dots j_n)$ numbers x such that $x < j$ and $x_k = 1$.

(2) There can be at most $2^{n-2} - binary(j_2 \dots j_{k-1} j_{k+1} \dots j_n)$ numbers x such that $j \leq X_\sigma(x)$ and $x_1 = x_k = 0$.

On the other hand, when $j_1 = 1$, it suffices to establish the following two statements.

(3) There can be at most binary$(j_2 \dots j_{k-1} j_{k+1} \dots j_n)$ numbers x such that $x < j$ and $x_1 = x_k = 1$.

(4) There can be at most $2^{n-2} - binary(j_2 \dots j_{k-1} j_{k+1} \dots j_n)$ numbers x such that $j \leq X_\sigma(x)$ and $x_1 = 0$.

First we prove (1) by assuming $j_1 = 0$. Let x be such that $x < j$ and

$x_k = 1$. Then $x_1 = 0$ and

$$\begin{aligned} & binary(0x_2 \ldots x_{k-1}1x_{k+1} \ldots x_n) \\ = \; & x \\ < \; & j \\ = \; & binary(0j_2 \ldots j_n) \\ \leq \; & binary(0j_2 \ldots j_{k-1}1j_{k+1} \ldots j_n). \end{aligned}$$

Hence binary$(x_2 \ldots x_{k-1}x_{k+1} \ldots x_n) < binary(j_2 \ldots j_{k-1}j_{k+1} \ldots j_n)$. The number of binary sequences $0x_2 \ldots x_{k-1}1x_{k+1} \ldots x_n$ subjected to this inequality is at most binary$(j_2 \ldots j_{k-1}j_{k+1} \ldots j_n)$.

Now we prove (2) by assuming $j_1 = 0$. Let x be such that $j \leq X_\sigma(x)$ and $x_1 = x_k = 0$. Then

$$\begin{aligned} & binary(0j_2 \ldots j_{k-1}0j_{k+1} \ldots j_n) \\ \leq \; & binary(0j_2 \ldots j_n) \\ = \; & j \\ \leq \; & X_\sigma(x) \\ = \; & binary(0y_2 \ldots y_{m-1}0y_{m+1} \ldots y_n) \\ \leq \; & binary(0y_2 \ldots y_{m-1}y_{m+1} \ldots y_k0y_{k+1} \ldots y_n), \end{aligned}$$

where each $y_{\sigma(i)} = x_i$. Hence

$$\text{binary}(j_2 \ldots j_{k-1}j_{k+1} \ldots j_n) \leq binary(y_2 \ldots y_{m-1}y_{m+1} \ldots y_n)$$

The number of binary sequences $0y_2 \ldots y_{m-1}0y_{m+1} \ldots y_n$ subjected to this inequality is at most $2^{n-2} - binary(j_2 \ldots j_{k-1}j_{k+1} \ldots j_n)$.

Next, we prove (3) by assuming $j_1 = 1$. Let x bc such that $x < j$ and $x_1 = x_k = 1$. Then

$$\begin{aligned} & binary(1x_2 \ldots x_{k-1}1x_{k+1} \ldots x_n) \\ = \; & x \\ < \; & j \\ = \; & binary(1j_2 \ldots j_n) \\ \leq \; & binary(1j_2 \ldots j_{k-1}1j_{k+1} \ldots j_n). \end{aligned}$$

Hence binary$(x_2 \ldots x_{k-1}x_{k+1} \ldots x_n) < binary(j_2 \ldots j_{k-1}j_{k+1} \ldots j_n)$. The number of binary sequences $1x_2 \ldots x_{k-1}1x_{k+1} \ldots x_n$ subjected to this inequality is at most binary$(j_2 \ldots j_{k-1}j_{k+1} \ldots j_n)$.

Finally we prove (4) by assuming $j_1 = 1$. Let x be such that $j \leq X_\sigma(x)$ and $x_1 = 0$. Then $x_k = 1$ and

$$\begin{aligned} & binary(1j_2 \dots j_{k-1}0j_{k+1} \dots j_n) \\ \leq\ & binary(1j_2 \dots j_n) \\ =\ & j \\ \leq\ & X_\sigma(x) \\ =\ & binary(1y_2 \dots y_{m-1}0y_{m+1} \dots y_n) \\ \leq\ & binary(1y_2 \dots y_{m-1}y_{m+1} \dots y_k0y_{k+1} \dots y_n), \end{aligned}$$

where each $y_{\sigma(i)} = x_i$. Hence

$$\text{binary}(j_2 \dots j_{k-1}j_{k+1} \dots j_n) \leq binary(y_2 \dots y_{m-1}y_{m+1} \dots y_n)$$

The number of binary sequences $1y_2 \dots y_{m-1}0y_{m+1} \dots y_n$ subjected to this inequality is at most $2^{n-2} - binary(j_2 \dots j_{k-1}j_{k+1} \dots j_n)$. □

4 Generalization of the layout optimality of divide-and-conquer networks

Notation The $2^n \times 2^n$ n stage network with exchanges induced by permutations $\sigma_1, \sigma_2, \dots, \sigma_{n-1}$, respectively, is denoted as $[: \sigma_1 : \sigma_2 : \dots : \sigma_{n-1} :]$.

Definition The network $[: \sigma_1 : \sigma_2 : \dots : \sigma_{n-1} :]$ is *routable* if every input is connectable to every output.

Theorem 4.1 *The network* $[: \sigma_1 : \sigma_2 : \dots : \sigma_{n-1} :]$ *is routable if and only if* n, $\sigma_1^{-1}(n)$, $(\sigma_1\sigma_2)^{-1}(n)$, $\dots$, $(\sigma_1\sigma_2 \dots \sigma_{n-1})^{-1}(n)$ *are n distinct numbers. In that case, at least one of the permutations* σ_j *is of rank 1 (Proposition 3.1 and Lemma 4.5 in [17]).*

Theorem 4.1 is a criterion for routability of bit-permuting networks. It will be used for the proof of Theorem 4.6. A $2^n \times 2^n$ divide-and-conquer network is a special banyan-type network defined as:

Definition Assume that $\lfloor n/2 \rfloor \leq$ k $\leq \lceil n/2 \rceil$. Recursively let $[: \pi_1 : \pi_2 : \dots : \pi_{k-1} :]$ and $[: \lambda_{k+1} : \lambda_{k+2} : \dots : \lambda_{n-1} :]$ be $2^k \times 2^k$ and $2^{n-k} \times 2^{n-k}$ *divide-and-conquer networks* [17], where each π_j is a permutation on integers 1 to k and each λ_j is a permutation on integers 1 to n-k. Then, a

$2^n \times 2^n$ *divide-and-conquer* network $[\ : \sigma_1 : \sigma_2 : \ldots : \sigma_{n-1} : \ ]$ can be defined by

• For $1 \leq j < k$, let $\sigma_j =$ (n n-1 $\ldots 1)^{n-k} \pi_j$ $(1\ 2 \ldots n)^{n-k}$. This $2^n \times 2^n$ exchange induced by σ_j is the vertical stacking of 2^{n-k} disjoint copies of the $2^k \times 2^k$ exchange induced by π_j.

• σ_k is the permutation on the set $1, 2, \ldots, n$ such that $\sigma_k(j) \equiv (j + k)$ mod(n).

• For $k < j < n$, let $\sigma_j =$ (n n-1 $\ldots 1)^k \lambda_j (1\ 2 \ldots n)^k$. Thus the $2^n \times 2^n$ exchange induced by σ_j is the vertical stacking of 2^k disjoint copies of the $2^{n-k} \times 2^{n-k}$ exchange induced by λ_j.

Theorem 4.2 *A $2^n \times 2^n$ divide-and-conquer network achieves the minimum layout complexity among all banyan-type networks (Theorem 7.3 in [17]).*

All $2^n \times 2^n$ banyan-type networks are equivalent in a very loose sense through intra-stage rearrangement of 2×2 elements ([11], [27]). In some applications two banyan-networks are exchangeable in use only when they are equivalent in some stronger sense or when they meet some ad hoc properties. Thus it is not always permissible to replace a banyan-type network by a divide-and-conquer network. Below we generalize the notion of divide-and-conquer networks. The layout optimality will be preserved through the generalization. A wider choice of optimal networks enhances the chance of meeting ad hoc requirements.

Definition Assume that $\lfloor n/2 \rfloor \leq$ k $\leq \lceil n/2 \rceil$. Recursively let $[\ : \pi_1 : \pi_2 : \ldots : \pi_{k-1} : \ ]$ and $[\ : \lambda_{k+1} : \lambda_{k+2} : \ldots : \lambda_{n-1} : \ ]$ be $2^k \times 2^k$ and $2^{n-k} \times 2^{n-k}$ *generalized divide-and-conquer networks* [17], where each π_j is a permutation on integers 1 to k and each λ_j is a permutation on integers 1 to n-k. Then, a $2^n \times 2^n$ *generalized divide-and-conquer* network $[\ : \sigma_1 : \sigma_2 : \ldots : \sigma_{n-1} : \ ]$ can be defined by

• For $1 \leq j < k$, let $\sigma_j =$ (n n-1 $\ldots 1)^{n-k} \pi_j$ $(1\ 2 \ldots n)^{n-k}$.

• Let σ_k be any permutation that maps the numbers 1, 2, $\ldots$, n-k into the set $\{k+1, k+2, \ldots, n\}$.

• For $k < j < n$, let $\sigma_j = (\text{n n-1} \ldots 1)^k \lambda_j (1\ 2 \ldots n)^k$.

By Theorem 3.2, the layout complexity of σ_k is equal to 2^{n-1}. For $j < k$ (*resp.* $j > k$), the induced $2^n \times 2^n$ exchange by σ_k is a stack of disjoint copies of the exchange induced by π_j (*resp.* λ_j) and hence share the same layout complexity with π_j (*resp.* λ_j). Figure 3 gives an example of a 64×64 generalized divide-and-conquer network.

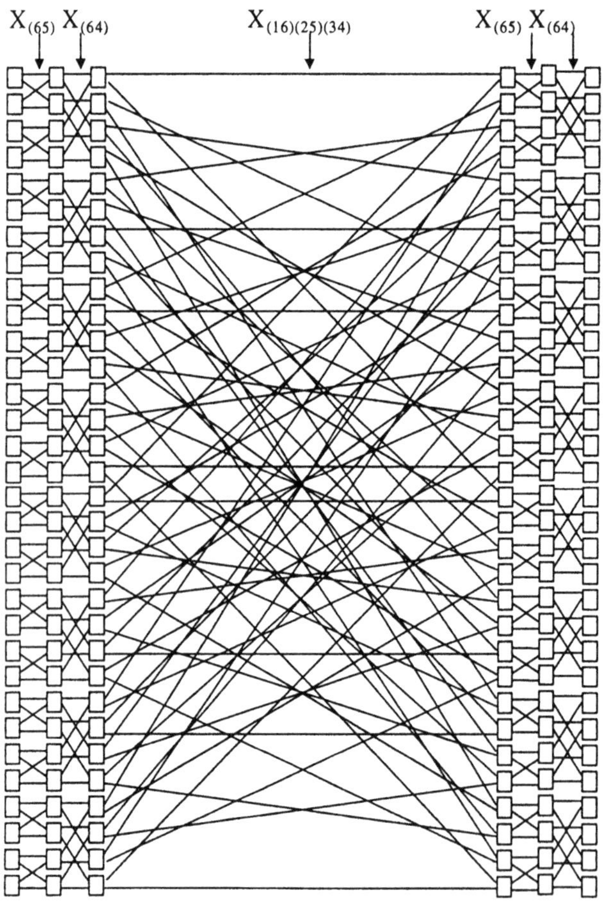

Figure 3: An example of 64×64 generalized divide-and-conquer network (courtesy of Academic Press [19]).

Five examples of 256×256 generalized divide-and-conquer networks are

listed in the follow:
(1) [: (87) : (86)(75) : (87) : (84)(73)(62)(51) : (87) : (86)(75) : (87) :],
the 256 × 256 divide-and-conquer network.
(2) [: (87) : (5867) : (87) : (17263845) : (87) : (57)(68) : (87) :].
(3) [: (87) : (58)(67) : (87) : (1826)(3547) : (87) : (5768) : (87) :].
(4) [: (87) : (57)(68) : (87) : (182538)(47) : (87) : (5867) : (87) :].
(5) [: (87) : (85)(76) : (87) : (81)(72)(63)(54) : (87) : (85)(76) : (87) :].

Although these five examples involve different sequences of exchanges, they all share the same layout complexity $2^1+2^3+2^1+2^7+2^1+2^3+2^1 = 152$ according to Theorem 3.2. Clearly every $2^n \times 2^n$ generalized divide-and-conquer network shares the same layout complexity as a $2^n \times 2^n$ divide-and-conquer network. Theorem 4.2 can be generalized as:

Theorem 4.3 *Every $2^n \times 2^n$ generalized divide-and-conquer network achieves the optimal layout complexity among all $2^n \times 2^n$ banyan-type networks.*

To prove the converse of Theorem 4.3, we shall need the following lemmas.

Notation Denote the layout complexity of a $2^n \times 2^n$ generalized divide-and-conquer network as $\lambda(n)$.

From Theorem 3.2 we have the recursive formula $\lambda(n) = \lambda(\lfloor n/2 \rfloor) + 2^{n-1} + \lambda(\lceil n/2 \rceil)$

Lemma 4.4 $\lambda(n) \leq 2^{n-1} + 2^{n-2}$.

Proof. When $n \leq 5$, the lemma is trivial. When $n \geq 6$, the proof is by induction on n in the sequel.

If $n = 2m \geq 6$, then

$$\begin{aligned} \lambda(n) = &\ 2^{n-1} + 2\lambda(m) \\ \leq &\ 2^{n-1} + 2^m + 2^{m-1} . \end{aligned}$$

If $n = 2m+1 \geq 7$, then

$$\begin{aligned} \lambda(n) = &\ 2^{n-1} + \lambda(m) + \lambda(m+1) \\ \leq &\ 2^{n-1} + 2^{m+1} + 2^{m-2} . \end{aligned}$$

In either case,

$$\begin{aligned}\lambda(n) &= 2^{n-1} + 2^{n-3} + 2^{n-3} \\ &\leq 2^{n-1} + 2^{n-2}.\end{aligned}$$

□

Lemma 4.5 *If $\lambda(k) + \lambda(n-k) = \lambda(\lfloor n/2 \rfloor) + \lambda(\lceil n/2 \rceil)$ then $k = \lambda(\lfloor n/2 \rfloor)$ or $\lambda(\lceil n/2 \rceil)$ [19].*

Theorem 4.6 *The $2^n \times 2^n$ generalized divide-and-conquer networks are the only $2^n \times 2^n$ banyan-type networks that achieve the optimal layout complexity.*

Proof. The proof is by induction on n. Let $[\,:\sigma_1:\sigma_2:\ldots:\sigma_{n-1}:\,]$ be a $2^n \times 2^n$ banyan-type network with the minimum layout complexity. From Theorem 4.1, one of the permutations, say, σ_k is of rank 1.

Remove the exchange induced by σ_k from network. The remainder is split into the two sides. On the left-hand side is the 2^{n-k} logically separate $2^k \times 2^k$ banyan-type networks. There are two possibilities depending whether these logically separate networks are physically separate in layout.

(1) It so, then they are identical $2^k \times 2^k$ banyan-type networks. Hence the layout complexity of the whole left-hand side is simply $\lambda(k)$.

(2) Else, the rank of at least one of the permutations $\sigma_1, \sigma_2, \ldots, \sigma_{k-1}$ is less than n-k+1. From Lemma 4.4, the layout complexity of the exchange induced by this permutation is at least $2^k > \lambda(k)$.

Similarly, if the right-hand side is the 2^k copies of a $2^{n-k} \times 2^{n-k}$ banyan-type network, then the layout complexity is $\lambda(n-k)$. Else, one of the permutations $\sigma_{k+1}, \sigma_{k+2}, \ldots, \sigma_{n-1}$ induces an exchange of layout complexity greater than $\lambda(n-k)$.Therefore,

$$\begin{aligned}& \mathit{layout\ complexity\ of\ the\ network}\ [:\sigma_1:\sigma_2:\ldots:\sigma_k:\ldots:\sigma_{n-1}:] \\ \geq\ & \lambda(k) + 2^{n-1} + \lambda(n-k) \\ =\ & \mathit{the\ layout\ complexity\ of\ some}\ 2^n \times 2^n\ \mathit{banyan-type\ network} \\ \geq\ & \lambda(n), \mathit{by\ Theorem 3} \\ =\ & \mathit{layout\ optimality\ of}\ [:\sigma_1:\sigma_2:\ldots:\sigma_k:\ldots:\sigma_{n-1}:], \\ & \mathit{by\ the\ initial\ assumption.}\end{aligned}$$

Thus the equality holds throughout. This can only be true when the aforementioned left-hand (resp. right-hand) side is the vertical stacking of 2^{n-k} (*resp.* 2^k) copies of a $2^k \times 2^k$ (*resp.* $2^{n-k} \times 2^{n-k}$) banyan-type network. Moreover, $\lambda(k) + \lambda(n-k) = \lambda(\lfloor n/2 \rfloor) + \lambda(\lceil n/2 \rceil)$. From Lemma 4.5, we must have $k = (\lfloor n/2 \rfloor)$ or $(\lceil n/2 \rceil)$.

It remains to show that

- σ_k maps the numbers 1, 2,..., n-k into the set {k+1, k+2,..., n}, or equivalently,
- σ_k^{-1} maps the numbers k+1, k+2, ..., n into the set {1, 2, ..., n-k}.

Assume to the contrary that σ_k^{-1} maps the numbers k+1, k+2, ..., n to the set {1, 2, ...,h-1,h+1,..., n-k, n-k+j}, where $1 \le h \le n-k < j \le n$. We shall derive a contradiction.

The numbers $n, \sigma_1^{-1}(n)$, $(\sigma_1\sigma_2)^{-1}(n)$, ..., $(\sigma_1\sigma_2\ldots\sigma_{k-1})^{-1}(n)$ is a sequence of k distinctive numbers n, n-1,..., n-k+1 by Theorem 4.1. Moreover, the permutation $\sigma_1\sigma_2\ldots\sigma_{k-1}$ does not change any number from 1 to n-k. Similarly, the sequence
$n, \sigma_{k+1}^{-1}(n)$, ..., $(\sigma_{k+1}\sigma_{k+2}\ldots\sigma_{n-1})^{-1}(n)$ is n-k distinctive numbers of n, n-1, ..., k+1.
Hence, we have the set-theoretic equality

$$\begin{aligned}
& \{(\sigma_1\sigma_2\ldots\sigma_k)^{-1}(n), (\sigma_1\sigma_2\ldots\sigma_k\sigma_{k+1})^{-1}(n), \ldots, (\sigma_1\sigma_2\ldots\sigma_{n-1})^{-1}(n)\} \\
= & \{(\sigma_1\sigma_2\ldots\sigma_k)^{-1}(k+1), (\sigma_1\sigma_2\ldots\sigma_k)^{-1}(k+2), \ldots, (\sigma_1\sigma_2\ldots\sigma_k)^{-1}(n)\} \\
= & \{(\sigma_1\sigma_2\ldots\sigma_{k-1})^{-1}(1), (\sigma_1\sigma_2\ldots\sigma_{k-1})^{-1}(2), \ldots, (\sigma_1\sigma_2\ldots\sigma_{k-1})^{-1}(h-1), \\
& (\sigma_1\sigma_2\ldots\sigma_{k-1})^{-1}(h+1), \ldots, (\sigma_1\sigma_2\ldots\sigma_{k-1})^{-1}(n), (\sigma_1\sigma_2\ldots\sigma_{k-1})^{-1}(j)\} \\
= & \{1, 2, \ldots, h-1, h+1, \ldots, n-k, (\sigma_1\sigma_2\ldots\sigma_{k-1})^{-1}(j)\},
\end{aligned}$$

here $n-k+1 \le (\sigma_1\sigma_2\ldots\sigma_{k-1})^{-1}(j)$. Thus,

$$\begin{aligned}
& \{n, \sigma_1^{-1}(n), \ldots, (\sigma_1\sigma_2\ldots\sigma_{n-1})^{-1}(n)\} \\
= & \{n, \sigma_1^{-1}(n), \ldots, (\sigma_1\sigma_2\ldots\sigma_{k-1})^{-1}(n)\} \cup \\
& \{(\sigma_1\sigma_2\ldots\sigma_k)^{-1}(n), (\sigma_1\sigma_2\ldots\sigma_k\sigma_{k+1})^{-1}(n), \ldots, (\sigma_1\sigma_2\ldots\sigma_{n-1})^{-1}(n)\} \\
= & \{n-k+1, n-k+2, \ldots, n\} \cup \\
& \{1, 2, \ldots, h-1, h+1, \ldots, n-k, (\sigma_1\sigma_2\ldots\sigma_{k-1})^{-1}(j)\}
\end{aligned}$$

Therefore, h is missing from the sequence $n, \sigma_1^{-1}(n)$, ..., $(\sigma_1\sigma_2\ldots\sigma_{n-1})^{-1}(n)$. This contradicts the routability of the network $[\,:\sigma_1:\sigma_2:\ldots:\sigma_k:\ldots:\sigma_{n-1}:\,]$ according to Theorem 4.1. □

5 Conclusion

We treat the VLSI layout of an inter-stage exchange pattern in a multi-stage interconnection network as a *channel routing problem* under the popular *two-layer Manhattan model.* The layout complexity of an exchange pattern is defined in terms of the *global density* of the channel routing problem so as to reflect the layout area. We prove that the layout complexity of the exchange induced by a permutation s on integers 1 to n is 2^{n-k}, where k is the smallest integer such that $\sigma(k) \neq k$.

Among all the $2^n \times 2^n$ banyan-type networks, the divide-and-conquer networks have been known to achieve the layout optimality. All $2^n \times 2^n$ banyan-type networks are equivalent in a very loose sense through intra-stage rearrangement of 2×2 elements. In some applications two banyan-networks are exchangeable in use only when they are equivalent in some stronger sense or when they meet some ad hoc properties. Thus it is not always permissible to replace a banyan-type network by a divide-and-conquer network. A wider choice of optimal networks would enhance the chance of meeting ad hoc requirements. We thus generalize the notion of divide-and-conquer networks. The generalized divide-and-conquer networks turn out to be the only banyan-type networks that achieve the layout optimality.

References

[1] D.P. Agrawal, Graph theoretical analysis and design of multistage interconnection Networks, *IEEE Trans. Computers*, Vol. C-32, no.7, pp. 637-648, July, 1983.

[2] K. E. Batcher, Sorting networks and their applications, *Proc. of AFIP 1968 Spring Joint Computer Conf.*, Vol. 32, pp. 307-314, 1968.

[3] K. E. Batcher, The flip network in STARAN, *Proc. 1976 Int. Conf. Parallel Processing*, pp. 65-71.

[4] H. Burchardt and L. C. Barbosa, Contributions to the Application of the Viterbi Algorithm, *IEEE Trans. Information Theory*, Vol.31, pp. 626-634, 1985.

[5] J. B. Dennis, Data Flow Supercomputer,*Computers*, Vol.13, No.11, pp. 48-56, Nov.1980.

[6] L. R. Goke and G. J. Lipovski, Banyan Networks for Partitioning Multiprocessing Systems, *Proc. First Annual Computer Architecture Conf.*, pp. 21-28, Dec. 1973.

[7] A. Huang and S. Knauer, Starlite: a wideband digital switch, *Proceedings of GLOBECOM'84*, pp. 121-125, 1984.

[8] H. S. Kim and A. Leon-Garcia, Nonblocking property of reverse banyan networks, *IEEE Trans. Commun.*, Vol. 40, No 3, pp. 472-476, March 1992.

[9] D. E. Knuth, *The Art of Computer Programming*, Volume 3, Addison-Wesley, 1973.

[10] D. H. Lawrie, Access and Alignment of Data in an Array Processor, *IEEE Trans. Computers*, Vol. C-24, No. 12, pp. 1145-1155, Dec. 1975.

[11] S.-Y. R. Li, Formalization of Self-route Networks and the Rotary Switch, *Proc.INFOCOM'94*, pp. 438-446, Toronto, June 1994.

[12] S.-Y. R. Li, Partial Sorting and Concentration by Parallel Networks, *Proc. of International Workshop on Discrete Mathematics and Algorithms*, pp. 27-43, 1994.

[13] S.-Y. R. Li, and C. M. Lau, Concentrators in ATM Switching, *Computer Systems Science and Engineering*, (11) (6) (1996), pp. 335-342.

[14] S.-Y. R. Li, and W. Lam, ATM Switching by Divide-and-conquer interconnection of Partial Sorters, *Microprocessors and Microsystems* 22 (1999), pp. 579-587.

[15] S.-Y. R. Li, G. M. Koo and H. Li, An algorithm for the construction of concentrators from 2 × 2 sorters, *American Mathematical Society, DIMACS series in Discrete Mathematics and Theoretical Computer Science*, Vol. 42, (1998), pp. 197-219.

[16] S.-Y. R. Li and H. Li, Optimization in the fast knockout algorithm for self-route concentration, *IEEE Proceedings of ICC'98*, pp. 630-634, Atlanta, June,1998.

[17] S.-Y. R. Li, Optimal multi-stage interconnection by divide-and-conquer networks, *Proc. of the 2nd IASTED International Conference on Parallel and Distributed Computing and Networks*, pp. 318-323, Brisbane, Australia, Dec.1998.

[18] S.-Y. R. Li, H. Li and G. M. Koo, Fast knockout algorithm for self-route concentration, *Computer Communications* 22 (1999), pp. 1574-1584.

[19] S.-Y. R. Li, *Switching Networks: Algebraic Principles and Broadband Applications*, book under preparing for Academic Press. Summer 2000.

[20] S. F. Lundstrom and G. Barnes, A Controllable MIMD Architecture, *Proc. Int. Conf. Parallel Processing*, pp. 19-27, 1980.

[21] J. H. Patel, Processor-Memory Interconnections for Multiprocessors, *Proc. Sixth Annual Symp. Computer Architecture*, pp. 168-177, April 1979.

[22] D. S. Parker, Note on Shuffle/Exchange-Type Switching Networks, *IEEE Trans. on Computers*, Vol. C-29, No. 3, pp. 213-222, March 1980.

[23] M. Sarrafzadeh and C. K. Wong, *An Introduction to VLSI physical design*, McGraw-Hill, 1996.

[24] B. J. Smith, A Pipelined, Shared Resource MIMD computer, *Proc. Int. Conf. Parallel Processing*, pp. 6-8, 1978.

[25] H. S. Stone, Parallel Processing with the Perfect Shuffle, *IEEE Trans. Computers*, Vol. 20, pp. 153-161, 1971.

[26] F. A. Togago and T.C. Kwok, The Tandem Banyan Switching Fabric: A Simple High-Performance Fast Packet Switch, *Proc. IEEE INFOCOM'91*, Vol. 3, 11A.2, pp. 1245-1253, 1991.

[27] C.-L. Wu and T. Y. Feng, On a Class of Multistage Interconnection Networks, *IEEE Trans. on Computers*, Vol. C-29, No. 8, pp. 694-702, Aug. 1980.

[28] C.-L. Wu and T. Y. Feng, On a Distributed-Processor Communication Architecture, *Proc. Compcon.*, pp. 599-605, 1980.

SWITCHING NETWORKS: RECENT ADVANCES
D.-Z. Du and H.Q. Ngo (Eds.) pp. 277-306

Multicast Switching Networks

Dongsoo S. Kim, PhD.
Electrical and Computer Engineering
Indiana University Purdue University Indianapolis
E-mail: dkim@engr.iupui.edu

Contents

In the previous chapters, we examined the principle switching architecture for peer-to-peer communications. We discussed how to build efficient switching networks employing intelligent control algorithms. In this chapter, we consider switching networks that can handle communication from single source to multiple destination, such as teleconferencing, commercial video distribution and distributed data processing.

A simple crossbar switch shown in Figure 1(a) is capable of multicasting data packets to any number of output ports. However, the switch requires $O(n^2)$ crosspoints, or switching elements, for n external ports. For this reason, the crossbar switch is inappropriate in constructing a large scale switching network. A knockout switch in Figure 1(b)is a way to implement the crossbar switch for packet switching [20, 4, 11]. Each input of the network

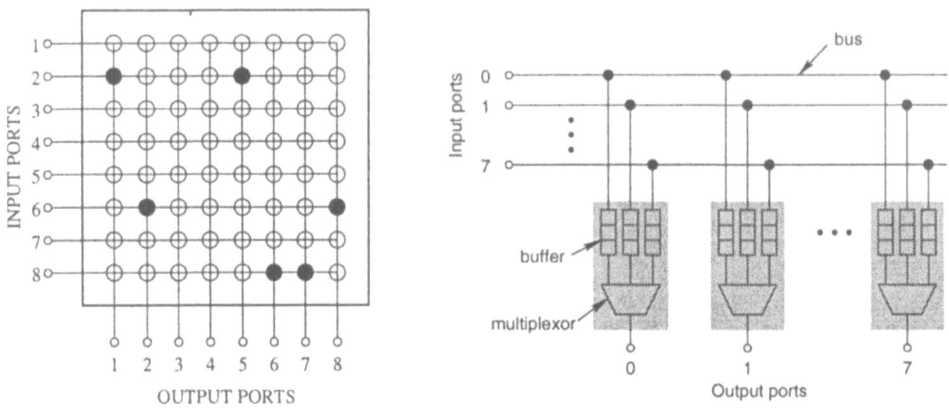

(a) (b)

Figure 1: Simple switches (a) Crossbar switch, (b) Knockout switch

broadcasts incoming packets to its own bus. Controllers at the output ports store packets into internal buffers by matching their destination addresses. When two or more packets have the same destination address, a multiplexer resolves the contention by knocking out low priority packets. Though the knockout switch is able to multicast data packets, it also has the scalability problem as the crossbar switch does. In this chapter, we will consider multi-stage switching networks which support multicast communications and is scalable.

1 Clos Switching Networks for Single Rate Multicast

A multicast connection request is defined as a triple (x, Y), where x is a source port and Y denotes a set of destination ports. The cardinality of Y, $|Y| = d$, is called a degree of fanout of the multicast request. Figure 1(a) shows three requests $(2, \{1,5\})$, $(6, \{2,8\})$, and $(8, \{6,7\})$. A set of connection requests are compatible if they have no common input port and no common output port (port compatibility), so those three requests are compatible. A new connection request is said to be compatible if the addition of the new request are compatible to the existing requests in the network. For example, a new request $(4, \{3,4\})$ is compatible to the network in Fig-

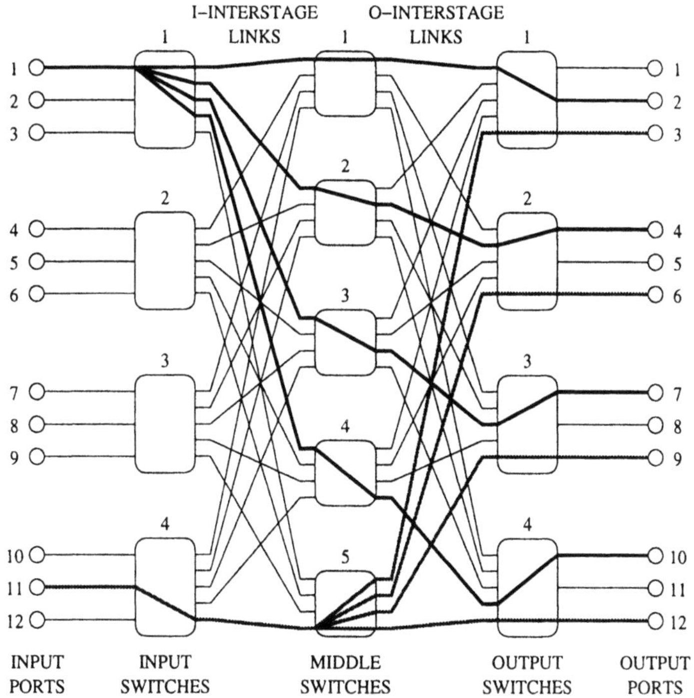

Figure 2: Clos switch $C(3,4,5)$. A connection $(1,\{1,2,3,4\})$ is routed with a pre-fanout scheme, and $(4,\{1,2,3,4\})$ with a lazy-fanout scheme.

ure 1(a), but $(4,\{4,5\})$ is not. A connection request is called to be satisfiable if a switch can provide resources from input port to output ports. A switch is non-blocking if it can satisfy any compatible set of connection requests. A knockout switch shown in Figure 1(b) makes the crossbar switch suitable for a packet switch architecture. A packet coming into the input port is broadcasted along the bus. Each bus interface has access to all the input packet and one output port. The bus interface observes the destination address in each input packet and stores packets into a first-in first-out buffer if their destination addresses are associated with its output port. The multiplexer selects a packet based on a knockout tournament strategy before forwarding them to the output port [4]. The crossbar switch and knockout switch are non-blocking not only for peer-to-peer communications, but for multicast connections. However, the complexity of both switches is $O(n^2)$ for n external ports, which prohibits them to be employed for a large scale switch network [16].

A three-stage interconnection network shown in Figure 2 is called a Clos switch [2]. The Clos switch, denoted by $C(n_1, r_1, n_2, r_2, m)$, consists of r_1

input switch modules ($n_1 \times m$), m middle switches ($r_1 \times r_2$), and r_2 output switches ($m \times n_2$). If $n_1 = n_2 = n$ and $r_1 = r_2 = r$, the Clos network is symmetric and denoted by $C(n, r, m)$. Let us consider a multicast request $(1, \{2, 3\})$. The input switch 1 can duplicate a packet in this request and route two copies through different middle switches to the output switch 1 as if they are point-to-point packets. However, it is to waste a middle and its links if the output switches have multicast capability [8, 9]. It is necessary for an input switch to connect to only one output switch if an input port needs to connect to two or more output ports in a same output switch. In this condition, the maximum fanout of a connection request is limited to r for a symmetric network, or r_2 for an asymmetric network. Without loss of generality, we can specify a connection request (x, Y) as an input switch and a set of output switches for $1 \leq |Y| \leq r$, instead of input and output ports. Figure 2 shows two multicast connections $(1, \{1, 2, 3, 4\})$ and $(4, \{1, 2, 3, 4\})$.

1.1 Pre-Fanout Multicast

A simple multicast routing is that the input switches perform all packet duplications and the middle switches route the copies of packets as if they are point-to-point connections such as a connection $(1, \{1, 2, 3, 4\})$ in Figure 2. To make the network non-blocking for any compatible connection requests, we need to count the number of middle switches required in the worst case. Let us consider a multicast request (x, Y) and $Y = (y_1, y_2, \cdots, y_k)$ for $1 \leq k \leq r$. The input switch x could contain up to $n - 1$ requests except the given request; and each request is fanned out up to r middle switches so that $(n - 1)r$ middle switches out of m can be occupied by other requests from the input switch x. For the compatible request, each output switch in Y can carry at most $n - 1$ connections, which means that at most $n - 1$ middle switches from the output switch cannot provide a path for the new request. As a result, $(n - 1)r + n - 1$ middle switches cannot be used for the request either from the input stage or the output stage in the worst case. If the network has r more middle switches than the $(n - 1)r + n - 1$ possible unavailable switches, we surely route the request by duplicating packets only at the input stage. In other word, the network is non-blocking for the pre-fanout route scheme if $m \geq nr + n - 1$ [14, 5, 6, 14].

1.2 Lazy-Fanout Multicast

Let us assume the input switches do not have the multicast capability for some reasons. Then, the middle switches should perform all necessary duplication. This routing scheme is called a lazy-fanout scheme because a packet is not duplicated until the last chance. Figure 2 illustrates this type of routing for a connection $(4, \{1, 2, 3, 4\})$. In this case, the maximum number of middle switches that can carry other connections from the input x is $n - 1$. However, once we select a middle switch, it should have available links to every output switches in Y. At the worst case, each middle switch contains one or none busy links to the output stage. There could be $(n - 1)r$ such middle switches. The network is non-blocking for this lazy-fanout routing scheme if the network has one more middle switch than $n - 1 + (n - 1)r$, or $m \geq nr + n - r$.

A crossbar switch requires $N_1 N_2$ crosspoints for N_1 input ports and N_2 output ports so that an equivalent crossbar switch to $\mathcal{C}(n, r, m)$ requires n^2r^2 crosspoints, or $O(n^2r^2)$. The number of crosspoints of a Clos network $\mathcal{C}(n, r, m)$, provided each switch module is a crossbar switch, is found as $2rnm + mr^2$. The point-to-point communication takes an advantage that the Clos network is strictly non-blocking if $m \geq 2n - 1$, or $O(nr(n + r))$ crosspoints. However, if we are limited to either the pre-fanout or the lazy-fanout required $O(nr)$ middle switches, the complexity of crosspoints falls into $O(n^2r^2 + nr^3)$, which gives no benefit of the multistage interconnection network over the simple crossbar switch.

1.3 Controlled Nonblocking Multicast

In the previous routing schemes, we assumed that either input switches or middle switches have no multicast capability. Supposed that every switch module is capable of multicasting packets. An input switch duplicates a packet to a small number of middle switches, which multicast to all output switches. The question in this scheme is what the minimum number of middle switches is in order to support non-blocking multicast [18].

Let us consider Figure 3 where four connection requests are routed on $\mathcal{C}(3, 4, 5)$. If we use a binary vector for the set of output switches, they can be represented as follows:

$$
\begin{aligned}
C_1 &= (x_1, Y_1) = (1, (1\,1\,1\,1)),\\
C_2 &= (x_2, Y_2) = (2, (0\,0\,0\,1)),\\
C_3 &= (x_3, Y_3) = (3, (1\,1\,1\,0)), \text{ and}\\
C_4 &= (x_4, Y_4) = (4, (1\,0\,0\,0)).
\end{aligned}
$$

A collection of the binary vectors well represents the characteristic of connection requests denoted by a $(k \times r)$ binary matrix $\mathbf{D} = (D_{ij})$ (demand

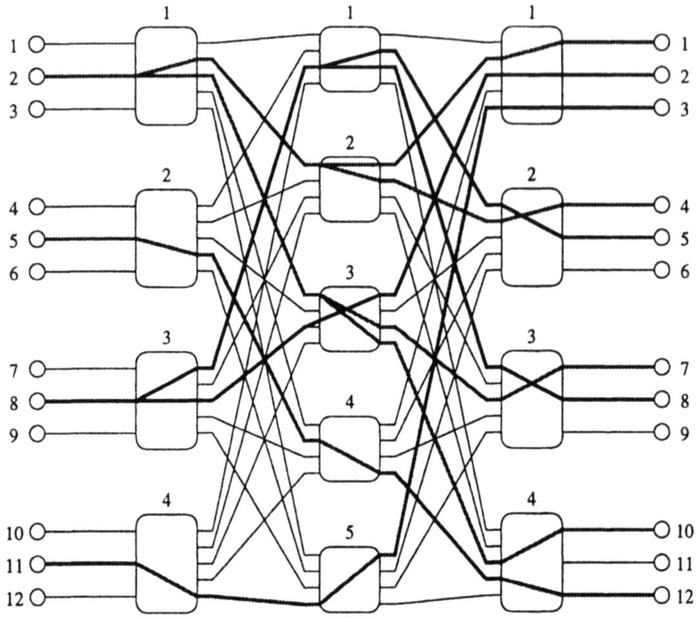

Figure 3: Clos switch $C(3, 4, 5)$. It configures four requests – $(1, (1\,1\,1\,1)$, $(2, (0\,0\,0\,1))$, $(3, (1\,1\,1\,0))$, and $(4, (1\,0\,0\,0))$.

matrix) where D_{ij} is 1 if the i-th request contains the output switch j as its destination, 0 otherwise. The previous 4 requests can be summarized with a request matrix as

$$\mathbf{D} = (D_{ij}) = \begin{pmatrix} 1 & 1 & 1 & 1 \\ 0 & 0 & 0 & 1 \\ 1 & 1 & 1 & 0 \\ 1 & 0 & 0 & 0 \end{pmatrix}$$

Let $\mathbf{1}$ be a unit row vector. Then, $\mathbf{1D}$ is a row vector summing up each column, where it indicates the total number of requests on each output switch so that it is not larger than $\mathbf{1}n$ that is the number of external ports in each output switch as

$$\mathbf{1D} = \begin{pmatrix} 1 \\ 1 \\ 1 \\ 1 \end{pmatrix}^T \begin{pmatrix} 1 & 1 & 1 & 1 \\ 0 & 0 & 0 & 1 \\ 1 & 1 & 1 & 0 \\ 1 & 0 & 0 & 0 \end{pmatrix} = (3\ 2\ 2\ 2) \leq (3\ 3\ 3\ 3).$$

We shall examine the state of the I-interstage links as an $(r \times m)$ binary matrix, $\mathbf{U} = (U_{ij})$ (access matrix), where U_{ij} is 1 if the link between the input switch i and the middle switch j carries no request, 0 otherwise. The network state in Figure 3 is denoted by

$$\mathbf{U} = \begin{pmatrix} 1 & 0 & 0 & 1 & 1 \\ 1 & 1 & 1 & 0 & 1 \\ 0 & 1 & 0 & 1 & 1 \\ 1 & 1 & 1 & 1 & 0 \end{pmatrix}.$$

Let $\mathbf{A_i}$ denote the i-th row of the matrix $\mathbf{A}$. For example, $\mathbf{U_4}$ is a row vector (1 1 1 1 0), which indicates the middle switches that can be utilized for a request of $(4, Y)$.

We also can specify the configuration of the O-interstage links as an $(m \times r)$ binary matrix, $\mathbf{Q} = (Q_{ij})$ (destination matrix), where Q_{ij} is 1 if the link between the middle switch i and the output switch j is occupied by some request, and 0 if the link carry no request. Likewise, $\mathbf{1Q}$ is a row vector summing up each column, where it indicates the total number of requests on each output switch so that it is not larger than n. Note that a Clos network expresses a relation of $\mathbf{1D} = \mathbf{1Q} \leq \mathbf{1}n$ in a stable state after routing all compatible request. For example, Figure 3 can be denoted by

$$\mathbf{1Q} = \begin{pmatrix} 1 \\ 1 \\ 1 \\ 1 \\ 1 \end{pmatrix}^T \begin{pmatrix} 0 & 1 & 1 & 0 \\ 1 & 1 & 0 & 0 \\ 1 & 0 & 1 & 1 \\ 0 & 0 & 0 & 1 \\ 1 & 0 & 0 & 0 \end{pmatrix} = (3\ 2\ 2\ 2) \leq (3\ 3\ 3\ 3).$$

The complement of $\mathbf{Q}$, $\mathbf{V} = \overline{\mathbf{Q}}$ (available matrix), indicates the set of O-interstage links which are currently available for other requests. The available matrix $\mathbf{V}$ for the previous example is referred as

$$\mathbf{V} = \begin{pmatrix} 1 & 0 & 0 & 1 \\ 0 & 0 & 1 & 1 \\ 0 & 1 & 0 & 0 \\ 1 & 1 & 1 & 0 \\ 0 & 1 & 1 & 1 \end{pmatrix}.$$

Consider a new multicast request (x, Y). Beside of the port compatibility, the request will be compatible to the current configuration of the network if $Y + \mathbf{1Q} \leq \mathbf{1}n$, meaning that the total number of requests in each output switch is not greater than the number of its external ports. The question is then raised which middle switch modules will be used for routing the new request, or which rows in the binary matrix $\mathbf{V}$ covers the vector Y. For the given request, some middle switches are not accessible from the input switch x. For a given vector $\mathbf{a}$ of size n, let us define an expanding function δ which produce an $n \times n$ diagonal matrix $\delta(\mathbf{a}) = (\delta_{ij})$ as

$$\delta_{ij} = \begin{cases} 1 & \text{if } i = j \text{ and } \mathbf{a}_i = 1 \\ 0 & \text{otherwise} \end{cases}$$

Then, $\delta(\mathbf{U_x})\mathbf{V}$ indicates the set of middle switches and O-interstage links that are not only available for the destination but also accessible from the input switch. In addition, An output switch that is not in the destination set Y has no effect in finding the route for the request. Such a output switch can be taken out by $\mathbf{V}\ \delta(Y)$ As a whole, $\mathbf{V}^r = \delta(\mathbf{U_x})\ \mathbf{V}\ \delta(Y)$ indicates the set of available O-interstage links effective to the given request. For example, assume we have a new request $r = (4, (0\,1\,1\,1))$ on the network in Figure 3. The effective available matrix $\mathbf{V^r}$ is defined by

$$\mathbf{V^r} = \delta(\mathbf{U_4})\ \mathbf{V}\ \delta(Y) = \delta(1\,1\,1\,1\,0)\ \mathbf{V}\ \delta(0\,1\,1\,1) = \begin{pmatrix} 0 & 0 & 0 & 1 \\ 0 & 0 & 1 & 1 \\ 0 & 1 & 0 & 0 \\ 0 & 1 & 1 & 0 \\ 0 & 0 & 0 & 0 \end{pmatrix}.$$

We can route the request through the middle switch 1, 2 and 3 or through the middle switches 1 and 4. The question is to minimize the number of middle switches for covering the request. The request-covering generalizes an well-known NP-complete set-covering problem and is therefore NP-hard [3]. We can find a near-optimal cover of the request by applying a simple greedy algorithm. The greedy method picks a middle switch that covers the most remaining unsatisfied destinations in each iteration.

Algorithm 1. (Greedy Request Cover.)

Input: $r = (x, Y)$ (A new request)
$\mathbf{V^r}$ (Effective Available matrix)
Output: C (Set of middle switches for routing r)
Method:
$R \leftarrow Y$
$C \leftarrow \emptyset$
while $R \neq \mathbf{0}$
do select a row m_i that maximizes $c(i) = \mathbf{V^r}(i)R^T$
$R \leftarrow R - \mathbf{V^r}(i)$

$C \leftarrow C \cup \{m_i\}$

return C

In Algorithm 1, the set R contains the set of remaining output switches to be covered, and the set C contains the set of middle switches to cover the output switches. In a greedy approach, $\mathbf{V^r}(i)$ is chosen to cover as many uncovered output switches as possible. The middle switch i is inserted into C and the output switches in $\mathbf{V^r}(i)$ are eliminated from Q.

Let C^* be an optimal set of middle switches covering a request (x, Y). The greedy algorithm returns $|C^*|(\ln|Y| + 1)$ middle switches at the worst case. The proof of this ratio bound remains to the reader. We assumed a general set-cover approach to obtain the ratio bound, but we can get a better result if we observe special properties in the Clos network.

Let us assume each output switch in $\mathcal{C}(n, r, m)$ carries at most n' connections for $1 \leq n' \leq n$, and select any x rows in the destination matrix $\mathcal{M}$ for $x \geq n'$. Then, the intersection of them becomes zeroes as

$$\cap_{i \in x} \mathcal{M}_i = 0. \tag{1}$$

The above formula describes a property of the destination matrix in which the number of middle switches for covering a request is bounded by the number of connection paths in each output switch as well as the degree of fanout of the request. The following lemma will further investigate the property of the destination matrix.

Lemma 1. For a $\mathcal{C}(n, r, m)$ and a request (x, Y), assume that each output switch in Y contains at most n' connections. If any k among m' middle switches cannot cover the request, then $m' \leq \lfloor n' r^{1/k} \rfloor$.

Proof. If $k > n'$, it is clear that the intersection of the k rows in $\mathcal{M}$ is not a null by the formula 1, so that we consider only the case $k \leq \min(n', r)$.

Suppose $\mathcal{M}'$ be these m' middle switches by extracting proper rows from $\mathcal{M}$ and $m' = |\mathcal{M}'|$. Let $c_1(i)$ be the number of ones in the i-th row, which is denoted by $\mathcal{M}' Y^T$. We know that each column contains at most n' ones and there are r columns, so we obtain

$$\sum_{i=1}^{m'} c_1(i) \leq n' r. \tag{2}$$

Let $c_1 = \min_i \{c_1(i)\}$ and we got

$$m' c_1 \leq \sum_{i=1}^{m'} c_1(i) \leq n' r. \tag{3}$$

If $c_1 = 0$, then we can satisfy the request with the middle switch that makes $c_1 = 0$, which is a contradiction to the assumption. Henceforth,

$$m' \leq \frac{n'r}{c_1}. \tag{4}$$

Without loss of generality, let us assume $\mathcal{M}'_1$ have the minimum c_1. Construct a new destination matrix $\mathcal{M}'_1$ by taking only the c_1 columns from $\mathcal{M}'$. In other words, we partially satisfy the request using the middle switch 1 but c_1 destinations are left to be routed.

Generally speaking, assume that the request needs to be routed to c_{j-1} destinations and we have $m' \times c_{j-1}$ destination matrix $\mathcal{M}^{j-1}$, where $1 \leq j < k$. Let $c_j(i)$ be the number of ones in i-th row of $\mathcal{M}^{j-1}$ and we have

$$\sum_{i=1}^{m'} c_j(i) \leq n'c_{j-1}. \tag{5}$$

Let $c_j = \min_i\{c_j(i)\}$ and we obtain

$$m'c_j \leq \sum_{i=1}^{m'} c_j(i) \leq n'c_{j-1} \tag{6}$$

Note that $c_k \neq 0$ because of the assumption that the intersection of any k rows in $\mathcal{M}'$ is not a null. Therefore, we get that

$$m' \leq \frac{n'c_{j-1}}{c_k}. \tag{7}$$

After performing the above operation $k-1$ times, we have an $m' \times c_{k-1}$ destination matrix $\mathcal{M}^{k-1}$, where each column has at most n' ones, so

$$m' \leq n'c_{k-1}. \tag{8}$$

From 4, 7, and 8, we have

$$m' \leq \min\left\{\frac{n'r}{c_1}, \cdots, \frac{n'c_{j-1}}{c_j}, \cdots, n'c_{k-1}\right\}. \tag{9}$$

The upper bound of 9 can be obtained from the geometric means as

$$\begin{aligned} m' &\leq \min\left\{\frac{n'r}{c_1}, \cdots, \frac{n'c_{j-1}}{c_j}, \cdots, n'c_{k-1}\right\} \\ &\leq \left(\frac{n'r}{c_1} \cdots \frac{n'c_{j-1}}{c_j} \cdots n'c_{k-1}\right)^{1/k} \\ &\leq n'r^{1/k} \end{aligned}$$

Clearly, m' is an integer so that $m' \leq \lfloor n'r^{1/k} \rfloor$. □

The contrapositive of Lemma 1 gives a very useful result – *If each output switch contains at most n' connections and there are more than $\lfloor n' r^{1/k} \rfloor$ available middle switches for an input switch x, then there is a set of k middle switches that can satisfy any multicast request (x, Y).* Supposed that we have a compatible request (x, Y) of $|Y| = r$ and each output switch and the input switch x contain at most $(n-1)$ connections because of the compatibility. The above statement instructs the routing algorithm to use no more than k middle switches to satisfy an multicast request. If all previous request followed this guideline, there are $(n-1)k$ unavailabe middle switches for the new requst at the worst case.

Theorem 1. A $C(n, r, m)$ is nonblocking for a multicast request if

$$m > \min_{1 \le k \le \min(n-1, r)} \left\{ (n-1)k + \lfloor (n-1) r^{1/k} \rfloor \right\}.$$

Proof. Supposed that any request can be routed using no more than k middle switches. It is not necessary to use more than r middle switches to route a request whose degree of fanout is not larger than r. In addition, each output switch contains at most $(n-1)$ connection so that the set of any $(n-1)$ middle switches can satisfy a request if they carry no connection from the input switch based on Eq 1. Therefore, we obtain $1 \le k \le \min(n-1, r)$. For a given input switch which has at most $(n-1)$ connections plus the new request, the switch can have at most $(n-1)k$ busy I-interstage links. We can route the new multicast request using at most k middle switches if we have $\lfloor (n-1) r^{1/k} \rfloor$ middle switches in addition to the $(n-1)k$ middle switches which have connections to the input switch. □

We now deal with the optimal k for minimizing $(n-1)k + \lfloor (n-1) r^{1/k} \rfloor$, where n and r are constants which are determined by the choice of a Clos network. It is very difficult to find its symbolic minimum. To observe a basic property of the formula, let $f(k) = k + r^{1/k}$ and then we obtain $(n-1)k + \lfloor (n-1) r^{1/k} \rfloor \le (n-1) f(k)$. Figure 4 illustrates the behavior of the function $f(k)$ for $r = 20$. If we select $k = \log r$, then $r^{1/k} = r^{1/\log r} = 2$. Therefore, $m > (n-1)(\log r + 2)$ is sufficient for noblocking for a multicast. Note that the routing algorithm is the same as the greedy scheme because finding the minumum $\min_i \{c_j(i)\}$ in Lemma 1 is a dual of finding the maximum of $\mathbf{V^r}(i) R^T$ in Algorithm 1.

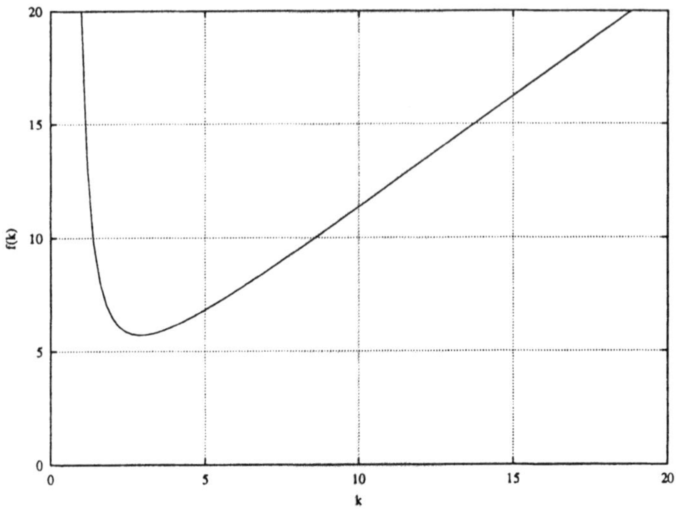

Figure 4: $f(k) = k + r^{1/k}$ for $r = 20$.

2 Multirate Multicast Switches

In the previous section, we assumed a traditional telephone network, where a physical path between a caller and a recipient is established during the entire conversation. In this *circuit switching (CS)* technique, the path is dedicated to the single connection and provides a reliable quality of service until the parties complete the conversation. Advances in digital technology have made it possible for several telephones and data terminals to share a link simultaneously as illustrated in Figure 5. An end-system formats its data into packets that contains meta-data to identify the source and destination and to control the characteristic of the data packets. The end-system transmits packets into a dedicated link. The packets are then transmitted in a shared link. Statistical multiplexers were emploited to improve the effi-

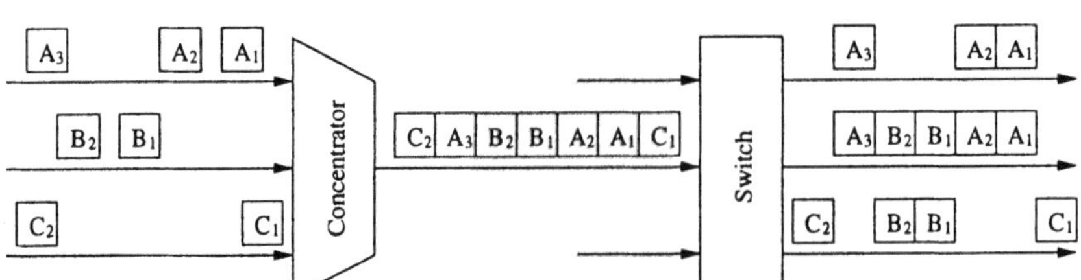

Figure 5: Statistical multiplexing of multicast requests

ciency of the shared communication lines. An asynchronous transfer mode (ATM) is an example of the statistical multiplexing, whose main goal is to integrate a broad range of communication applications, including telephone services, bursty data traffic, video/audio streams and so forth. They demand different criteria for service quality.

A *multirate* models the maximum bandwidth requirement, in which each connection takes a portion of the link capacity. One or more more virtual channels are established between two terminating systems before their conversation, but not a dedcated physical link. While seting up the virtual channels, a terminating system negotiates parameters of service quality including a peak bandwidth, packet loss ratio and delay allowance.

A multicast request in the multirate environment is defined as a triple (x, Y, ω), where x and Y are same as in the single rate, and ω is a normalized bandwidth requirement(weight) of the connection request. To take the advantage of high-end technology and to reduce the cost of a switching system, the internal link of the switch runs faster than the external link. Let β be the normalized capacity of the external link to the internal links. The reciprocal of β is called the *speed advantage* [15], which is utilized to optimize the internal structure of a switching system. The weight of each request (ω) belongs to a closed interval $[b, B]$, i.e. $0 \leq b \leq \omega \leq B \leq \beta$ (continuous bandwidth) [1]. In many networks, communication channels are classified on bandwidth and other service criteria to simplify hardware devices and resource management. A weight of a request in this case is chosen from a given finite set $\{b, 2b, \cdots, \lambda b = B\}$, where λ is a positive integer (discrete bandwidth).

If an internal link blocks a connection request of weight ω, then the link already carries more than $1 - \omega$ weights. Furthermore, the blocking link cannot carry weight less than b because b is the minimum weight of all connection requests. Thus, internal blocking links should carry weights $s(\omega)$ as

$$s(\omega) = \max\{1 - \omega + \epsilon, b\},$$

where $\epsilon \approx 0$ for a continuous bandwidth case and $\epsilon = b$ for a discrete bandwidth case.

Let us define a *connection vector* as $\mathbf{I} = \omega Y$ for a request (x, Y, ω). The connection vectors of four requests shown in Figure 6 are $\mathbf{I}_1 = (.2\ .2\ .2\ .2)$, $\mathbf{I}_2 = (0\ 0\ .3\ 0)$, $\mathbf{I}_3 = (.4\ .4\ .4\ .4)$ and $\mathbf{I}_4 = (0\ .5\ 0\ 0)$. We need some binary operators on the vectors. For two given vectors, $\mathbf{x} = (x_1\ x_2 \cdots x_n)$

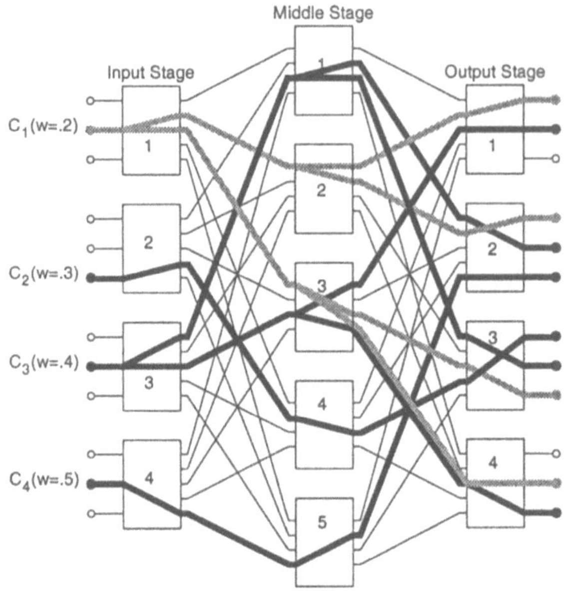

Figure 6: $\mathcal{C}(3,4,5)$ Clos Network. Connection requests are $C_1 = (1,(1\ 1\ 1\ 1),.2)$, $C_2 = (2,(0\ 0\ 1\ 0),.3)$, $C_3 = (3,(1\ 1\ 1\ 1),.4)$ and $C_4 = (4,(0\ 1\ 0\ 0),.5)$.

and $\mathbf{y} = (y_1\ y_2 \cdots y_n)$ of length n,

$$\mathbf{x} \leq \mathbf{y} \text{ iff } x_i \leq y_i \text{ for all } i = 1,2,\cdots,n,$$
$$\mathbf{x} < \mathbf{y} \text{ iff } \mathbf{x} \leq \mathbf{y} \text{ but } \mathbf{x} \neq \mathbf{y}.$$

We also can define $\geq$ and $>$ operators in the similar manner. For example, if $\mathbf{x} = (.3\ .2\ .5)$, $\mathbf{y} = (.4\ .3\ .9)$ and $\mathbf{z} = (.4\ .5\ .3)$ then $\mathbf{x} \leq \mathbf{y}$ but $\mathbf{x} \not\leq \mathbf{z}$ because $x_3 \not\leq z_3$. The k-th element of a connection vector $\mathbf{I}$, $\mathbf{I}(k)$ denotes the weight of the request on output switch k so that the sum of connection vectors, $\sum_{\text{all } i} \mathbf{I}_i$, represents the sum of weights loaded on each output switch. In the previous example, $\sum \mathbf{I}_i = (.6\ 1.1\ .9\ .6)$ and thus the loads of output switches 1 and 4 are .6, the load of output switch 2 is 1.1, and the load of output switch 3 is .9.

We can describe the configuration of O-interstage links defining *destination vector* $\mathbf{M}_j$ for each middle switch j. The k-th element of $\mathbf{M}_j$, $\mathbf{M}_j(k)$, is the sum of weights loaded on the link between middle switch j and output switch k. In Figure 6, for example, $\mathbf{M}_1 = (0\ .4\ .4\ 0)$, $\mathbf{M}_2 = (.2\ .2\ 0\ 0)$, $\mathbf{M}_3 = (.4\ 0\ .2\ .6)$, $\mathbf{M}_4 = (0\ 0\ .3\ 0)$ and $\mathbf{M}_5 = (0\ .5\ 0\ 0)$. Note that the link between middle switch 3 and output switch 4 is shared by 2 requests. The sum of destination vectors, $\sum_{\text{all } j} \mathbf{M}_j$, represents the sum of weights loaded on each output switch as the connection vectors. Because each output switch has n output ports, its total load is at most βn. It is easy to obtain the relationship between connection vectors and destination vectors after routing all requests as $\sum_{\text{all } i} \mathbf{I}_i = \sum_{\text{all } j} \mathbf{M}_j \leq \beta \mathbf{n}$.

Note that the configuration of middle stage switches is described as $m \times r$ matrix, called a *destination matrix*. For instance, the network shown in Figure 6 has the destination matrix as

$$\mathbf{M} = \begin{bmatrix} 0 & .4 & .4 & 0 \\ .2 & .2 & 0 & 0 \\ .4 & 0 & .2 & .6 \\ 0 & 0 & .3 & 0 \\ 0 & .5 & 0 & 0 \end{bmatrix}$$

The sum of row i in the destination matrix $\mathbf{M}$ expresses the total loads on middle switch i and the sum of column j describes the total loads on output switch j. For two vectors, $\mathbf{x} = (x_1\ x_2 \cdots x_n)$ and $\mathbf{y} = (y_1\ y_2 \cdots y_n)$, we also need operators min and max defined as follows:

$$\begin{aligned} \mathsf{min}(\mathbf{x}, \mathbf{y}) &= (\min(x_1, y_1)\ \ \min(x_2, y_2) \cdots \min(x_n, y_n)) \\ \mathsf{max}(\mathbf{x}, \mathbf{y}) &= (\max(x_1, y_1)\ \ \max(x_2, y_2) \cdots \max(x_n, y_n)) \end{aligned}$$

We can generally extend the operators for any number of vectors, $\mathbf{x}_j = (x_{j,1}\ x_{j,2} \cdots x_{j,n})$ for $1 \leq j \leq k$,

$$\begin{aligned} \mathop{\mathsf{min}}_{1 \leq j < k} (\mathbf{x}_j) &= (\min_{1 \leq j \leq k} (x_{j,1})\ \ \min_{1 \leq j \leq k} (x_{j,2}) \cdots \min_{1 \leq j \leq k} (x_{j,n})) \\ \mathop{\mathsf{max}}_{1 \leq j \leq k} (\mathbf{x}_j) &= (\max_{1 \leq j \leq k} (x_{j,1})\ \ \max_{1 \leq j \leq k} (x_{j,2}) \ldots, \max_{1 \leq j \leq k} (x_{j,n})) \end{aligned}$$

For the previous destination vectors, $\mathsf{min}_{1 \leq j \leq 5}(\mathbf{M}_j) = (0\,0\,0\,0)$ and $\mathsf{max}_{1 \leq j \leq 5}(\mathbf{M}_j) = (.4\ .5\ .4\ .6)$.

2.1 Nonblocking Multirate Multicast

Let us consider a multirate Clos network $C(n, r, m)$ and a multicast request (x, Y, ω). How many middle switches are needed for routing the request? It is certain that no more than r middle switches are needed, because r is the maximum degree of fanout. At the worst case, each middle switch can provide a path to a single output switch without replicating the message at the middle switch. In addition to the maximum fanout r, another bound can be found by realizing the characteristic of weight distribution on the O-interstage link as shown in the following lemma.

Lemma 2. If a multicast request (x, Y, ω) is compatible to $C(n,r,m)$, the min of more than $\lceil \frac{\beta n-\omega}{s(\omega)} \rceil$ destination vectors $\mathbf{M}_i$'s is less than $\mathsf{s}(\omega)$.

Proof. This is proven by contradiction. Suppose that

$$m' > \lceil \frac{\beta n - \omega}{s(\omega)} \rceil \quad \text{and} \quad \mathbf{M}_{\min} = \min_{1 \le i \le m'} (\mathbf{M}_{j_i}) \not< \mathsf{s}(\omega). \tag{10}$$

Consider $m' \times r$ destination matrix $\mathbf{M}'$ consisting of the destination vector $\mathbf{M}_{j_i}$ for $1 \le i \le m'$ in question. From the assumption, the minimum vector $\mathbf{M}_{\min}$ contains at least one element whose value is no less than $s(\omega)$. Thus, the destination matrix $\mathbf{M}'$ has at least one column, say column h, all of whose values are no less than $s(\omega)$. It means that output switch h has m' O-interstage links, each of which weighs no less than $s(\omega)$. Due to compatibility, the total weights on the output switch should be less than or equal to $\beta n - \omega$. Therefore, $m's(\omega) \le \beta n - \omega$ is obtained. With (10), we obtain $\lceil \frac{\beta n-\omega}{s(\omega)} \rceil < m' \le \frac{\beta n-\omega}{s(\omega)}$. It is a contradiction because of $\lceil x \rceil \ge x$ for any x. □

A routing algorithm scan *available* middle switches whose I-interstage links to the input switch x carry weights less than $s(\omega)$ to route a new multicast request (x, Y, ω). Should a middle switch j be found, whose *residual capacity vector* $\mathbf{1} - \mathbf{M}_j$ is greater than or equal to the connection vector ωY, the request can be routed through the middle switch. If there is no such middle switch, try to find two middle switches j_1 and j_2 whose *maximum capacity vector* $\mathbf{1} - \min(\mathbf{M}_{j_1}, \mathbf{M}_{j_2})$ is not less than the connection vector to satisfy the request through these two middle switches. Generally, when it is possible to find k middle switches whose maximum capacity vector $\mathbf{1} - \min_{1 \le i \le k}(\mathbf{M}_{j_i})$ is greater than or equal to the connection vector, we can route the request through these k available middle switches. The value k does not need to be greater than r since it is the maximum degree of fanout for any requests. In addition, Lemma 2 instructs that any multicast requests can be satisfied with no more than $\lceil \frac{\beta n-\omega}{s(\omega)} \rceil$ middle switches, which are maximized when $\omega \to B$ and $\epsilon \to 0$. The goal of our routing scheme is to find an optimal k so that any compatible requests are routed using no more than k middle switches. Let m' be the maximumber of middle switches blocking from an input switches. The relationship between m' and k is revealed in the following lemma.

Lemma 3 (number of middle switches blocking from an input switch). For a multirate Clos network $C(n,r,m)$ whose external links operate on β for $0 < \beta \leq 1$, if each connection request replicates its message to at most k middle switches in any configuration, the number of middle switches blocking a new connection request (x, Y, ω) from the input switch is no more than $\lfloor \frac{\beta n - \omega}{s(\omega)} k \rfloor$.

Proof. The sum of weights coming to input switch x, excluding the new request, is at most $\beta n - \omega$ by compatibility. Because each request replicates its message at most k times in the input switch, the sum of weights from the input switch to the middle stage is no more than $(\beta n - \omega)k$. Let m' be the number of links between the input switch and middle switches which are blocking the connection request, i.e. the number of links carrying at least $s(\omega)$. Then, $m's(\omega) \leq (\beta n - \omega)k$. For an integer m', $m' \leq \lfloor \frac{\beta n - \omega}{s(\omega)} k \rfloor$. □

We can restrict the number k to a small number in order to keep the number of blocking I-interstage links small. However, it doen't necessarily enhance the routability of the network because it results in that middle switches are needed be fanned out to a large number of output switches. Let us take a look to the following lemma that shows the relation between k and the number of available O-interstage links.

Lemma 4 (number of middle switches blocking to output stage). For a multirate 3-stage Clos network $C(n,r,m)$ whose external links operate on β for $0 < \beta \leq 1$, assume a new multicast request (x, Y, ω) is compatible to the existing configuration. When given $m'' > \lfloor \frac{\beta n - \omega}{s(\omega)} r^{1/k} \rfloor$ accessible middle switches, there exist some k middle switches among m'' such that they satisfy the new connection request.

Proof. This lemma can be proven by its contrapositive—If any k middle switches among m'' block the request to output stage, then $m'' \leq \lfloor \frac{\beta n - \omega}{s(\omega)} r^{1/k} \rfloor$. Assume the request has the degree of fanout d, and the input switch of x has m'' accessible middle switches. Consider the corresponding O-interstage as an $m'' \times d$ destination matrix as illustrated in Table 1(a). The sum of each column is less than or equal to $\beta n - \omega$ because of the compatibility, and thus the total sum of the matrix is at most $(\beta n - \omega)d$. For each row j, let $c_1(j)$ be the number of elements whose values are at least $s(\omega)$, or the number of links connected to the output switch j but which cannot be used to route the given request of weight ω. Then, the sum of the j-th row is greater than or equal to $c_1(j)s(\omega)$. Therefore, $\sum_{j=1}^{m''} c_1(j)s(\omega) \leq (\beta n - \omega)d$. Let c_1 be the minimum among $c_1(j)$, and thus $m''c_1 s(\omega) \leq (\beta n - \omega)d$. Note

ω_{11}	$\cdots$	ω_{1d}
ω_{21}	$\cdots$	ω_{2d}
$\vdots$	$\ddots$	$\vdots$
$\omega_{m''1}$	$\cdots$	$\omega_{m''d}$

(a) Initial destination matrix

ω_{11}	$\cdots$	ω_{1c_1}
$\min(\omega_{11}, \omega_{21})$	$\cdots$	$\min(\omega_{1c_1}, \omega_{2d})$
$\vdots$	$\ddots$	$\vdots$
$\min(\omega_{11}, \omega_{m''1})$	$\cdots$	$\min(\omega_{1c_1}, \omega_{m''d})$

(b) Contracted destination matrix

Table 1: Destination matrices

that c_1 should not be zero. If it is, the single middle switch can provide the request a routing path. It means that the network does not block the request. This is a contradiction to the assumption. So,

$$m'' \leq \frac{\beta n - \omega}{s(\omega)} \frac{d}{c_1}. \tag{11}$$

Without loss of generality, let the destination vector of middle switch 1 have the minimum, c_1. By using middle switch 1, $d - c_1$ destination output switches among d are satisfied. Only c_1 output switches are still blocked after using the first middle switch. The number of columns in the destination matrix is reduced to c_1 by discarding $d - c_1$ columns which were already satisfied by using the first middle switch. The contracted matrix is denoted by $\mathbf{M}^*$. Then, replace each row vector with the minimum of the row vector itself and the first row vector and denote it as $\mathbf{M}^1$, which is $\forall_{1 \leq j \leq m''} : \mathbf{M}^1_j = \min(\mathbf{M}^*_1, \mathbf{M}^*_j)$ as shown in Table 1(b). In general, for any $m'' \times c_{i-1}$ destination matrix $\mathbf{M}^{i-1}$ for $i < k$, it is said that there exist only c_{i-1} output switches which are needed to be satisfied. Let $c_i(j)$ be the number of elements whose values are not less than $s(\omega)$, i.e. the number of links which still block the new connection request. Similarly, $\sum_{j=1}^{m''} c_i(j)s(\omega) \leq (\beta n - \omega)c_{i-1}$. Let c_i be $\min_{1 \leq j \leq m''}\{c_i(j)\}$, and $m'' c_i s(\omega) \leq \sum_{j=1}^{m''} c_i(j)s(\omega) \leq (\beta n - \omega)c_{i-1}$. Note also that c_i shouldn't be zero. If it were, i middle switches for $i < k$ could be used to satisfy the request. This is a contradiction to the assumption. It is certain that

$$m'' \leq \frac{\beta n - \omega}{s(\omega)} \frac{c_{i-1}}{c_i}. \tag{12}$$

When $i = k$, each row vector in the destination matrix $\mathbf{M}^{k-1}$ has at least one element that is not less than $s(\omega)$. Thus,

$$m''s(\omega) \leq (\beta n - \omega)c_{k-1} \qquad \text{and} \qquad m'' \leq \frac{\beta n - \omega}{s(\omega)} c_{k-1}. \tag{13}$$

The minimum m'' is found from (11), (12) and (13). A minimum of a sequence is less than or equal to its geometric mean so that

$$m'' \leq \frac{\beta n - \omega}{s(\omega)} d^{1/k},$$

which is maximized at $d = r$. Since m'' is an integer, $m'' \leq \lfloor \frac{\beta n - \omega}{s(\omega)} r^{1/k} \rfloor$. □

In contrast to Lemma 3, Lemma 4 shows that the smaller the value of k, the more middle switches block the request to the output stage, because the algorithm needs to find a fewer middle switches for satisfying the same number of destination output switches. We can combine these two blocking conditions to find the minimum number of middle switches in the multirate environment.

Theorem 2. For a multirate $C(n, r, m)$ whose external links operate on β for $0 < \beta \leq 1$, it is nonblocking for multicast if

$$m > \lim_{\epsilon \to 0} \left(\min_{1 \leq k \leq \min(\lceil \frac{\beta n - B}{s(B)} \rceil, r)} \left(\left\lfloor \frac{\beta n - B}{s(B)} k \right\rfloor + \left\lfloor \frac{\beta n - B}{s(B)} r^{1/k} \right\rfloor \right) \right),$$

where k is an arbitrary constant.

Proof. By Lemmas 2, 3 and 4, the number of middle switches blocking from an input stage is at most $\lfloor \frac{\beta n - \omega}{s(\omega)} k \rfloor$ and the number of middle switches blocking at an output stage is at most $\lfloor \frac{\beta n - \omega}{s(\omega)} r^{1/k} \rfloor$. Moreover, the min of more than $\lceil \frac{\beta n - \omega}{s(\omega)} \rceil$ destination vectors is less than $\mathbf{s}(\omega)$ and the maximum fanout is r. This obtains the maximum number of blocking middle switches when $\omega = B$ and $\epsilon \to 0$. □

We can find an optimal k approximately using the result in the previous section, as finding the minimum of $f(r) = k + r^{1/k}$. Or, we can find the minimum m experimentally for fixed other constants.

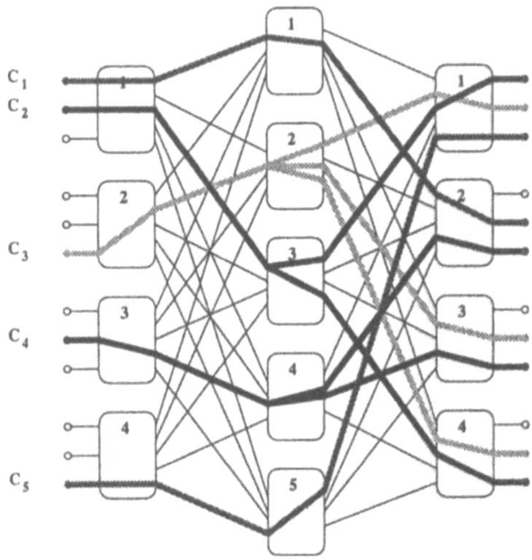

$C_1 = (1, \{2\})$
$C_2 = (1, \{1, 4\})$
$C_3 = (2, \{1, 3, 4\})$
$C_4 = (3, \{2, 3\})$
$C_5 = (4, \{1\})$

Figure 7: Clos switching network $\mathcal{C}(3, 4, 5)$.

3 Blocking Multicast Switches

We have seen the requirements of Clos networks for nonblocking any multicast request that results in a large number of middle switches. In this section, we consider a blocking probability of a multicast request for a given Clos network. We will investigate a blocking model of the lazy fanout algorithm and extend the principle to optimize the blocking property and the complexity of the network [7, 10, 12, 13, 17, 19, 21].

To describe the state of the I-interstage and O-interstage links, let U_x denote the set of *accessible* middle switches whose links to input switch x do not carry any requests. In Figure 7 for example, $U_1 = \{2, 4, 5\}$, $U_2 = \{1, 3, 4, 5\}$, $U_3 = \{1, 2, 3, 5\}$ and $U_4 = \{1, 2, 3, 4\}$. Let V_y denote the set of *available* middle switches whose links to output switch y carry no requests. In the same figure, $V_1 = \{1, 4\}$, $V_2 = \{2, 3, 5\}$, $V_3 = \{1, 3, 5\}$ and $V_4 = \{1, 4, 5\}$. In order to describe the available middle switch set for all output switches, we define V_Y as

$$V_Y = \bigcap_{i=1}^{d} V_{y_i} \text{ for } Y = \{y_1, y_2, \cdots, y_d\}.$$

The lazy fanout algorithm routes a multicast request (x, Y) by using a single middle switch. To route the request successfully, there should exist at least one middle switch in $U_x \cap V_Y$. When $U_x \cap V_Y = \emptyset$, it cannot route the request unless it is allowed to rearrange the existing requests. For instance, consider a new request $C_{\text{new}} = (1, \{2, 4\})$ in Figure 7. Since $U_1 \cap V_{\{2,4\}} = \{2, 4, 5\} \cap \{5\} = \{5\}$, the middle switch 5 is accessible and available for the new request; therefore the request C_{new} can be successfully routed through the middle switch 5 under the lazy-fanout algorithm. For another new request $C'_{\text{new}} = (4, \{2, 4\})$, $U_4 \cap V_{\{2,4\}} = \{1, 2, 3, 4\} \cap \{5\} = \emptyset$, thus the request cannot be routed the request through any middle switch.

The input switch x has at most $n - 1$ requests plus the new compatible request (x, Y). Since each request is allowed to use only one middle switch under the lazy-fanout strategy, among m links from the input switch x are occupied at most $(n-1)$ I-interstage links. In a meanwhile, an output switch y in Y has at most $n-1$ requests due to the compatibility, and consequently has at most $n - 1$ occupied O-interstage links.

The blocking probability of a request with the degree of fanout d under a given network load, denoted by $P_B(d)$, is a function of two probabilities p_1 and p_2, where p_1 is the probability that an I-interstage link is occupied by another request, and p_2 is the probability that an O-interstage link is occupied by other request. All these parameters are related to other independent variables such as traffic patterns, network load and network configurations. Assume that an input port has a probability φ_d of requests for a fanout d, and their blocking probabilities are $P_B(d)$ for $C(n, r, m)$. Then, the *input network utilization*, i.e. the probability of occupancy of the input port, is $a = \sum_d \varphi_d[1 - P_B(d)]$ for $1 \leq d \leq r$. The expected number of occupied input ports is anr. Let $f(d)$ denote the probability density function(p.d.f.) of incoming requests with a fanout of d. Then, $f(d) = \varphi_d / \sum_i^r \varphi_i$. Suppose that requests granted in the switching system are distributed according to a p.d.f. of $g(d)$. This p.d.f can be obtained by

$$g(d) = \frac{f(d)[1 - P_B(d)]}{\sum_k f(k)[1 - P_B(k)]}$$

If a routing algorithm allows a multicast request to use a single middle switch and requests are randomly routed over the m middle stage switches, we obtain $p_1 = an/m$.

The expected number of occupied output ports is $anr \sum_k kg(k) = anr\bar{d}$, where $\bar{d}$ is the average degree of fanout. Let b be the *output network utilization*, i.e. the probability that an output port is occupied. Note that $anr\bar{d} = bnr$. Provided that requests are randomly routed over the mr O-interstage links, $p_2 = bn/m$ is acquired.

Unlike a point-to-point Clos network, the input utilization in a multicast network may differ from the output utilization. It is safe to choose a larger utilization to indicate a general network utilization. Note that $a\bar{d} = b$ for a symmetric network and $\bar{d}$ is no less than 1, so the general network utilization becomes b.

3.1 Blocking Probability of Lazy-Fanout Algorithm

A multicast request (x, Y) is blocked if and only if $U_x \cap V_Y = \emptyset$. Find the probability for when the input switch x has j occupied I-interstage links, or $|U_x| = m - j$. The distribution of busy I-interstage links from an input switch is assumed to be random among m links and independent of the distribution of busy links from other input switches. This is approximated by a binomial (m, p_1) distribution. The number of occupied links is not larger than $n - 1$ under the lazy fanout algorithm.

$$\Pr(x \text{ has } j \text{ occupied links}) = \Pr(|U_x^c| = j \mid |U_x^c| \leq n - 1) \tag{14}$$

$$= \frac{\binom{m}{j} {p_1}^j {q_1}^{m-j}}{\sum_{i=0}^{n-1} \binom{m}{i} {p_1}^i {q_1}^{m-i}} \tag{15}$$

where $q_1 = 1 - p_1$ and $0 \leq j \leq n - 1$.

Consider $m \times |Y|$ O-interstage links for a compatible multicast request (x, Y). Exploring an occupancy problem in an $m \times d$ rectangular matrix of cells makes it possible to explain the property of the interstage links A cell can be occupied by a ball with probability p_2 under the condition that each column has at most $n - 1$ balls. The distribution of balls in a column is assumed to be independent of the distribution in other columns. They are approximated by binomial (m, p_2) distributions. Let E_d be the number of empty rows in the $m \times d$ matrix, and A_j the number of balls in the j-th column. Let $P_d(m, p_2, n, k)$ be the probability of k empty rows, then $\Pr(A_j = h) = P_1(m, p_2, n, m - h)$ for $1 \leq j \leq d$.

The probability of k empty rows can be found by inducing on d. First of all, consider the case that a column has k empty cells, or $m - k$ balls.

$$P_1(m, p_2, n, k) = \Pr(A_1 = m - k \mid A_1 < n) = \frac{\binom{m}{m-k} {p_2}^{m-k} {q_2}^k}{\sum_{i=0}^{n-1} \binom{m}{i} {p_2}^i {q_2}^{m-i}} \tag{16}$$

Given two independent events $E_{d-1} = j$ and $A_d = h$, there are $\binom{m}{h}$ ways to put h balls into m cells of the d-th column, $\binom{j}{j-k}$ ways to put $j - k$ balls

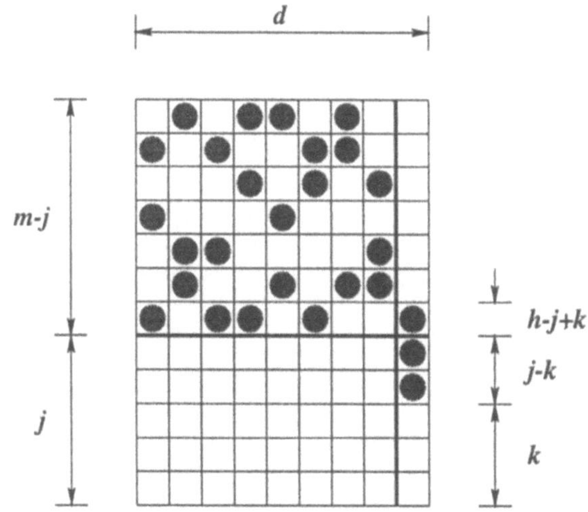

Figure 8: k empty rows in $m \times d$ matrix

from h into j cells which are the empty rows in $m \times (d-1)$ sub-matrix, and $\binom{m-j}{h-j+k}$ ways to put the remaining $h-j+k$ balls into $m-j$ cells as illustrated in Figure 8. Therefore, the conditional probability of k empty rows, $\Pr(E_d = k \mid A_d = h,\ E_{d-1} = j)$, is obtained through

$$\Pr(E_d = k \mid A_d = h,\ E_{d-1} = j) = \frac{\binom{j}{j-k}\binom{m-j}{h-j+k}}{\binom{m}{h}} \tag{17}$$

We are now in the position to develop the recursive function $P_d(m, p_2, n, k)$, or $\Pr(E_d = k)$, which is the probability of k empty rows in an $m \times d$ matrix under the condition that each cell has the occupancy probatility p_2 and a column is occupied by at most $n-1$ balls as follows from (16) and (17):

$$\Pr(E_d{=}k) = \sum_{\text{all } h,j} \Pr(E_d{=}k \mid A_d{=}h,\ E_{d-1}{=}j) \Pr(A_d{=}h) \Pr(E_{d-1}{=}j) \tag{18}$$

$$= \sum_{\text{all } h,j} \frac{\binom{j}{j-k}\binom{m-j}{h-j+k}}{\binom{m}{h}} \frac{\binom{m}{h} {p_2}^h {q_2}^{m-h}}{\sum_{i=0}^{n-1} \binom{m}{i} {p_2}^i {q_2}^{m-i}} P_{d-1}(m, p_2, n, j), \tag{19}$$

where $0 \le h \le \min(n-1, m-k)$ and $k \le j \le \min(m, k+h)$.

A compatible new request (x, Y) will be blocked if $U_x \cap V_Y = \emptyset$, or $V_Y \subseteq V_x^c$. Under the condition that $|U_x^c| = j$ and $|V_Y| = k$, the blocking probability is to choose none of the $m-j$ available middle switches with sample size k

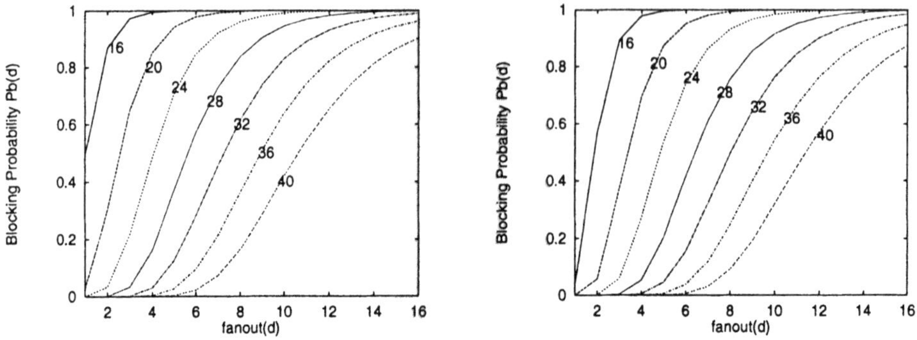

(a) small average degree of fanout (b) large average degree of fanout

Figure 9: Analysis of blocking probabilities versus degree of fanout for multicast Clos networks $C(16,16,m)$ at 80% reference network utilization. The digits on the lines indicate the number of middle stage switches.

and population m in a hyper-geometric $(k, m, m-j)$ distribution.

$$\Pr(V_Y \subseteq V_x^c \mid |U_x^c| = j,\ |V_Y| = k) = \frac{\binom{j}{k}\binom{m-j}{0}}{\binom{m}{k}} = \frac{\binom{j}{k}}{\binom{m}{k}} \tag{20}$$

The blocking probability of a request (x, Y) for $|Y| = d$, $P_B(d)$, is computed from (19) and (20) as

$$P_B(d) = \sum_{\text{all } j,k} \Pr(m-j \text{ accessible}, k \text{ available}, \text{no common middle}) \tag{21}$$

$$= \sum_{\text{all } j,k} \frac{\binom{j}{k}}{\binom{m}{k}} \frac{\binom{m}{j} {p_1}^j {q_1}^{m-j}}{\sum_{i=0}^{n-1} \binom{m}{i} {p_1}^i {q_1}^{m-i}} P_d(m, p_2, n, k), \tag{22}$$

where $p_1 = bn/m\bar{d}$, $q_1 = 1 - p_1$ and $p_2 = bn/m$.

The portion of high degree fanout requests increases, so does the dependencies between the events of $E_{d-1}{=}j$ and $E_d{=}k$. Because a multicast request is routed through only one middle switch under the algorithm no matter which degree of fanout, the balls in the matrix tend to form clusters along the rows as shown in Figure 9(a) and 9(b). This tendency makes other rows to have more empty cells. As a result, the dependent events have a positive covariance. Ignoring the dependency results in overestimating the blocking probability.

The blocking probability $P_B(d)$ is zero if $m > (r+1)(n-1)$, which is consistent with the nonblocking condition in Section 1.2. It can be proved using the following lemma recursively.

Lemma 5. If $m > (r+1)(n-1)$, then $P_d(m, p_2, n, k) = 0$ for $k \leq (r+1-d)(n-1)$.

Proof. When $d = 1$, for the induction base, $k \leq r(n-1)$, and $m - k > n-1$, it contradicts the compatibility. Assume that $P_{d-1}(m, p_2, n, k) = 0$ for $k \leq (r+1-(d-1))(n-1) = (r+2-d)(n-1)$. Find the value of index j from Equation 19 when $k \leq (r+1-d)(n-1)$. The index of the first summation gives $0 \leq h \leq \min(n-1, m-k)$ and thus $k \leq h+k \leq \min(n-1+k, m)$. If $k \leq (r+1-d)(n-1)$ and $m > (r+1)(n-1)$, then $n-1+k < m$ when $d \geq 2$ so that $h+k \leq n-1+k \leq (r+2-d)(n-1)$. From the index of the second summation in the lemma, it is obtained that $k \leq j \leq \min(m, h+k)$ so $k \leq j \leq (r+2-d)(n-1)$. By the assumption, $P_{d-1}(m, p_2, n, j) = 0$ for all j, where $j \leq (r+2-d)(n-1)$. Therefore, $P_d(m, p_2, n, k)$ is also equal to zero. □

3.2 Split Routing Algorithm

A problem in the lazy fanout algorithm is that the blocking probability for a large fanout request is close to 1 at the given reference network utilization, if m is not large. Therefore, the large fanout request will rarely get routed through the switching system or has to wait until the network utilization becomes low enough to route the request. Although large fanout requests are infrequent, it is not desirable for large fanout requests to have the high blocking probabilities. To meliorate this problem, we need to develop another routing algorithm providing a durable blocking probability.

A *split routing* allows more than one middle switches to be used for a large fanout request. It replicates the request at the input switch and restricts the replicated requests to having the same number of destinations, except possibly for the last one. This algorithm requires a few more middle switches in a nonblocking fashion than the CS nonblocking network algorithm [18]. However, estimating the blocking probability on the unsized configuration, the results show that the probability becomes close to zero when the number of middle switches is far less than the nonblocking condition. In addition, it is a merit of the simple routing scheme that it can be parallelized and pipelined to enhance time complexity and make it easy to implement in VLSI hardware.

n	r	m		
		pre-fanout	lazy fanout	optimal split
16	16	271	256	124(4)
32	32	1055	1024	376(8)
64	64	4159	4096	1016(8)

Table 2: Required number of middle switches of nonblocking networks $\mathcal{C}(n,r,m)$ for the split routing algorithm. Numbers enclosed by parenthesis are the split sizes

Theorem 3. $\mathcal{C}(n,r,m)$ is nonblocking under the split routing strategy of split size s if $m \geq \lceil r/s \rceil n + sn - s$

Proof. Divide N elements into groups of size s (possibly the last group has less than s) and there are $\lceil N/s \rceil$ groups. Suppose that a request (x, Y) for $|Y| = d$ is compatible with the existing network. The input switch x can already carry $n-1$ requests, each of which can already be connected to at most $\lceil r/s \rceil$ distinct middle switches. Thus, at most $\lceil r/s \rceil(n-1)$ middle switches can block the new request from the input switch. Partition Y into $Y_1, Y_2, \cdots$ such that $|Y_i| = s$ but the size of the last subset may be less than s if $|Y|$ is not divisible by s. Consider one of the subsets Y_i. An output switch $y \in Y_i$ has at most $n-1$ calls and thus $n-1$ O-interstage busy links, or blocking middle switches. In the worst case, the blocking middle switches are disjointed so that the number of the blocking middle switches is at most $s(n-1)$. The new request is partitioned into $\lceil d/s \rceil$ sub-requests. If there are no less than $\lceil r/s \rceil(n-1) + s(n-1) + \lceil d/s \rceil$ middle switches, then $\lceil d/s \rceil$ middle switches do not block the request from both the input switch and the set of output switches, which is maximized at $d = r$. □

Note that, for $s = 1$, Theorem 3 is reduced to the nonblocking condition for the pre-fanout algorithm and, for $s = r$, it is reduced to the condition for the lazy fanout algorithm. For a three-stage $\mathcal{C}(n,r,m)$ with $s = c\sqrt{r}$ for some constant c, the multicast Clos network is nonblocking under the split strategy if $m = O(n\sqrt{r})$ and thus has $O(n^2 r^{3/2})$ crosspoints, which is an advantage over the previous strategies. Table 2 exhibits the required number of middle switches to support nonblocking paths depending on the routing strategies. The optimal number of middle switches is obtained by choosing proper split sizes.

Although the nonblocking theorem assumed to use $\lceil d/s \rceil$ middle switches to route a request of fanout d, it is not mandatory in this algorithm, furthermore, routing requests through less middle switches, if possible, gains the effect of packing without causing harm to other requests. To analyze the blocking behavior of the split routing algorithm, let N be the total number of requests granted into the system and N_s the number of splitted sub-requests. $\bar{d}_s$ denote the expected degree of fanout of the splitted sub-requests. Then,

$$N_s \leq N \sum_{d=1}^{r} \lceil d/s \rceil g(d) \quad \text{and} \quad \bar{d}_s = \frac{N\bar{d}}{N_s} \geq \frac{\bar{d}}{\sum_{d=1}^{r} \lceil d/s \rceil g(d)}$$

An I-interstage link in this split routing algorithm has the probability to be occupied as $p_1 = bn/m\bar{d}_s$, which is greater than when the lazy-fanout algorithm is applied. Nevertheless, the O-interstage link has the same occupancy probability $p_2 = bn/m$ as the other algorithm does, because the occupancy of the O-interstage link is independent of routing algorithms. The probability of j occupied links is obtained as

$$\Pr(x \text{ has } j \text{ occupied links}) = \frac{\binom{m}{j} {p_1}^j {q_1}^{m-j}}{\sum_{i=0}^{\min(m,(n-1)\lceil r/s \rceil)} \binom{m}{i} {p_1}^i {q_1}^{m-i}}$$

The blocking probability of the splitted request of fanout d_s, $P_s(d_s)$, is computed by using approaches similar to those in the previous section.

$$P_s(d_s) = \frac{\sum_{\text{all } j} \sum_{\text{all } k} p_1^j q_1^{m-j} P_{d_s}(m, p_2, n, k) \binom{m-k}{j-k}}{\sum_{\text{all } i} \binom{m}{i} {p_1}^i {q_1}^{m-i}},$$

where Y_s is the set of output switches in a splitted request $0 \leq i \leq \min(m, (n-1)\lceil r/s \rceil)$, $0 \leq j \leq \min(m, (n-1)\lceil r/s \rceil)$ and $0 \leq k \leq \min(j, m)$.

Splitting a request of fanout d into requests of size s makes $\lfloor d/s \rfloor$ requests of size s and one request of size $d - s\lfloor d/s \rfloor$. For example, if the split size is 4 and the fanout of a request is 15, then $\lfloor 15/4 \rfloor = 3$ splitted requests of size 4 and one splitted request of size $15 - 4\lfloor 15/4 \rfloor = 3$. If d is divisible by s, then there will be only $\lfloor d/s \rfloor$ splitted requests of size s. The same argument can be applied by assigning $P_s(0) = 0$. A request of fanout d is blocked if any splitted requests are blocked. Thus, its blocking probability is,

$$P_B(d) = 1 - (1 - P_s(s))^{\lfloor d/s \rfloor}(1 - P_s(d - s\lfloor d/s \rfloor)),$$

where $B(x)$ is the event to block a splitted request of fanout x and $N(x)$ the event not to block a splitted request of fanout x.

Algorithm 2. (Split Routing).

Input: $r = (x, Y)$ (A new request)
Output: return true if routed, otherwise false
Method:

1 $n \leftarrow \lceil |Y| / s \rceil$
2 partition Y into $Y_1, Y_2, \cdots, Y_n$
3 **for** $i \leftarrow 1$ **to** n
4 $\quad w \leftarrow (\bigcap_{\forall y \in Y_i} V_y) \cap U_x$
5 **if** $w = \emptyset$ **then return false**
6 $m_i \leftarrow$ select arbitrarily one among w
7 **for** $i \leftarrow 1$ **to** n
8 $\quad U_x \leftarrow U_x \setminus \{m_i\}$
9 $\quad V_y \leftarrow V_y \setminus \{m_i\}$ for $\forall y \in Y_i$
10 route (x, Y_i) using m_i for $1 \leq i \leq n$
11 **return true**

It is simple to parallelize the network control algorithm by maintaining the states of middle switches, U_x and V_y for all input switch x and output switch y. Line 2 partitions the set of destinations into n subsets, each having the same cardinality, except the last subset, if the degree of fanout of the request is not divisible by the split size s. The first iteration in line 3–4 computes intersection of U_x and V_Y. The intersections can be computed independently from other subsets. By privatizing the local variable w, the loop can be parallelized. The second iteration in line 7–9 updates the global set variables U_x and V_y to make the next connection requests use them consistently. Line 9 can be parallelized since U_y's are independent from each other. Line 10 routes the request through the networks.

We can observe that the complexity of the algorithm is $O(mr)$ in a single processor. If a proper split size s is chosen, for example, $s = c_1\sqrt{r}$ and $n = c_2\sqrt{r}$ for some constants c_1 and c_2, a request can be routed in $O(m\sqrt{r})$ time complexity with $O(\sqrt{r})$ processors. In addition to the parallelization, the effect of pipelining can be obtained if there are many requests routed in a short period. The critical sections are the first and the second loops. Once the sets U_x and V_y are successfully updated, a computation waiting on the first loop can be immediately initiated.

References

[1] Shun-Ping Chung and Keith W. Ross. On nonblocking multirate interconnection networks. *SIAM J. on Comp.*, 20(4):726–726, August

1991.

[2] C. Clos. A study of non-blocking switching networks. *Bell Syst. Tech. J.*, 32:406–424, March 1953.

[3] T. Cormen, C. Leiserson, and R. Rivest. *Introduction to Algorithms.* McGraw Hill, 1990.

[4] K. Y. Eng, M. G. Hluchyj, and Y. S. Yeh. Multicast and broadcast services in a knockout packet switch. In *Proc. on IEEE INFOCOM*, pages 29–34, 1988.

[5] F. K. Hwang. Rearrangeability of multiconnection three-stage networks. *Networks*, 2:301–306, 1972.

[6] F. K. Hwang. Three-stage multi-connection networks which are non-blocking in the wide sense. *Bell Syst. Tech. J.*, 58(10):1283–1287, December 1979.

[7] F. K. Hwang and C. H. Lin. Broadcasting in a three-stage point-to-point nonblocking network. *International Journal of Reliability, Quality and Safety Engineering*, 2(3):299–307, 1995.

[8] Dongsoo S. Kim and Ding-Zhu Du. Multirate broadcast switching networks nonblocking in a wide-s ense. In *DIMACS Workshop on Network Switching*, July 1997.

[9] Dongsoo S. Kim and Ding-Zhu Du. Multirate multicast switching networks. In *The Fourth Computing and Combinatorics Conferences*, August 1998.

[10] Dongsoo S. Kim and Ding-Zhu Du. Performance of split routing algorithm for three-stage multicast networks. *IEEE/ACM Trans on Networking*, 8(4):526–534, August 2000.

[11] Young Man Kim and Kyungsook Y. Lee. KSMIN's : Knockout switch-based multistage interconnection networks for high-speed packet switching. *IEEE Trans. on Comm.*, 43(8):2391–2398, August 1995.

[12] C. Y. Lee. Analysis of switching networks. *Bell Syst. Tech. J.*, 34(6):1287–1315, November 1955.

[13] Marco Listanti and Luca Veltri. Blocking probability of three-stage multicast switches. In *International Conference of Communications(ICC'98)*, 1998.

[14] G. M. Masson and B. W. Jordan, Jr. Generalized multistage connection networks. *Networks*, 2:192–209, 1972.

[15] Riccardo Melen and Jonathan S. Turner. Nonblocking multirate networks. *SIAM J. on Comp.*, 18(2):301–313, April 1989.

[16] Andrew S. Tanenbaum. *Computer Networks.* Prentice Hall, 3rd edition, 1996.

[17] Yuanyuan Yang and Neil H. Kessler. Modeling the blocking behavior of clos networks. In *DIMACS Workshop on Network Switching*, 1997.

[18] Yuanyuan Yang and Gerald M. Masson. Nonblocking broadcast switching networks. *IEEE Trans. on Comp.*, 40(9):1005–1015, September 1991.

[19] Yuanyuan Yang and Jianchao Wang. Performance analysis and simulation of multicast networks. In *Proceedings of International Conference on Parallel Processing(ICPP'97)*, 1997.

[20] Y. S. Yeh, M. G. Hluchyj, and A. S. Acampora. The knockout switch: A simple, modular architecture for high performance packet switching. *IEEE J. on Sel. Areas in Comm.*, 5(8):1274–1283, October 1987.

[21] Ellen Witte Zegura. An improved model for evaluating blocking probability in generalized connectors. In *Infocom*, pages 455–463, 1994.

SWITCHING NETWORKS: RECENT ADVANCES
D.-Z. Du and H.Q. Ngo (Eds.) pp. 307-357

Notes on the Complexity of Switching Networks

Hung Quang Ngo
Department of Computer Science and Engineering
University of Minnesota, Minneapolis, MN 55455
E-mail: hngo@cs.umn.edu
Ding-Zhu Du
Department of Computer Science and Engineering
University of Minnesota, Minneapolis, MN 55455
E-mail: dzd@cs.umn.edu

Contents

1 Introduction

1.1 Overview

There are various complexity measures for switching networks and communication networks in general. These measures include, but not limited to, the number of switching components, the delay time of signal propagating through the network, the complexity of path selection algorithms, and the complexity of physically designing the network. This chapter surveys the study of the first measure, and partially the second measure. It is conceivable that the number of switching components, or the "size" of a network, affects directly the third and fourth measures.

The most common and most intuitively obvious way to study a particular type of switching networks is to model the network by a graph. It is customary to use a directed acyclic graph with a degsinated set of vertices called *inputs* and another disjoint set of vertices called *outputs*. Depending on the functionality of the network under consideration, this graph has to satisfy certain conditions. Our job is then to determine the minimum number of edges of graphs satisfying these conditions, and construct an optimal one if possible. For example, a typical functionality of a network is *rearrangeability*. The corresponding graph model is the *connectors*, also referred to as *rearrangeable networks*. For different types of networks and their desired functionalities, the reader is referred to [16, 72, 40]. This chapter assumes that the reader is familiar with concepts such as rearrangeability, strictly nonblockingness or wide sense nonblockingness of multistage interconnection networks.

There have been several articles and books discussing aspects of switching network complexity or their associated graphical models. For example, the articles by Pippenger (1990, [72]), and Bien (1989, [17]); the classic book by Beneš (1965, [16]), a book by Hui (1990, [39]), and a recent book by Hwang (1998, [40]). We shall see that this chapter's material, although has some small overlaps with the articles and books above, have been collected in one place for the first time. Many open questions shall also be presented along the way.

We shall attempt to make this chapter as self-contained as possible, requiring mostly elementary background on probability and linear algebra. This objective will certainly affect our choice of which results to be covered. Moreover, we shall also try to select results that are more intuitive, which might not be the best results known on the problems under discussion. Pointers to more advanced results shall be given as needed.

1.2 Basic models

We now give definitions of several important graphs and their components arising in studying the complexity of switching networks, settling down the main terminologies for the rest of the chapters.

Definition 1.1. A bipartite graph $G = (I, O; E)$, where I and O are the bipartitions, is called an (n, d, c)*-expander* if $|I| = |O| = n$, $\Delta(G) = d$, and for every $X \subset I$ such that $|X| \leq n/2$, we have

$$|\Gamma(X)| \geq \left(1 + c\left(1 - \frac{|X|}{n}\right)\right)|X|$$

where $\Gamma(X)$ is the set of neighbors of all vertices X, $\Delta(G)$ denotes the maximum degree of vertices in G, d is called the *density*, and c is called the *expansion rate* of the expander G. It is a *strong* (n, d, c)*-expander* if the above inequality holds for all $X \subseteq I$. A *family of linear expanders of density* d *and expansion* c is a sequence $\{G_i\}_{i=1}^{\infty}$, where G_i is an (n_i, d, c)-expander and $n_i \to \infty$, $n_{i+1}/n_i \to 1$ as $i \to \infty$.

Definition 1.2. An (n, m)*-network* is an directed acyclic graph (DAG) with n distinguished vertices called *inputs* and a disjoint set of m distinguished vertices called *outputs*. When $n = m$, we call it an n*-network* for short. The *size* of a network is its number of edges. The *depth* of a network is the maximum length of a path from an input to an output.

An (n, m)-network is our main model for a switching network with n inputs and m outputs. With respect to physical switching networks, the vertices of our DAG represent *links* between switching components of the underlying physical networks, and there is an edge between two links if these two links are an input and an output link of some switch. Moreover, the depth of the DAG implicitly represent the delay of signal propagating from inputs to outputs.

A classical objective of studying switching networks of various types was to design these networks with as few switching components as possible. The main hope was that reducing this number also reduces several other complexity measures. Hence, one of our objectives is to find the minimum size (n, m)-network (or n-network for that matter) satisfying certain properties, and construct one with as small a size as possible. We shall see that this objective is usually a trade-off with another objective which is to have a network with small depth, which presumably has small signal propagation delay. Determining this trade-off in some precise manner is thus another research topic.

Definition 1.3. An (n, m)-*concentrator* is an (n, m)-network where $n \geq m$, such that for any subset S of m inputs there exists a set of m vertex disjoint paths connecting S to the outputs. An n-*concentrator* is an (n, n)-concentrator. An (n, m, d)-*concentrator* is an (n, m)-concentrator with at most dn edges. An (n, d)-*concentrator* is an (n, n, d)-concentrator. A *family of linear concentrators of density d* is a sequence $\{G_i\}_{i=1}^{\infty}$ such that each G_i is an (n_i, d)-concentrator, where $n_i \to \infty$ as $i \to \infty$.

Definition 1.4. An n-*superconcentrator* is an n-network with inputs I and outputs O such that for any $S_1 \subseteq I$ and $S_2 \subseteq O$ with $|S_1| = |S_2| = c$, there exist a set of c vertex disjoint paths connecting S_1 to S_2. An (n, k)-*superconcentrator* is an n-superconcentrator with at most kn edges. A *family of linear superconcentrators of density k* is a sequence $\{G_i\}_{i=1}^{\infty}$ such that each G_i is an (n_i, k)-superconcentrator, where $n_i \to \infty$ as $i \to \infty$.

Concentrators and superconcentrators are models for various communication networks and parallel architectures [72, 39]. For example, we can think of concentrators as a model for switching networks in which a set of processors communicate with a set of identical memories. While if a set of processors can request a particular set of memories, we need superconcentrators.

The graphs we just introduced are only part of the set of models we shall discuss. They were defined here to facilitate the history discussion in the next section. Other models shall be defined in associated sections.

1.3 A historical perspective

In this section, we briefly summarize the study of superconcentrators, concentrators and expanders. An objective is to give the reader a feel of why certain results were mentioned or covered in this chapter. Another reason for discussing these graphs' history is that they have rich and deep connections to many other areas of Computer Science and Mathematics, motivating many beautiful and difficult problems. Other complexity models, although shall be discussed later on in the chapter, will not be mentioned here. We shall briefly summarize their development in the corresponding section.

Concentrators were first introduced by Pinsker (1973 [65]) in the context of telephone switching networks. The notion of a superconcentrator, according to Pippenger (1996, [73]), was first introduced by Aho, Hopcroft and Ullman (1975, [1]), who attributed it to conversations with W. Floyd. They wanted to use superconcentrator as a tool to establish non-linear lower bounds for the complexity of circuits computing Boolean functions. Valiant (1975, [85, 84]) constructed linear size superconcentrators using Pinsker's concentrators, thus negating the intention of Aho, Hopcroft and Ullman. Paul, Tarjan and Celoni, continuing the theme, applied superconcentrators to demonstrate the optimality of certain algorithms (1977, [63, 64]). In 1973, Pippenger [66] also independently obtained some initial results on concentrators. From 1977 to 1979, Pippenger [69, 68, 67] and Fan Chung [24] studied the size bounds of minimum sized superconcentrators and concentrators. The concepts of connectors, generalizers, generalized connectors were also introduced by Pippenger in these papers as models for different switching networks.

Pinsker, Valiant, Pippenger and Chung works were roughly based on the same line of reasoning: the probabilistic method. Their proofs of the existence of good concentrators and superconcentrators were based on the existence of good bounded concentrators. The proofs were not constructive and thus of little practical value, although certainly very interesting. In 1973, Margulis [53] gave the first explicit construction of a family of expanders. As we shall see, expanders can be used to construct bounded concentrators; hence, Margulis construction yields an explicit construction of superconcentrators. Margulis construction was fairly technical using deep results from the Representation Theory of finite groups. He showed that there exists a constant $k > 0$ such that for $n = m^2, m \in \mathbb{N}$, each graph $\hat{G}_n$ of his constructed family of graphs is an $(n, 5, k)$-expander. Extending this work, Gabber and Galil (1981, [35]) showed that $k = k_0 = \frac{2-\sqrt{3}}{4}$ works. Their proof was a bit less technical, involving relatively elementary analysis. Moreover, they specified a way to construct a family of $(n, 7, 2k_0)$-expanders (larger density). They also specify how to use linear expanders to construct linear bounded concentrators and then to use the later to construct linear superconcentrators.

As we can see, the explicit constructions were too specialized to the given parameters, and thus not entirely satisfactory. Ron Rivest and S. Bhatt suggested another method which reuse earlier results on probabilistic construction: randomly choose a graph in a clever way, then check to see if the graph is a concentrator or superconcentrator. The main question is obviously that how hard the checking procedure is. Blum, Karp, Vornberger, Papadimitriou and Yannakakis (1981, [19]) gave a negative answer: the check(s) is coNP-complete.

In the meantime, researchers keep working on bounding the best possible size of the graphs. They also limit attention to graphs with a fixed depth. We will discuss more on this later. In 1983, so much excitement were ignited when expanders were used to explicitly construct a "parallel sorting network" which sorts n numbers in $O(\log n)$ time using n parallel processors. The work was done by M. Ajtai, J. Komlós, and E. Szemerédi [3], settling a long standing unsolved problem. This problem looked so impossible that, in fact, Knuth [48] in his previous edition of "The Art of Computer Programming" Vol. 3, stated it as a 50 exercise to show that constructing such a sorting network is impossible.

In 1984, an important idea comes into place as Tanner [83] used association schemes (see [21]), in particular certain class of distance regular graphs arising from finite geometry, to construct better expanders. The basic idea is to characterize the expansion rate using the graph's eigenvalue. Immediately after that, Noga Alon and partially Milman took the idea to its peak by a series of papers ([9, 5, 6, 7, 8]). Alon characterized the expansion rate using the second smallest eigenvalue $\mu(G)$ of the Laplacian of a graph G. Basically, as $\mu(G)$ gets larger the expansion rate is larger. As eigenvalues of real symmetric matrices (as the case with a graph's Laplacian) can be computed easily in polynomial time, we now have another way to "check" if a graph has certain expansion rate. The suggestion by Rivest and Bhatt can now be used. Alon did just that. He showed how to randomly generate expanders. Notice that since the problem is coNP-complete, quite a bit of information is lost in the characterization. (See [46], for example, for a sample research on this loss.) However, this method is fairly practical in certain applications.

Obviously a very natural question to ask is: "how large can $\mu(G)$ be?". Alon and Boppana (mentioned in [7]) proved that for any fixed d and for any infinite family of graphs G with maximum degree d,

$$\limsup \mu(G) \leq d - 2\sqrt{d-1}$$

Nilli (from Israel, 1991, [61]) got an upper bound that does not have any hidden constant, but the term $d - 2\sqrt{d-1}$ is still dominating. It is obviously interesting to construct families of expanders which achieve this eigenvalue bound. The special case when $d-1$ is a prime congruent to 1 modulo 4 was done by Lubotzky, Phillips and Sarnak (1988, [51]), and independently by Margulis (1988, [53]). They constructed a family of d-regular graphs with $\lambda \leq 2\sqrt{d-1}$, where $\lambda(G)$ is the second largest eigenvalue of G. Formally, we define

Definition 1.5. A d-regular graph G with n vertices is called a *Ramanujan graph* if $\lambda(G) \leq 2\sqrt{d-1}$. (Or equivalently $\mu(G) \geq d - 2\sqrt{d-1}$.)

Ramanujan graphs are optimal in the sense of the eigenvalue bound. Why were the graphs named after *Ramanujan* ? Basically, the construction was the Cayley graph of certain group, where the generators were chosen to be the solutions to certain representation of integers in quaternary quadratic form. The general formula is unknown, but Ramanujan had a conjecture in 1916 about the formula (well, obviously these kinds of conjectures can only come from Ramanujan), which was proved in several special cases by Eichler (1954, [30]) and Igusa (1959, [43]).

From this point on, a lot more research papers were published on various aspects of these graphs, several of which will be discussed in later section. One big open problem (mentioned in [34]) is

Open Problem 1.6. Find an elementary (e.g. purely combinatorial) proof that certain family of graphs is expanders. For example the ones constructed in [34] or in [51].

Solving this problem would lead to an entirely new wave of research on expanders. The spectral characterization of expansion rate loses a lot of information. The proofs of known explicit constructions were either too complicated, or used the spectral characterization which already reduces the power of the graphs. Calculating eigenvalues, although gives the quantitative measure, sheds no light as to why certain set of graphs are good expanders. A combinatorial proof would be much more satisfactory and would certainly lead to new development in the area.

It is worth mentioning that these graphs also play important roles in areas other than switching networks. To mention a few, for example, Noga Alon (1986, [7]) used geometric expanders (expanders constructed from finite geometry) to deduce a certain strengthening of a theorem of de Bruijn and Erdős on the number of lines determined by a set of points in a finite projective plane. He also obtained some results on Hadamard matrices and constructed some graphs relevant to Ramsey theory. Sipser and Spielman (1996, [81] – see also [82]) constructed asymptotically good linear codes from expanders.

2 The complexity of checking whether a graph is a superconcentrator or concentrator

As we have mentioned in the introduction, if it is possible to check in polynomial time if a graph is a (super)concentrator, then it might be possible to devise a randomized algorithm to generate these graphs. The results in this section are from [19], which says that the checks are both coNP-complete.

Definition 2.1. A *matcher* is a bipartite graph $B = (V_1, V_2, E)$ such that $|V_1| = |V_2| = 2m$ and that for any m-subset S of V_1, there exists a matching from S into

V_2.

We shall show that deciding whether a bipartite graph is a matcher is coNP-complete, and then deduce as corollaries that checking whether a graph is a concentrator or superconcentrator is also coNP-complete. In this section and throughout the chapter, we will use $[n]$ to denote the set of integers from 1 to n, and $deg(v)$ to denote the degree of a vertex v in some graph.

Theorem 2.2 (Blum et al., 1981 [19]). *Deciding whether a bipartite graph $B = (V_1, V_2, E)$ is a matcher is coNP-complete.*

Proof. We will reduce the complement of the following NP-complete problem to an instance of the matcher problem.

Big-clique. Given a graph (N, A) (N for nodes and A for arcs), with $|N| = 2k$, does is have a clique of size k ?

We can assume $|A| \geq \binom{k}{2}$, since otherwise the answer is clearly NO. Given (N, A), we construct a bipartite graph $B = (V_1, V_2, E)$ as follows.

- $V_1 = N \times [2k]$.
- $V_2 = A \cup \{u_1, \dots, u_{|A|}\} \cup \bigcup_{n \in N} \{v_1^n, \dots, v_{c(n)}^n\}$, where $c(n) = 2k - deg(n)$.
- To get E, each vertex $(n, i) \in V_1$, where $n \in N$ and $i \in [2k]$, is connected to the following vertices in V_2:
  - All arcs $a_j \in A$ which are incident to n.
  - All the u-nodes $u_1, u_2, \dots, u_{\binom{k}{2}-1}$.
  - All nodes v_j^n for $j = 1, \dots, c(n)$.

Our proof is complete if we can show that (N, A) does not have a clique of size k iff B is a matcher. Notice that

$$|V_2| = 2|A| + (2k)^2 - \sum_{n \in N} deg(n) = 4k^2 = |V_1|.$$

Moreover, by Hall's theorem B is a matcher iff $|\Gamma(S)| \geq |S|$ for all $S \subset V_1, |S| \leq \frac{1}{2}|V_1| = 2k^2$. Thus, it suffices to prove the following claim.

Claim. There exists $S \subset V_1$ such that $|S| \leq 2k^2$ and $|\Gamma(S)| < |S|$ iff (N, A) has a clique of size k.

($\Rightarrow$). We first make some observation as follows. Let

$$C = \{n \in N \mid (n, i) \in S \text{ for some } i\}.$$

Let $x = |C|$ and $y = |\{a \in A \mid a = (n_1, n_2) \text{ for some } n_1, n_2 \in C\}|$. Notice that $1 \leq x \leq 2k$, and that

$$\begin{aligned} |\Gamma(S)| &= |\{a \in A \mid a \text{ is incident to some } n \in C\}| + \\ &\quad \binom{k}{2} - 1 + \\ &\quad 2k|C| - \sum_{n \in C} deg(n) \\ &= \binom{k}{2} - 1 + 2kx - y \end{aligned}$$

Now, $|\Gamma(S)| < |S| \leq 2kx$ implies $y > \binom{k}{2} - 1$. However, y is exactly the number of edges of a graph with x vertices, making $y \leq \binom{x}{2}$. Hence, $k \leq x \leq 2k$. Moreover, if $x > k$ then as $\binom{k}{2} - 1 + 2kx - \binom{x}{2}$ is a strictly increasing function in $[1, 2k]$, we get a contradiction:

$$2k^2 \geq |S| > \binom{k}{2} - 1 + 2kx - \binom{x}{2} \geq 1 + \binom{k}{2} - 1 + 2k^2 - \binom{k}{2} = 2k^2$$

Consequently, $x = k$. This implies $y = \binom{k}{2}$, so that C is a k-clique. $(\Leftarrow)$. Let C be a clique of size k. Let $S = C \times [2k]$, then

$$|\Gamma(S)| = \binom{k}{2} - 1 + 2k^2 - \binom{k}{2} < |S|$$

□

It is easy to see that the problems of determining if a graph is a "concentrator" or "superconcentrator" are in coNP. To prove in polynomial time that a graph is not a concentrator, we only need to present a vertex cut C which separates a set of more than $|C|$ inputs to the outputs. Menger's Theorem ensures that this cut must exists. The proof that a graph is not a super concentrator is similar. Consequently, to show that they are coNP-complete, we can just reduce them to the "matcher" problem.

Corollary 2.3. *Deciding whether a graph G is a concentrator is coNP-complete*

Proof. We reduce "matcher" to "concentrator". Given a bipartite graph $B = (V_1, V_2, E)$ with $|V_i| = 2m$, we construct G by orienting all edges from V_1 to V_2, add a set V_3 of m vertices, and then connect all vertices of V_2 to all vertices of V_3. Clearly B is a matcher iff G is a concentrator. □

Corollary 2.4. *Deciding whether a graph G is a superconcentrator is coNP-complete.*

Proof. In the previous proof, change the size of V_3 to be $2m$. □

Corollary 2.5. *Deciding whether a graph G is an $(n, d, 0)$-expander is coNP-complete.*

3 Expanders

The main objective of this section is to present several results on the eigenvalue characterization of the expansion rate of expanders. Moreover, we shall give a partial answer to the question: "how large can the second smallest eigenvalue of the Laplacian of a graph be?", and give some pointers to results on direct constructions of expanders.

3.1 Algebraic graph theory

We first fix notations and terminologies from algebraic graph theory needed for the rest of the section. The reader is referred to the standard books by Biggs (1993, [18]), Godsil (1993, [36]), Cvetković, Doob, and Sachs (1995, [27]), and Chung (1997, [25]) for more information.

Let $G = (V, E)$ be a simple graph with n vertices, we will always assume the eigenvalues of G are $\lambda_1 \geq \lambda_2 \geq \cdots \geq \lambda_n$. Those are the eigenvalues of the adjacency matrix A of G. Moreover, $\lambda(G)$ will be used to denote λ_2. Let D be the $n \times n$ diagonal matrix indexed by vertices of G with $(D)_{vv} = deg(v)$; then, the matrix $L := D - A$ is called the *Laplacian* matrix of G. We shall use $\mu_1 \geq \mu_2 \geq \cdots \geq \mu_n$ to denote the eigenvalues of L. In contrast to the λ's, we use $\mu(G)$ to denote μ_{n-1}. The reason for this is that when G is d-regular then $\lambda_i = d - \mu_{n-i+1}$.

Let N be the incident matrix of any orientation H of $G(V, E)$. Let $L^2(V)$ ($L^2(E)$) be the space of real valued functions on V (E), with the usual inner product $\langle f, g\rangle$ and the usual norm $\|f\| = \sqrt{\langle f, f\rangle}$. Note that $L^2(V)$ is isomorphic to $\mathbb{R}^n$ and the Rayleigh quotient for f is $R_A(f) = \frac{\langle Lf, f\rangle}{\|f\|^2}$. Also note that

$$\begin{aligned}
\langle Lf, f\rangle &= \langle N^T N f, f\rangle = \langle Nf, Nf\rangle \\
&= \sum_{(u,v)\in E(H)} (f(u) - f(v))^2 \\
&= \sum_{u\sim v} (f(u) - f(v))^2
\end{aligned}$$

Here $u \sim v$ is used in lieu of $(u, v) \in E$ for convenience. The previous equation implies that L is non-negative definite, thus $\mu_i \geq 0, \ \forall i$. Moreover, it is not hard to see that μ_n is always 0 and that $\mu_{n-1} = 0$ iff G is not connected.

The following bounds on μ_{n-1} will be used quite often.

Proposition 3.1. *Let $f \in L^2(V)$ such that $\sum_v f(v) = 0$, then*

$$\mu(G) \leq \frac{\langle Lf, f\rangle}{\|f\|} = \frac{\sum_{u\sim v}(f(u) - f(v))^2}{\sum_v f^2(v)}$$

In fact, a stronger statement holds

$$\mu(G) = \min_{f\neq 0} \frac{\langle Lf, f\rangle}{\|f\|}$$

with the min *runs over all f satisfying $\sum_v f(v) = 0$.*

Note. $\sum_{u\sim v}(f(u) - f(v))^2$ is sometime called the *Dirichlet sum* of G.

Proof. Let $u_n = \mathbf{1}/\sqrt{n}$ be a unit μ_n-eigenvector of L, then the variational characterization of the eigenvalues (see [37], e.g.) gives

$$\mu_{n-1} = \min_{\substack{0\neq f\in\mathbb{C}^n\\ f\perp u_n}} R_L(f) = \min_{\substack{0\neq f\in\mathbb{C}^n\\ f\perp \mathbf{1}}} \frac{\langle Lf, f\rangle}{\|f\|}$$

The condition $f \perp \mathbf{1}$ is the same as $\sum_u f(u) = 0$. □

Proposition 3.2. *Suppose G is connected, and $f \in L^2(V)$ is a $\mu(G)$-eigenvector. Let $V^+ := \{v \in V \mid f(v) > 0\}$ and $V^- := V - V^+$. Let $g \in L^2(V)$ be defined by*

$$g(v) = \begin{cases} f(v) & \text{if } v \in V^+ \\ 0 & \text{otherwise.} \end{cases}$$

Then, we have

$$\mu \geq \frac{\sum_{u\sim v}(g(u) - g(v))^2}{\sum_v g^2(v)}$$

Proof. Note that since G is connected, $\mu \neq 0$, making $f \neq 0$. Hence, $V^+ \neq \emptyset$. By definition, we have $(Lf)(v) = \mu f(v), \forall v \in V$. Thus,

$$\mu = \frac{\sum_{v \in V^+} (Lf)(v) f(v)}{\sum_{v \in V^+} f^2(v)}$$

But,

$$\sum_{v \in V^+} f^2(v) = \sum_{v \in V} g^2(v)$$

and,

$$\begin{aligned}
\sum_{v \in V^+} (Lf)(v) f(v) &= \sum_{v \in V^+} \left(d(v) f^2(v) - \sum_{u \in \Gamma(v)} f(v) f(u) \right) \\
&= \sum_{uv \in E(V^+)} (f(u) - f(v))^2 + \sum_{uv \in E(V^+, V^-)} f(u)(f(u) - f(v)) \\
&\geq \sum_{u \sim v} (g(u) - g(v))^2
\end{aligned}$$

which completes our proof. □

3.2 The eigenvalue characterization of expansion rate for regular strong expanders

The results in this section are from Tanner (1984, [83]), Alon and Milman (1985, [9]), and Alon (1986, [6]). As the title of the section indicated, our goal is to show that the larger the expansion rate of a regular strong expander G is, the larger $\mu(G)$ has to be, and vice versa.

We again need to define several types of graphs closely related to expanders: enlargers, magnifiers, and bounded concentrators.

Definition 3.3. An (n, d, ϵ)*-enlarger* is a graph G on n vertices with maximum degree d and $\mu_{n-1}(G) \geq \epsilon$.

Definition 3.4. An (n, d, c)*-magnifier* is a graph $G = (V, E)$ on n vertices, $\Delta(G) = d$ and for every $X \subset V$ with $|X| \leq n/2$, $|\Gamma(X) - X| \geq c|X|$ holds. The *extended double cover* of a graph $G = (V, E)$ with $V = [n]$ is the bipartite graph H on the sets of inputs $X = \{x_1, \ldots, x_n\}$ and outputs $Y = \{y_1, \ldots, y_n\}$ so that $(x_i, y_j) \in E(H)$ iff $i = j$ or $(i, j) \in E(G)$.

Definition 3.5. An $(n, \theta, d, \alpha, c)$*-bounded strong concentrator* is a bipartite graph $G = (I, O; E)$ with n inputs, θn outputs, $\theta < 1$, and at most dn edges, such that if $X \subseteq I$ with $|X| \leq \alpha n$, then $|\Gamma(X)| \geq c|X|$. An (n, θ, d, α)*-bounded concentrator* is an $(n, \theta, d, \alpha, 1)$*-bounded strong concentrator.*

For the purpose of this section, enlargers were introduced just to shorten the sentence: "let G be a graph with maximum degree d and $\mu(G)$ large enough." On the other hand, we need magnifiers because their extended double covers are expanders. Magnifiers are, in a sense, the non-bipartite version of expanders. It is intuitively clear, and will be made precise later, that bounded concentrators are closely related to expanders. The proof of the following lemma is straightforward, hence omitted.

Lemma 3.6. *The double cover of an* (n, d, c)*-magnifier is an* $(n, d+1, c)$*-expander.*

The following theorem, which is an improved version of Theorem 3.4 in [6], makes our goal precise.

Theorem 3.7. *Let* $G = (I, O; E)$ *be a d-regular bipartite graph, where* $|I| = |O| = n$ *and* $\mu = \mu(G)$ $(= d - \lambda(G))$. *The following hold:*

(i) If G is an (n, d, c) *strong expander then* $\mu \geq \frac{c^2}{576-48c+2c^2}$.

(ii) If $\mu \geq \epsilon$, *then G is an* (n, d, c)*-expander, where* $c = \frac{2d\epsilon - \epsilon^2}{d^2}$.

We shall prove this theorem using a sequence of lemmas, with the following plan in mind. To show (i), we will show

G is a strong expander $\rightarrow$ G is a magnifier $\rightarrow$ G is an enlarger (i.e. $\mu(G)$ is large)

Similarly, to prove (ii) we shall show the reverse sequence

G is an enlarger $\rightarrow$ G is a strong expander

Let us now proceed to show that a strong expander is a magnifier.

Lemma 3.8. *Let* $G = (I, O; E)$ *be an* (n, d, c)*-strong-expander. Then G is a* $(2n, d, \frac{c}{12-c})$*-magnifier.*

Proof. Be definition, every $X_1 \subset I$ satisfies

$$|\Gamma(X_1)| \geq \left(1 + c(1 - \frac{|X_1|}{n})\right) |X_1|. \tag{1}$$

In particular,

$$|\Gamma(X_1)| \geq |X_1| \tag{2}$$

When $|X_1| = n/2$, (1) implies $c \leq 2$. Moreover, for every $X_2 \subseteq O$, setting $X_1 = I - \Gamma(X_2)$ in (2) yields $\Gamma(X_2) \geq |X_2|$.

We are to show that for every $X \subset I \cup O$ with size at most n, it must be the case that

$$|\Gamma(X) - X| \geq \frac{c}{12-c}|X|$$

Let $X_1 = X \cap I$ and $X_2 = X \cap O$. The intuition is that when $|X_1|$ is very small comparing to $|X_2|$, then there will be a lot of neighbors of X_2 lying outside of X_1 in I, making $|\Gamma(X) - X|$ large. On the other hand, when $|X_1|$ is relatively large comparing to $|X_2|$, then there will be "many" neighbors of X_1 not in X_2. We now turn this intuition into mathematical rigor.

When $|X_1| \leq |X_2|(1 - \frac{c}{6})$, we have

$$|X| \leq \frac{12-c}{6}|X_2|.$$

Hence,

$$|\Gamma(X) - X| \geq |\Gamma(X_2)| - |X_1| \geq |X_2| - |X_1| \geq \frac{c}{6}|X_2| \geq \frac{c}{12-c}|X|.$$

When $|X_2|(1 - \frac{c}{6}) < |X_1| \leq \frac{n}{2}$, we get

$$|X| < \left(1 + \frac{1}{1-\frac{c}{6}}\right)|X_1| = \frac{12-c}{6-c}|X_1|.$$

This relation, the fact that $c \leq 2$, and (1) yield

$$\begin{aligned}
|\Gamma(X) - X| &\geq |\Gamma(X_1)| - |X_2| \\
&\geq (1 + \frac{c}{2} - \frac{1}{1-\frac{c}{6}})|X_1| \\
&= \frac{4-c}{2}\frac{c}{6-c}|X_1| \\
&> \frac{4-c}{2}\frac{c}{12-c}|X| \\
&\geq \frac{c}{12-c}|X|
\end{aligned}$$

Lastly, when $|X_1| \geq \frac{n}{2}$ it follows that

$$|\Gamma(X) - X| \geq |\Gamma(X_1)| - |X_2| \geq (1 + \frac{c}{2})\frac{n}{2} - \frac{n}{2} = \frac{c}{4}n \geq \frac{c}{4}|X| \geq \frac{c}{12 - c}|X|.$$

Here, we use the fact that $c \leq 2$ and that the function $f(x) = (1 + c(1 - \frac{x}{n}))x$ is increasing when $n \geq 2$. □

Next, to complete part (i) of Theorem 3.7 we show that every magnifier has "large" μ.

Lemma 3.9. *Let $G = (V, E)$ be an (n, d, c)-magnifier, then G is an (n, d, ϵ)-enlarger where $\epsilon = \frac{c^2}{2+2(c+1)^2}$.*

Proof. We apply and use notations of Proposition 3.2. Since we have to show that $\mu \geq \epsilon = \frac{c^2}{2+2(c+1)^2}$, it suffices to show that

$$\frac{\sum\limits_{u \sim v} (g(u) - g(v))^2}{\sum\limits_{v} g^2(v)} \geq \frac{c^2}{2 + 2(c+1)^2}$$

This is done by using the maxflow-mincut theorem. Consider a network N with vertex set $\{s\} \uplus V^+ \uplus V \uplus \{t\}$, where s is the source, t is the sink, and $\uplus$ denotes the disjoint union. The edges and their capacities are defined as follows.

(a) For each $u \in V^+$, (s, u) has capacity $cap(s, u) = 1 + c$.

(b) For each pair $(u, v) \in V^+ \times V$, $cap(u, v) = \begin{cases} 1 & \text{if } uv \in E \text{ or } u = v \\ 0 & \text{otherwise.} \end{cases}$

(c) For each $v \in V$, $cap(v, t) = 1$.

We claim that the min-cut of N has capacity $(1 + c)|V^+|$. Consider any cut C. Let $X := \{u \in V^+ \mid (s, u) \notin C\}$. As G has magnifying rate c and $\Gamma_N(X) = X \cup \Gamma_G(X)$, it is easy to see that $|\Gamma_N(X)| \geq (1 + c)|X|$. Moreover, for every $v \in \Gamma_N(X)$ there must be at least one edge with capacity one in C incident to it. All these edges are disjoint, thus the capacity of C is at least $(1+c)(|V^+|-|X|)+|\Gamma_N(X)| \geq (1 + c)|V^+|$. Lastly, the cut $\{(s, u) \mid u \in V^+\}$ has capacity exactly $(1 + c)|V^+|$, proving the claim. By the maxflow-mincut theorem, there exists an orientation $\bar{E}$ of edges in G and a function $h : \bar{E} \to \mathbb{R}$ such that

$$0 \leq h(u, v) \leq 1 \text{ for all } (u, v) \in \bar{E}$$

$$\sum_{v:(u,v)\in \bar{E}} h(u,v) = \begin{cases} 1+c & \text{if } u \in V^+ \\ 0 & \text{otherwise.} \end{cases}$$

$$\sum_{u:(u,v)\in \bar{E}} h(u,v) \leq 1 \text{ for all } v \in V$$

Now, the following is straightforward:

$$\begin{aligned} \sum_{(u,v)\in \bar{E}} h^2(u,v)\big(g(u)+g(v)\big)^2 &\leq 2 \sum_{(u,v)\in \bar{E}} h^2(u,v)\big(g^2(u)+g^2(v)\big) \\ &= 2\sum_{v\in V} g^2(v)\Bigg(\sum_{u:(u,v)\in \bar{E}} h^2(u,v) + \sum_{u:(v,u)\in \bar{E}} h^2(v,u)\Bigg) \\ &\leq 2(1+(1+c)^2)\sum_{v\in V} g^2(v) \end{aligned}$$

and

$$\begin{aligned} \sum_{(u,v)\in \bar{E}} h(u,v)\big(g^2(u)-g^2(v)\big) &= \sum_{u\in V} g^2(u)\Bigg(\sum_{v:(u,v)\in \bar{E}} h(u,v) - \sum_{v:(v,u)\in \bar{E}} h(v,u)\Bigg) \\ &\geq c\sum_{v\in V} g^2(v) \end{aligned}$$

Now, combining all inequalities above along with Cauchy-Schwarz inequality we get

$$\begin{aligned}
\mu &\geq \frac{\sum_{u\sim v}\left(g(u)-g(v)\right)^2}{\sum_{v} g^2(v)} \\
&= \frac{\sum_{u\sim v}\left(g(u)-g(v)\right)^2 \sum_{(u,v)\in\bar{E}} h^2(u,v)\left(g(u)+g(v)\right)^2}{\sum_{v} g^2(v) \sum_{(u,v)\in\bar{E}} h^2(u,v)\left(g(u)+g(v)\right)^2} \\
&\geq \frac{\left(\sum_{(u,v)\in\bar{E}} h(u,v)|g^2(u)-g^2(v)|\right)^2}{2(1+(c+1)^2)\left(\sum_{v\in V} g^2(v)\right)^2} \\
&\geq \frac{1}{2+2(c+1)^2}\left(\frac{\sum_{(u,v)\in\bar{E}} h(u,v)(g^2(u)-g^2(v))}{\sum_{v\in V} g^2(v)}\right)^2 \\
&\geq \frac{c^2}{2+2(c+1)^2}
\end{aligned}$$

□

The previous two lemmas trivially imply part (i) of Theorem 3.7. Now we are ready to complete part (ii) of Theorem 3.7. We first need a lemma from [83]. Let $G = (I, O; E)$ be a bipartite graph such that $|I| = n$, $|O| = m$, and that $deg(i) = a, \forall i \in I$, $deg(o) = b$, $\forall o \in O$. Let M be an $n \times m$ 01-matrix whose rows are indexed by I and whose columns are indexed by O such that $m_{io} = 1$ if $io \in E$ and $m_{io} = 0$ otherwise. Let $B = MM^T$, then clearly B is real, symmetric, and non-negative definite. As usual, we let $\theta_1 \geq \theta_2 \geq \cdots \geq \theta_n$ be the eigenvalues of B and $u_1, u_2, \ldots, u_n$ be a set of corresponding orthonormal eigenvectors. Notice that $(B)_{ij}$ is the number of common neighbors of i and j, hence $\sum_j (B)_{ij} = ab, \forall i$. (Each neighbor of i is counted b times in the sum.) This implies $\mathbf{1}/\sqrt{n}$ is an eigenvector of B with corresponding eigenvalue ab. Suppose θ is any eigenvalue of B and e is a θ-eigenvector. Let e_i be a coordinate of e with highest absolute value, then $(Be)_i = \theta e_i$ implies

$$\theta|e_i| = |\sum_j (B)_{ij} e_j| \leq |e_i| \sum_j (B)_{ij} = ab|e_i|$$

Consequently, ab is also the largest eigenvalue of B, i.e. $\theta_1 = ab$. We may thus assume that $u_1 = \mathbf{1}/\sqrt{n}$.

Lemma 3.10. *If* $\theta_1 > \theta_2$, *then* G *is an* $(n, m/n, a, \alpha, c(\alpha))$*-bounded strong concentrator, where*

$$c(\alpha) \geq \frac{a^2}{\alpha(ab - \theta_2) + \theta_2}$$

Proof. Let β be a positive real number not exceeding α. For any $X \subseteq I$ with $|X| = \beta n$, let $x \in \{0,1\}^n$ be X's characteristic vector. Clearly $xx^T = \|x\|^2 = \beta n$. Similarly, let $Y = \Gamma(X)$ and y be its characteristic vector. As for any $o \in O$, $(xM)_o$ is the number of vertices in X adjacent to o, the sum of entries in xM is exactly $\beta n a$. Hence,

$$\|xM\|^2 = \sum_{o \in O} (xM)_o^2 \geq \left(\frac{\sum_{o \in O}(xM)_o}{|Y|}\right)^2 |Y| = \frac{\beta^2 n^2 a^2}{|Y|} \tag{3}$$

Now, write $x = \gamma_1 u_1 + \gamma_2 u_2 + \cdots + \gamma_n u_n$ we get

$$xB = \gamma_1 \theta_1 u_1 + \gamma_2 \theta_2 u_2 + \cdots + \gamma_n \theta_n u_n.$$

Thus, by orthonormality we have

$$\|xM\|^2 = (xB)x^T = \theta_1 \gamma_1^2 + \theta_2 \gamma_2^2 + \cdots + \theta_n \gamma_n^2 \tag{4}$$

Now,

$$\gamma_1 = x u_1^T = \frac{\sum_i x_i}{\sqrt{n}} = \beta \sqrt{n}$$

Hence, (4) gives

$$\begin{aligned} \|xM\|^2 &= \theta_1 \gamma_1^2 + \sum_{j=2}^{n} \theta_j \gamma_j^2 \\ &\leq ab\beta^2 n + \theta_2(\|x\|^2 - \gamma_1^2) \\ &= \beta^2 n(ab - \theta_2) + \theta_2 \beta n \end{aligned} \tag{5}$$

Lastly, combining (3) and (5) yield

$$\frac{|\Gamma(X)|}{|X|} = \frac{|Y|}{\beta n} \geq \frac{a^2}{\beta(ab - \theta_2) + \theta_2}$$

As this inequality is true for all $\beta \leq \alpha$, the proof is completed. □

Part (ii) of Theorem 3.7 could now be derived as a corollary of this lemma.

Corollary 3.11. *If $G = (I, O; E)$ is a d-regular bipartite graph with $|I| = |O| = n$ and $\mu = \mu(G)$, then*

(i) G is an (n, d, c)-strong expander, where

$$c = \frac{2d\mu - \mu^2}{d^2}$$

(ii) G is an (n, d, c)-expander, where

$$c = \frac{2d\mu - \mu^2}{d^2 - d\mu + \mu^2/2}$$

Proof. Let M and B be the matrices for G as in Lemma 3.10, and let A be G's adjacency matrix. Also let $\theta_1 \geq \cdots \geq \theta_n$ be the eigenvalues of B and $\lambda_1 \geq \cdots \geq \lambda_{2n}$ be the eigenvalues of A as usual. Since $(B)_{ij} = (MM^T)_{ij}$ counts the number of common neighbors of i and j, while $(A^2)_{ij}$ counts the number of length-2 paths from i to j, it is obvious that

$$\begin{pmatrix} B & 0 \\ 0 & B \end{pmatrix} = A^2$$

It is standard that as G is bipartite, whenever λ is an eigenvalue of A, then so is $-\lambda$. Moreover, $\lambda_1 = d$ and $\lambda_2 = d - \mu$ since G is d-regular. Thus, $\theta_1 = d^2$ and $\theta_2 = (d - \mu)^2$. Now, for any $X \subseteq I$ with $\alpha = |X|/n$, applying Lemma 3.10 gives

$$\begin{aligned} |\Gamma(X)| &\geq \frac{d^2}{\alpha\big(d^2 - (d-\mu)^2\big) + (d-\mu)^2}|X| \\ &= \left(1 + \frac{(2d\mu - \mu^2)(1-\alpha)}{d^2 - (2d\mu - \mu^2)(1-\alpha)}\right)|X| \\ &\geq \left(1 + \frac{(2d\mu - \mu^2)}{d^2}\left(1 - \frac{|X|}{n}\right)\right)|X| \end{aligned}$$

When $|X| \leq n/2$, we get

$$|\Gamma(X)| \geq \left(1 + \frac{(2d\mu - \mu^2)}{d^2 - d\mu + \mu^2/2}\left(1 - \frac{|X|}{n}\right)\right)|X|$$

□

3.3 On the second eigenvalue of a graph

As we have seen, the expanding rate of a graph increases as its second smallest Laplacian eigenvalue increases. Thus, it makes sense to study how large μ_{n-1} can be. Through out this section, we shall consider only d-regular graphs and let $\mu = \mu(G) = \mu_{n-1}(G) = d - \lambda_2(G)$. Alon and Boppana (mentioned in [7]) proved that for any fixed d and for any infinite family of graphs G with maximum degree d,

$$\limsup \mu(G) \leq d - 2\sqrt{d-1}$$

The bound is sharp when $d - 1$ is a prime congruent to 1 modulo 4, as shown by the explicit constructions of Lubotzky, Phillips and Sarnak [51], and independently by Margulis [53]. Alon [7] conjectured the following

Conjecture 3.12 (Alon [7]). As $n \to \infty$, the probability that $\mu(G) \leq d - 2\sqrt{d-1} - o(1)$ goes to 1.

In other words, as n gets large, with very high probability we have $\mu(G) \leq d - 2\sqrt{d-1}$. The conjecture is still open as far as we know. Friedman, Kahn and Szemeédi (1989, [33]) showed that

$$\mu(G) \geq d - 2\sqrt{d-1} - \log d - o(1)$$

as $n \to \infty$. Nilli (1991, [61]) got an upper bound that does not have any hidden constant:

Theorem 3.13 (Nilli, 1991 [61]). *Let G be a graph in which there are two edges with distance at least $2k+2$, and let d be the maximum degree of G. Then,*

$$\mu(G) \leq d - 2\sqrt{d-1} + \frac{2\sqrt{d-1} - 1}{k+1}$$

Proof. Let (u_1, u_2) and (w_1, w_2) be two edges with distance at least $2k+2$. Let L be the Laplacian matrix of G as usual. Let $U_0 = \{u_1, u_2\}$ and $W_0 = \{w_1, w_2\}$. For $1 \leq i \leq k$, let U_i (resp. W_i) be the set of vertices of distance i from U_0 (resp. W_0). Clearly $U := \cup_{0 \leq i \leq k} U_i$ has distance at least 2 from $W := \cup_{0 \leq i \leq k} W_i$, so that there is no edge joining the two unions. Moreover, it is easy to see that $|U_i| \leq (d-1)|U_{i-1}|$ and $|W_i| \leq (d-1)|W_{i-1}|$.

Let $f : V(G) \to \mathbb{R}$ be defined as

$$f(v) = \begin{cases} a(d-1)^{-i/2} & \text{for } v \in U_i, 0 \leq i \leq k \\ b(d-1)^{-i/2} & \text{for } v \in W_i, 0 \leq i \leq k \\ 0 & \text{otherwise.} \end{cases}$$

where $a > 0$, $b < 0$ are real numbers such that $\sum_v f(v) = 0$. The variational characterization yields

$$\mu(G) = \mu_{n-1} \le \frac{\langle Lf, f\rangle}{\langle f, f\rangle}$$

We also have

$$\begin{aligned}
\langle f, f\rangle &= \sum_{u\in U} f^2(u) + \sum_{w\in W} f^2(w) \\
&= \sum_{i=0}^{k} |U_i|\frac{a^2}{(d-1)^i} + \sum_{i=0}^{k} |W_i|\frac{b^2}{(d-1)^i} \\
&= A_1 + B_1
\end{aligned}$$

where A_1 and B_1 are the first and second sum, respectively. Moreover, as we have mentioned: (a) there are no edges joining U and W; (b) there are no edges connecting U_i or W_i to $V(G) - U \cup W$ if $i \le k-1$; and (c) there are at most $d-1$ edges joining a vertex of U_i (W_i) to a vertex of U_{i+1} (W_{i+1}), we have

$$\begin{aligned}
\langle Lf, f\rangle &= \sum_{u\sim v}(f(u) - f(v))^2 \\
&= \sum_{\substack{u\sim v\\ \{u,v\}\cap U\ne\emptyset}} (f(u) - f(v))^2 + \sum_{\substack{u\sim v\\ \{u,v\}\cap W\ne\emptyset}} (f(u) - f(v))^2 \\
&= \sum_{i=0}^{k-1} \sum_{\substack{u\sim v\\ u\in U_i, v\in U_{i+1}}} (f(u) - f(v))^2 + \sum_{i=0}^{k-1} \sum_{\substack{u\sim v\\ u\in W_i, v\in W_{i+1}}} (f(u) - f(v))^2 \\
&\quad + \sum_{\substack{u\sim v\\ u\in U_k, v\in V-U\cup W}} (f(u))^2 + \sum_{\substack{u\sim v\\ u\in W_k, v\in V-U\cup W}} (f(u))^2 \\
&\le a^2\left(\sum_{i=0}^{k-1} |U_i|(d-1)\left(\frac{1}{(d-1)^{i/2}} - \frac{1}{(d-1)^{(i+1)/2}}\right)^2 + |U_k|\frac{d-1}{(d-1)^k}\right) + \\
&\quad b^2\left(\sum_{i=0}^{k-1} |W_i|(d-1)\left(\frac{1}{(d-1)^{i/2}} - \frac{1}{(d-1)^{(i+1)/2}}\right)^2 + |W_k|\frac{d-1}{(d-1)^k}\right) \\
&= a^2\left(\sum_{i=0}^{k} \frac{|U_i|}{(d-1)^i}(d - 2\sqrt{d-1}) + (2\sqrt{d-1} - 1)\frac{|U_k|}{(d-1)^k}\right) + \\
&\quad b^2\left(\sum_{i=0}^{k} \frac{|W_i|}{(d-1)^i}(d - 2\sqrt{d-1}) + (2\sqrt{d-1} - 1)\frac{|W_k|}{(d-1)^k}\right) \\
&= A_2 + B_2
\end{aligned}$$

with A_2 and B_2 defined in the obvious way. To this end, we are left to show that

$$\frac{A_2 + B_2}{A_1 + B_1} \leq d - 2\sqrt{d-1} + \frac{2\sqrt{d-1} - 1}{k+1} =: C$$

We shall show that $A_2/A_1 \leq C$ and $B_2/B_1 \leq C$ instead. Notice that $\frac{|U_i|}{(d-1)^i} \geq \frac{|U_{i+1}|}{(d-1)^{i+1}}$, clearly

$$A_1 = \sum_{i=0}^{k} |U_i| \frac{a^2}{(d-1)^i} \geq a^2(k+1)\frac{|U_k|}{(d-1)^k}$$

hence,

$$\frac{A_2}{A_1} = d - 2\sqrt{d-1} + a^2 \frac{2\sqrt{d-1} - 1}{A_1} \frac{|W_k|}{(d-1)^k} \leq C$$

$B_2/B_1 \leq C$ is proved similarly.

□

Corollary 3.14. *Let G, d and k be defined as in the previous theorem, If G is d regular, then*

$$\lambda_2(G) \geq 2\sqrt{d-1}\left(1 - \frac{1}{k+1}\right) + \frac{1}{k+1}$$

3.4 Explicit Constructions of Expanders

Many practical applications require explicit constructions of expander graphs. Explicit constructions turn out to be a lot more difficult than showing existence. In 1973, Margulis [52] gave the first explicit construction of (strong) expanders. He explicitly constructed a family of bipartite graphs $\{\hat{G}_n\}$ for $n = m^2, m = 1, 2, \ldots,$ and show that there is a constant $k > 0$ such that for each $n = m^2$, $m \in \mathbb{N}$, $\hat{G}_n$ is an $(n, 5, k)$-strong expander. This construction was certainly undesirable as the constant k was not known. Moreover, his proof used deep results from Representation Theory. Angluin (1979, [12]) pointed out that Margulis' method could be used to construct $(n, 3, k')$-strong expanders but the constant k' is also not known. Gabber and Galil (1981, [34]) slightly modified Margulis' construction and used relatively elementary Taylor analysis to show how to construct a family of $(m^2, 5, (2-\sqrt{3})/4)$-strong expanders for each $m \in \mathbb{N}$. They also constructed a family of $(m^2, 7, (2-\sqrt{3})/2)$-strong expanders. Let us mention here their first construction. For each $m \in \mathbb{N}$, construct an $m^2 \times m^2$ bipartite graph

$\bar{G}_{m^2} = (I_m, O_m; E)$ where $I_m = O_m = \mathbb{Z}_m \times \mathbb{Z}_m$ and each vertex $(i, j) \in I_m$ is connected to 5 vertices in O_m defined by the following permutations:

$$\begin{aligned}
\sigma_1(i,j) &= (i,j) \\
\sigma_2(i,j) &= (i,i+j) \\
\sigma_3(i,j) &= (i,i+j+1) \\
\sigma_4(i,j) &= (i+j,j) \\
\sigma_5(i,j) &= (i+j+1,j)
\end{aligned}$$

Here, the additions are done modulo m. As we have mentioned, there is no known elementary proof that this family are expanders with the prescribed expansion rate. This construction was later modified slightly by Jimbo and Maruoka (1987, [44]) and Alon, Galil and Milman (1987, [8]) to obtain better superconcentrators.

As we have discussed in the introduction, after the eigenvalue characterization of expansion rate, the main problem was to construct Ramanujan graphs, which are optimal expanders in the eigenvalue sense. The special case where $d-1$ is a prime congruent to 1 modulo 4 was "solved" by Lubotzky, Phillips, and Sarnak (1988, [51]), and independently by Margulis (1988, [53]). Later, in 1994 Morgenstern [58, 57] constructed for every prime power q many families of $(q+1)$-regular Ramanujan graphs. All these constructions were Cayley graphs of certain groups.

Other works and information on expanders could be found in [4, 50, 11, 2, 46, 79, 17, 25].

4 Concentrators and Superconcentrators

4.1 Linear Concentrators and Superconcentrators

Valiant (1975, [84]) showed that there exists n-superconcentrators of size at most $238n$ and depth $O(\log^2 n)$. Valiant's proof was based on a recursive construction using Pinsker's concentrator (1973, [65]). Pinsker showed that there exist n-concentrators with at most $29n$ edges. A somewhat weaker version of this result was also obtained independently by Pippenger (1973, [66]). In 1977, Pippenger [67] gave a simpler construction of n-superconcentrators with size at most $40n$, maximum degree 16, and depth $O(\log n)$, while Valiant and Pinsker's graphs did not have $O(1)$ degree bound. This was certainly a big improvement. On the same line of reasoning, with finer analysis Fan Chung (1979, [24]) showed that there exists n-concentrators of size at most $27n$ and n-superconcentrators of size at most $38.5n$. Bassalygo (1981, [13]) improved the n-concentrator bound to $20n$ and n-superconcentrator bound to $36n$.

We present here the results of Pippenger [67] and a small generalization given by Gabber and Galil [34], since although it is not the best result, it is fairly intuitive.

Lemma 4.1 (Pippenger, 1977). *For every m, there is a bipartite graph with $6m$ inputs and $4m$ outputs, in which every input has outdegree at most 6, every output has indegree at most 9, and, for every $k \leq 3m$ and every set S of k inputs, there exists a matching from S into some k-subset of the outputs.*

Proof. Let $\mathcal{M} := \{0, 1, \ldots, 36m - 1\}$, and π be any permutation on $\mathcal{M}$. Let $G(\pi)$ be the bipartite graph obtained from π by taking $\{0, 1, \ldots, 6m - 1\}$ as inputs, $\{0, 1, \ldots, 4m - 1\}$ as outputs, and $E(G) := \{((x \bmod 6m), (\pi(x) \bmod 4m)) \mid x \in \mathcal{M}\}$.

We way that $G(\pi)$ is *good* if there are no $k \leq 3m$, a k-subset A of the inputs, and a k-subset B of the outputs such that $\Gamma(A) \subseteq B$. We would like to look for a good graph, which will satisfy our criteria. The existence of good graphs is shown by proving that the probability of $G(\pi)$ being *bad* (i.e. not good) is strictly less than 1.

Any k-subset A (B) of the inputs (outputs) corresponds uniquely to a $6k$-subset ($9k$-subset) $\mathcal{A}$ ($\mathcal{B}$) of $\mathcal{M}$ (but not vice versa.) Moreover, $\Gamma(A) \subseteq B$ iff $\pi(x) \in \mathcal{B}, \forall x \in \mathcal{A}$. For fixed $\mathcal{A}$ and $\mathcal{B}$, there are $(9k)_{6k}(36m - 6k)!$ permutations π which satisfy this condition. Hence, let P_m be the probability of $G(\pi)$ being *bad*, we have

$$\begin{aligned} P_m &\leq \sum_{k=1}^{3m} \binom{6m}{k}\binom{4m}{k}\frac{(9k)_{6k}(36m-6k)!}{(36m)!} \\ &= \sum_{k=1}^{3m} \frac{\binom{6m}{k}\binom{4m}{k}\binom{9k}{6k}}{\binom{36m}{6k}} \\ &=: I_m \end{aligned}$$

What we want to show is $I_m < 1$. As I_m has $3m$ terms, we first want to bound the largest term. Combinatorially, it's easy to see that

$$\binom{36m}{6k} \geq \binom{6m}{k}\binom{4m}{k}\binom{26m}{4k}$$

Hence,

$$I_m \leq \sum_{k=1}^{3m} \frac{\binom{9k}{6k}}{\binom{26m}{4k}} = \sum_{k=1}^{3m} L_k, \text{ where } L_k := \frac{\binom{9k}{6k}}{\binom{26m}{4k}}$$

It is not difficult to check that the sequence $1/L_k$ is unimodal, thus the largest L_k is either L_1 or L_{3m}.

As $3mL_1 < 1$ trivially, we can assume $L_{3m} > L_1$, so that I_m is at most

$$3mL_{3m} = 3m\frac{\binom{27m}{18m}}{\binom{26m}{12m}} = \frac{(27m)!(12m)!(14m)!}{(18m)!(9m)!(26m)!}$$

To this end, we use the following two inequalities. The first one is a good version of Stirling's formula (see Robbins [78]).

$$\sqrt{2\pi n}e^{-n}n^n \leq n! \leq e^{\frac{1}{12n}}\sqrt{2\pi n}e^{-n}n^n$$

and

$$e^x \leq \frac{1}{1-x}$$

Applying these inequalities yields $3mL_{3m} < 1$ for all $m \geq 3$. The cases where $m \leq 2$ are trivial. □

Corollary 4.2. *For every m, there is a bipartite graph with $4m$ inputs and $6m$ outputs, in which every input has outdegree at most 9, every output has indegree at most 6, and, for every $k \leq 3m$ and every set S of k outputs, there exists a matching from S into some k-subset of the inputs.*

Let $s(n)$ be the minimum size of an n-superconcentrator, and $\theta(n) := 4\lceil\frac{n}{6}\rceil$. We first obtain a recursive inequality for $s(n)$.

Lemma 4.3. *For any n, $s(n) \leq 13n + s(\theta(n))$.*

Proof. Let $m = \lceil\frac{n}{6}\rceil$, and G and G' be the graphs of Lemma 4.1 and Corollary 4.2 respectively. Let S' be a $\theta(n)$-superconcentrator with $s(\theta(n))$ edges. Construct an n-superconcentrator as shown in Figure 1. Clearly S has size at most $13n + s(\theta(n))$.

□

Now we are ready for the main result:

Theorem 4.4 (Pippenger, 1977). *We have $s(n) \leq 39n + O(\log n)$. In fact, $s(n) \leq 40n$.*

Proof. The ternary Beneš network (see, e.g. [16]) gives

$$s(n) \leq 3n(2\lceil\log_3 n\rceil - 1)$$

because the Beneš network is rearrangeable, it is certainly a superconcentrator. This gives $s(n) \leq 39n$ for $n \leq 3^7 = 2187$.

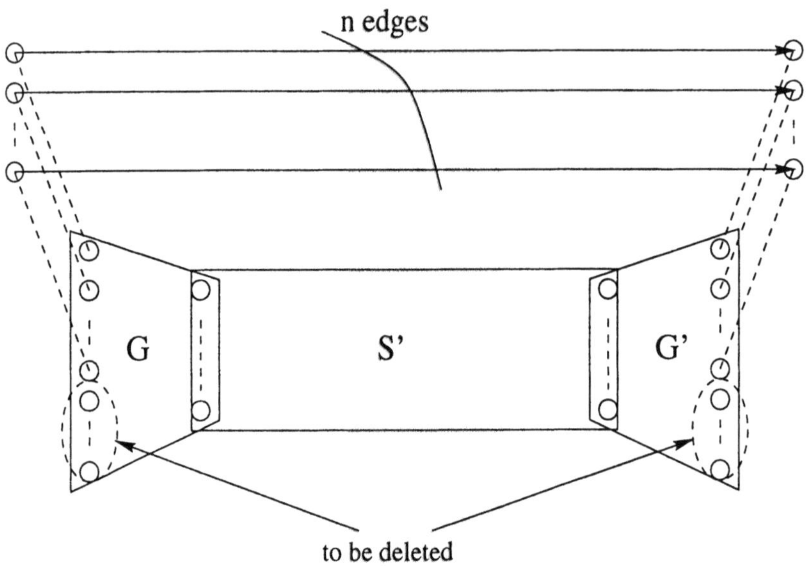

Figure 1: Recursive construction of an n-superconcentrator from a $\theta(n)$-superconcentrator

For larger values of n, we shall use the previous lemma. Define $\theta^0(n) = n$, $\theta^{t+1}(n) = \theta(\theta^t(n))$. Pick t such that $\theta^t(n) > 3^7 \geq \theta^{t+1}(n)$, then apply Lemma 4.3 $t+1$ times, we get

$$s(n) \leq 13(\theta^0(n) + \theta^1(n) + \cdots + \theta^t(n)) + s(\theta^{t+1}(n))$$

It is easy to show by induction on t that $\theta^t(n) \leq \left(\frac{2}{3}\right)^t n + 8$, which implies

$$s(n) \leq 39n + 104(t+3)$$

Moreover, as $\theta(n) \leq \frac{4384}{6561}n$, we have $3^7 < \theta^t(n) \leq \left(\frac{4384}{6561}\right)^t n$. Hence, $t = O(\log n)$ because $\frac{4384}{6561} < 1$. Finer analysis on t shows that $s(n) \leq 40n$. □

Notice that the graph G of Lemma 4.1 is nothing but a $(6m, 2/3, 6, 1/2)$-bounded concentrator. The above construction can be straightforwardly generalized as follows.

Lemma 4.5 (Gabber-Galil, 1981 [34]). *A family of n-superconcentrator of density $\frac{2k+1}{1-\theta}$ can be constructed if for each n an $(n, \theta, k, 1/2)$ bounded concentrator is given.*

As we have already mentioned, Bassalygo (1981, [13]) improved this bound further to $20n$, however we shall not discuss that result here. One might wonder what is known about the lower bound of $s(n)$. Lev and Valiant (1983, [49]) provided the best known so far, whose proof is omitted here.

Theorem 4.6 (Lev and Valiant, 1983). *An n-superconcentrator whose inputs have indegree 0 and outputs have outdegree 0 has size at least $5n - o(n)$.*

The proof of this lower bound is quite involved, and certainly more difficult than the upper bound proof. It is fairly disturbing that the gap between the lower and upper bounds remain quite large. Let us put it as an open problem.

Open Problem 4.7. Close the gap of $5n - o(n) \leq s(n) \leq 20n + o(n)$.

4.2 Superconcentrators with a given depth

The following functions often show up in probabilistic arguments concerning the problem.

Definition 4.8. Let $O^*(f(n))$ denote $f(n)n^{o(1)}$.

Definition 4.9. Let

$$\log^* n := \min\{l \geq 0 \mid \underbrace{\log \ldots \log n}_{l} \leq 1\}$$

where the logarithms are to base 2. By induction on k, define

$$\log^{*^{(k)}} n := \log^{\overbrace{* \cdots *}^{k}} n := \min\{l \geq 0 \mid \underbrace{\log^{\overbrace{* \cdots *}^{k-1}} \ldots \log^{\overbrace{* \cdots *}^{k-1}}}_{l} n \leq 1\}$$

The question of finding the minimum size of an n-superconcentrator with a given depth k was raised by Pippenger (1982, [71]). Let us denote this function by $s(n,k)$. Clearly $s(n,1) = n^2$. In the same paper, Pippenger showed that $\Omega(n \log n) = s(n,2) = O(n \log^2 n)$. Dolev, Dwork, Pippenger and Wigderson (1983, [29]) found that for even depth at least 4,

$$s(n, 2k) = \Theta\big(n \log^{*^{(k-1)}} n\big). \tag{6}$$

Rather surprisingly, Pudlák (1994, [76]) showed that for $k \geq 2$, it is also true that

$$s(n, 2k+1) = \Theta\big(n \log^{*^{(k-1)}} n\big). \tag{7}$$

Hence, when the depth is at least 4 the extra "odd" level does not help improve the superconcentrator size. Alon and Pudlák (1994, [10]) filled part of the gap by determining the minimum size for depth 3, proving

$$s(n,3) = \Theta(n \log n \log n) \tag{8}$$

They also improved Pippenger's lower bound for depth 2 superconcentrators to $s(n,2) = \Omega\big(n(\log n)^{\frac{3}{2}}\big)$. The only one case left where $s(n,k)$ has not been determined (up to a constant factor) is when $k = 2$, rather weird. One would think that the depth-2 case would be easier than larger depth cases. Finally, Radhakrishnan and Ta-Shma (2000, [77]) have determined the last value:

$$s(n,2) = \Theta\left(\frac{n \log^2 n}{\log \log n}\right) \tag{9}$$

We describe here only the initial results of Pippenger on $s(n,2)$ to get a feel of what is going on. In superconcentrators of depth 2, every path has length at most 2. By adding a vertex into every path of length 1, we increase the size of the superconcentrator by at most a factor of 2, which is irrelevant for our purposes. Hence, we may assume that the set of vertices V of our n-superconcentrator can be partitioned into three disjoint subset $V = I \uplus U \uplus O$, where U is the set of middle vertices called *links*.

Let $\mathcal{N}$ be an n-superconcentrator of size N and depth 2. Let X (Y) be a random set of inputs (outputs) where each $v \in X$ ($v \in Y$) appears independently with probability p. Let the random variable x (resp. y) denote the cardinality of X (resp. Y). As $\mathcal{N}$ is a superconcentrator, there exists a set of $m = \min\{x, y\}$ vertex disjoint paths joining X and Y. We first get a lower bound for $E[m]$.

Lemma 4.10. *With the notations just introduced, we have*

$$E[m] \geq np + O\big((np)^{\frac{2}{3}}\big)$$

Proof. Applying Markov's inequality, with $\epsilon \in [0,1]$ to be chosen, we get

$$\begin{aligned} E[m] &\geq np(1-\epsilon)P[m \geq np(1-\epsilon)] \\ &= np(1-\epsilon)P[x \geq np(1-\epsilon), y \geq np(1-\epsilon)] \\ &= np(1-\epsilon)P^2[x \geq np(1-\epsilon)] \end{aligned} \tag{10}$$

As x is the sum of n random indicators, $E[x] = np$ and $Var[x] = np(1-p)$. Now, Chebyshev's inequality gives

$$\begin{aligned} P[x < np(1-\epsilon)] &\leq P[|x - np| \geq np\epsilon] \\ &\leq \frac{Var[x]}{(np\epsilon)^2} \\ &\leq \frac{1}{np\epsilon^2} \end{aligned} \tag{11}$$

Combining inequalities (10) and (11), then set $\epsilon = \left(\frac{2}{np}\right)^{\frac{1}{3}}$ gives

$$\begin{aligned} E[m] &\geq np(1-\epsilon)\left(1 - \frac{1}{np\epsilon^2}\right)^2 \\ &\geq np(1-\epsilon)\left(1 - \frac{2}{np\epsilon^2}\right) \\ &= np\left(1 - (\frac{2}{np})^{\frac{1}{3}}\right)^2 \\ &\geq np\left(1 - 2(\frac{2}{np})^{\frac{1}{3}}\right) \\ &= np + O\left((np)^{\frac{2}{3}}\right) \end{aligned}$$

□

For each $u \in U$, let f_u and g_u be the number of edges directed into and out of u respectively. Then clearly

$$N = \sum_u (f_u + g_u)$$

We say that a link u is *useful* if it is on a path from an input of X to an output of Y. Let Z be the set of useful links and $z = |Z|$. The useful links are "useful" in the proof of the following theorem.

Theorem 4.11.

$$N \geq \frac{2}{3} n \log n + O(n)$$

Proof. The probability that a link u is connected from X is $1-(1-p)^{f_u} \leq f_u p$. Similarly, the probability that u is connected to Y is $\leq g_u p$. Hence,

$$P[u \in Z] \leq \min\{1, f_u g_u p^2\} \leq \min\{1, (f_u+g_u)^2 p^2/4\}$$

This implies

$$E[z] = \sum_{u \in U} P[u \in Z] \leq \sum_{u \in U} \min\{1, (f_u+g_u)^2 p^2/4\}$$

As it is trivial that $E[z] \geq E[m]$, we have

$$np + O\big((np)^{\frac{2}{3}}\big) \leq E[m] \leq E[z] \leq \sum_{u \in U} \min\{1, (f_u+g_u)^2 p^2/4\}$$

To this end, let $k = \lfloor \log n \rfloor$, set $p = 2^{-i}$, multiply both sides of the above equation by 2^i, and then sum it over $1 \leq i \leq k$, we get

$$n \log n + O(n) \leq \sum_{i=1}^{k} \sum_{u} \min\{2^i, (f_u+g_u)^2 2^{-i}/4\} \tag{12}$$

For a fixed $u \in U$, let $j = \lfloor \log(f_u+g_u) \rfloor$ and $t = \log(f_u+g_u) - j$, then $0 \leq t < 1$, and $f_u + g_u = 2^{j+t}$. As the function $2^t + 2^{1-t}$ is convex, we get

$$\begin{aligned} \sum_{i=1}^{k} \min\{2^i, (f_u+g_u)^2 2^{-i}/4\} &\leq \sum_{i=1}^{j} 2^i + \sum_{i=j+1}^{k} 2^{2j+2t-i} \\ &\leq (f_u+g_u)(2^{1-t}+2^t) \\ &\leq 3(f_u+g_u) \end{aligned} \tag{13}$$

Consequently, equations (12) and (13) yield

$$n \log n + O(n) \leq \sum_{u} \frac{3}{2}(f_u+g_u) = \frac{3}{2}N$$

□

4.3 Explicit Constructions and other results

We have seen how to use bounded concentrators to construct superconcentrators. Moreover, it is possible to construct bounded concentrators from expanders, as the following lemma shows. Consequently, explicit constructions of expanders induce explicit constructions of superconcentrators.

Lemma 4.12 (Gabber-Galil, 1981 [34]). *Let $p > 1$ be a fixed integer, and $\theta = \frac{p}{p+1}$. Suppose for each $n \in \mathbb{N}$ such that θn is an even integer we can construct an $(\theta n, k, \frac{2}{p-1})$-expander, then we can construct a family of superconcentrators with density $(2k+3)p + 1 + \epsilon$, where $\epsilon \to 0$ as $n \to \infty$.*

Proof. Firstly, for each n divisible by $(p+1)$ such that $np/(p+1)$ is even, we construct an $(n, \theta, (k+1)\theta, 1/2)$-bounded concentrator $G = (I, O; E)$. By definition, G has n inputs and θn outputs. The inputs are partitioned into two parts: a large part L containing θn vertices and a small part S containing $n/(p+1)$ vertices. The large part is connected to the output by a $(\theta n, k, 2/(p-1))$-expander. Each vertex in S is connected to exactly p consecutive outputs, so that the neighbors of vertices in S are completely disjoint. To show that G is an $(n, \theta, (k+1)\theta, 1/2)$-bounded concentrator, let X be any subset of I with $|X| \leq |I|/2$. Let $l = |X \cap L|$ and $s = |X \cap S|$. We need to show that $|\Gamma(X)| \geq |X|$. When $s \geq |X|/p$, S is connected to at least $|X|$ outputs, so that the inequality holds trivially. When $s < |X|/p$, we must have $l \geq r = \lceil \frac{p-1}{p}|X| \rceil$. Since $|X| \leq |I|/2$, we have

$$r = \left\lceil \frac{p-1}{p}|X| \right\rceil \leq \left\lceil \frac{p-1}{2p}|I| \right\rceil = \left\lceil \frac{p^2-1}{2p^2}|L| \right\rceil \leq \frac{1}{2}|L|$$

Thus, let R be a subset of L with r elements we can now use the fact that (L, O) is an $(\theta n, k, 2/(p-1))$-expander to get $|\Gamma(R)| \geq |X|$. The simple details are omitted.

Secondly, given any n, let n' be the smallest integer $> n$ such that $\theta n'$ is even and construct an $(n', \theta, (k+1)\theta, 1/2)$-bounded concentrator as above. Delete from the small part of this concentrator's inputs $n' - n$ vertices, turning it into an $(n, \theta + \epsilon_n, (k+1)\theta, 1/2)$-bounded concentrator, where $\epsilon_n = O(\frac{n'-n}{n}) = o(1)$. Lemma 4.5 now completes the proof. □

Gabber and Galil used this lemma and their explicitly constructed expanders to construct families of superconcentrators with density 271.8. This was improved by Chung to 261.5 and later by Buck (1986, [22]) to 190. Jimbo and Maruoka (1987, [44]) and Alon, Galil and Milman (1987, [8]) improved Gabber-Galil expanders slightly to obtained superconcentrators of density 122.74. The Ramanujan graphs allow Lubotzky et al. to reduce this to 78. Pippenger pointed out that density 64 is possible using double covers of 8-regular Ramanujan graphs, whose explicit constructions could be found in Morgenstern (1994, [57]). Morgenstern (1995, [59]) also explicitly constructed a family of bounded concentrators not using expanders. This construction yields density 66. All linear superconcentrators constructed above have logarithmic depths. Wigderson and Zuckerman (1999, [88]) constructed a linear-sized superconcentrator with sub-logarithmic depth: $(\log n)^{2/3+o(1)}$.

Open Problem 4.13. The probabilistic bound of $20n$ remains quite far apart from the best explicit construction bound of $64n$. Thus, an open question is to construct n-superconcentrators of size $< 64n$, and as close to $20n$ as possible.

As for limited depth, Meshulam (1984, [55]) constructed depth 2 superconcentrators of size $3n^{3/2} + O(n^{17/12})$, while Wigderson and Zuckerman [88] constructed one with size $O^*(n)$. Since connectors (defined in the next section) are also superconcentrators, the explicit constructions of n-connectors of depths k yield explicit constructions of n-superconcentrators of the same depth. For depth $k = 2j + 1$, n-connectors were constructed with size $O(n^{1+\frac{1}{(j+1)}})$, and depth $k = 2j$, $j \geq 2$ with size $O(n^{1+\frac{2}{3j-1}})$. More information can be found in the next section.

It is worthwhile to mention that there are several variations of concentrators and superconcentrators, which are also models for different types of switching networks. These include self-routing superconcentrators [73], partial concentrators [41, 42, 38], and natural bounded concentrators [59].

5 Connectors

In this section, We discuss the graphical models for rearrangeable, strictly nonblocking and wide-sense nonblocking networks.

Definition 5.1. An *n-connector* is an n-network with inputs I and outputs O so that for any one-to-one correspondence φ between I and O, there exist a set of n vertex disjoint paths joining i to $\varphi(i)$ for each input $i \in I$. When φ is restricted to those of the form $i_k \to o_{k+j \pmod n)+1}$, for some j, the n-connector is called an *n-shifter*.

Definition 5.2. An *generalized n-connector* is an n-network so that: given any one-to-many correspondence ϕ between inputs and disjoint sets of outputs, there exists a set of vertex disjoint trees joining i to the set $\phi(i)$ for all i with $\phi(i)$ defined.

Definition 5.3. An *generalized n-concentrator* is an n-network satisfying the condition that given any correspondence between inputs a nonnegative integers summing up to at most n, there exists a set of vertex disjoint trees that join each input to the corresponding number of distinct outputs.

Generalized concentrators used to be widely referred to as *generalizer*. Here we adopt this terminology from [31] because it is consistent with our overall use of terminologies. Clearly when each corresponding integer is 0 or 1, generalized concentrators are concentrators.

The n-connectors are the graphical version of a rearrangeable networks, while generalized n-connectors are the version for multicast rearrangeable networks. We shall also use the term *rearrangeable n-connector* for n-connector, to emphasize the difference of this network with strictly nonblocking connector and wide-sense nonblocking connector, which are to be defined shortly. Rearrangeable n-connector used to be called *rearrangeable n-network.* As the names suggested, these connectors are graphical versions of strictly nonblocking networks and wide-sense nonblocking networks, respectively.

Definition 5.4. A *strictly non-blocking n-connector* (SNB n-connector) is an n-connector with input set I and output set O, such that for any $i \in I$, $o \in O$ and a set $\mathcal{P}$ of vertex disjoint paths from $I - \{i\}$ to $O - \{o\}$, there is a path from i to o which is vertex disjoint from $\mathcal{P}$.

To formally define wide-sense nonblockingness (WSNBness), we need to settle several other technical concepts, whose counter parts in switching network are clear from the names.

Let $\mathcal{N}$ be a network with input set I and output set O. A *route* in $\mathcal{N}$ is a directed path from an input to an output. Two routes are *compatible* if they share only an initial segment, which could be empty. A *state* of $\mathcal{N}$ is a set of pairwise compatible routes. The set of all states of $\mathcal{N}$ could be partially ordered by inclusion. Hence, we can speak of a state S_1 being *contained in* a state S_2. A vertex or an edge in a state is *busy* if it is in some route of the state, and *idle* otherwise.

A *connection request* is an element (i, o) in $I \times O$. A connection request (i, o) is said to be *satisfied* by a route R if R originates from i and ends at o. A *generalized connection assignment* (GCA) is a set of connection requests, whose outputs are disjoint. A GCA is *realized* by a state if each request of the GCA is satisfied by some route of the state.

A GCA is (a, f)-limited if it contains at most a requests, of which at most f have a common input. *A state is (a, f)-limited* if the maximal GCA it realizes is (a, f)-limited. Often we speak of the maximal GCA realized by a state S as *the* GCA realized by S. A connection request (i, o) is *(a, f)-limited in a state S* if o is idle, and the GCA obtained by adjoining (i, o) into the GCA realized by S is (a, f)-limited.

Now we are ready to place the formal definition of WSNBness.

Definition 5.5. A *WSNB (a, f)-limited generalized connector* is a network for which there exists a collection $\mathcal{S}$ of states called the *safe states*, such that:

(i) The empty state $\emptyset$ is in in $\mathcal{S}$.

(ii) If $S \in \mathcal{S}$, then any state contained in S is also in $\mathcal{S}$.

(iii) Given $S \in \mathcal{S}$ and any (a, f)-limited connection request (i, o) in S, there exists $S' \in \mathcal{S}$ such that $S' \supset S$ and that S' has a route satisfying (i, o).

A *WSNB a-limited connector* is a WSNB $(a, 1)$-limited generalized connector. A *WSBN connector* is a ∞-limited connector. The prefixes "(n, m)-" and "n-" could be appended with the obvious meaning.

5.1 Rearrangeable connectors

As we have mentioned, connectors are models for rearrangeable networks. We put "rearrangeable" in front of "connectors" to emphasize the difference with SNB connectors and WSNB connectors. Let $c(n)$ denote the minimum size of an n-connector, and $c(n, k)$ be the minimum size of n-connectors with depth k. Pippenger and Valiant (1976, [74]) showed that

$$3n \log_3 n \le c(n) \le 6n \log_3 n + O(n)$$

Pippenger (1980, [70]) improved the lower bound of $c(n)$ to be $c(n) \ge 6 \log_6 n + O(n)$, and adopted a comment from the referee James Shearer of his paper to get

$$c(n) \ge \frac{45}{7} n \log_6 n + O(n) \tag{14}$$

which is the best lower bound known so far for $c(n)$.

In the case of connectors with a given depth, it is clear that $c(n, 1) = n^2$. When $k = 2$, de Bruijn, Erdős and Spencer (1974, [28]), while solving a problem of van Lint (1973, [86]), used a probabilistic argument to show $c(n, 2) = O(n^{\frac{3}{2}} \sqrt{\log n})$. Pippenger and Yao (1982, [75]) used an argument from Pippenger and Valiant (1976, [74]) to show

$$c(n, k) = \Omega(n^{1+\frac{1}{k}}) \tag{15}$$

and another probabilistic argument to prove

$$c(n, k) = O(n^{1+\frac{1}{k}} (\log n)^{\frac{1}{k}}) \tag{16}$$

which implies the results by de Bruijn et al.

In this section, we shall present the proofs of the two relations (14) and (15). First, we need a famous result which was conjectured by Minc (1963, [56]) and proved by Brègman (1973, [20]). Let A be a 01-square matrix of order n, and S_n the symmetric group of order n as usual. The *permanent* of A: per A is defined to be

$$\text{per } A = \sum_{\pi \in S_n} \prod_{i=1}^{n} (A)_{i\pi(i)}$$

Theorem 5.6 (Brègman-Minc Theorem, 1973). *Let A be an $n \times n$ 01-matrix where row i sums to r_i, $i = 1, \ldots, n$. Then*

$$per\ A \leq \prod_{i=1}^{n} (r_i!)^{\frac{1}{r_i}}$$

Theorem 5.7 (Pippenger, 1980 [70]).

$$c(n) \geq \frac{45}{7} n \log_6 n + O(n)$$

Proof. Let $G = (I, O; V, E)$ be an n-connector with input set i, output set O, vertex set V and edge set E. We can assume each $v \in V$ has indegree at least 2 and outdegree at least 2. Now, if some vertex v has two inarcs u_1v, u_2v and 2 outarcs vw_1 and vw_2, then we could delete v and add edges u_1w_1, u_1w_2, u_2w_1 and u_2w_2 without changing the rearrangeability of the graph and without increasing the number of edges. Consequently, we may assume each vertex in $V - I \cup O$ has total degree at least 5. Let π be any one to one correspondence between I and O. Let $G' = (V', E')$ be the graph obtained from G by gluing together v and $\pi(v)$ for every $v \in I$, and add a loop to every vertex $v \in V - I \cup O$. Let $d_G(v)$ be the total degree of vertex v in G, then

$$|E| = \frac{1}{2}\left(\sum_{v \in I \cup O} d_G(v) + \sum_{v \notin I \cup O} d_G(v) \right)$$

and,

$$\begin{aligned} |E'| &= \frac{1}{2}\left(\sum_{v \in I \cup O} d_G(v) + \sum_{v \notin I \cup O} d_G(v) + 2|V - I \cup O| \right) \\ &\leq \frac{1}{2}\left(\sum_{v \in I \cup O} d_G(v) + \sum_{v \notin I \cup O} d_G(v) \right) + \frac{1}{5} \sum_{v \notin I \cup O} d_G(v) \\ &\leq \frac{7}{10}\left(\sum_{v \in I \cup O} d_G(v) + \sum_{v \notin I \cup O} d_G(v) \right) \\ &= \frac{7}{5}|E| \end{aligned}$$

Let C be the set of cycle decompositions of G', then since G is rearrangeable, each I to O matching results in a different cycle decomposition in C. Hence $|C| \geq$

$n!$. Clearly, $|C| = \text{per } A$ where A is the adjacency matrix of G', i.e. $A_{uv} = 1$ iff $(u, v) \in E'$ and 0 otherwise. Let r_u be the sum of row u of A, then $\sum_v r_v = |E'|$. It is easy to see that the function $\log(x!)/x^2$ over the positive integers is decreasing if $x \geq 3$ and increasing when $x \leq 3$, namely $\log(x!)/x^2$ gets its maximum at $x = 3$. Theorem 5.6 gives

$$\begin{aligned} |E'| &= \sum_v r_v \geq \frac{3^2}{\log 6} \sum_{v \in V'} \frac{\log(r_v!)}{r_v} \\ &= 9 \sum_{v \in V'} \frac{\log_6(r_v!)}{r_v} \\ &\geq 9 \log_6(\text{per } A). \end{aligned}$$

Thus,

$$\begin{aligned} |E| &\geq \frac{5}{7}|E'| \\ &\geq \frac{45}{7} \log_6(\text{per } A) \\ &= \frac{45}{7} \log_6 |C| \\ &\geq \frac{45}{7} \log_6(n!) \\ &\geq \frac{45}{7} n \log_6 n + O(n) \end{aligned}$$

□

It should be noted that using the same trick for the case where G is undirected gives a lower bound for the size of an undirected n-connector:

$$|E| \geq \frac{18}{5} n \log_6(n) + O(n)$$

Theorem 5.8 (Pippenger-Yao, 1982 [75]). *An n-shifter of depth k has at least $kn^{1+\frac{1}{k}}$ edges.*

Proof. Let $T_k(n)$ be a directed rooted tree with n leaves and depth k where all edges directed to the direction of the leaves. Let $P_1, \ldots, P_n$ be the n paths from the root to the leaves of $T_k(n)$. Let

$$\Delta(T_k(n)) := \sum_{j=1}^{n} \sum_{v \in P_j} outdeg(v)$$

We first show that $\Delta(T_k(n)) \geq kn^{1+\frac{1}{k}}$ by induction on k.

As $\Delta(T_1(n)) = n^2$, the case $k = 1$ is trivial. For $k \geq 2$, supposed the root has degree d, which is connected to d subtrees $T^i_{k-1}(n_i)$, where the tree T_i has n_i leaves, for $1 \leq i \leq d$. Then, since the function $x^{1+\frac{1}{k-1}}$ is convex in x, we have

$$\begin{aligned}
\Delta(T_k(n)) &= dn + \sum_{i=1}^{d} \Delta(T^i_{k-1}(n_i)) \\
&\geq dn + \sum_{i=1}^{d} (k-1) n_i^{1+\frac{1}{k-1}} \\
&\geq dn + (k-1)d \left(\frac{\sum_{i=1}^{d} n_i}{d} \right)^{1+\frac{1}{k-1}} \\
&\geq dn + d(k-1) \left(\frac{n}{d} \right)^{1+\frac{1}{k-1}} .
\end{aligned}$$

Lastly, straightforward calculus completes the induction:

$$dn + d(k-1)\left(\frac{n}{d}\right)^{1+\frac{1}{k-1}} \geq kn^{1+\frac{1}{k}}.$$

Now, let G be an n-shifter. By definition, for each $j = 1, \ldots, n$ there are n vertex disjoint paths P_{ij} joining each input $i \in I$ to output $o \in O$, where $o = i + j \pmod{n} + 1$. Fix i, vary j from 1 to n and assemble the P_{ij} into a tree, keeping only the initial common segments. Call the resulting tree T_i, then T_i has n leaves and depth k. We thus have $\Delta(T_i) \geq kn^{1+\frac{1}{k}}$. Let

$$\mu(i,j,e) := \begin{cases} 1 & \text{if } e \text{ is an arc from a node of } P_{ij} \\ 0 & \text{otherwise} \end{cases}$$

then

$$\sum_{e \in E} \mu(i,j,e) \geq \sum_{v \in P_{ij}} d_{T_i}(v)$$

with strict inequality when some v along P_{ij} got split in T_i. Consequently,

$$\begin{aligned}\sum_{i=1}^{n}\sum_{j=1}^{n}\sum_{e\in E}\mu(i,j,e) &\geq \sum_{i=1}^{n}\sum_{j=1}^{n}\sum_{v\in P_{ij}} d_{T_i}(v)\\ &= \sum_{i=1}^{n}\Delta(T_i)\\ &\geq kn^{2+\frac{1}{k}}\end{aligned}$$

Lastly, as G is rearrangeable, $P_{1j},\dots,P_{nj}$ are vertex disjoint. Thus,

$$\sum_{i=1}^{n}\mu(i,j,e)\leq 1$$

This implies

$$kn^{2+\frac{1}{k}} \leq \sum_{i=1}^{n}\sum_{j=1}^{n}\sum_{e\in E}\mu(i,j,e) \leq \sum_{j=1}^{n}\sum_{e\in E} 1 \leq n|E|$$

which completes the proof. □

Corollary 5.9.

$$c(n,k) = \Omega(n^{1+\frac{1}{k}})$$

Proof. Any n-connector is an n-shifter □

5.2 Non-blocking connectors

In this section, we discuss results on SNB and WSNB connectors. Let $t(n,k)$ be the minimum size of a SNB n-connector of depth k. Let $w(n,k)$ be the minimum size of a WSNB n-connector of depth k. When there is no limitation on the depth, we use $t(n)$ and $w(n)$ respectively.

The multistage Clos network (see [16]) is strictly nonblocking, and thus it gives an upper bound of $O(n^{1+\frac{1}{k}})$ for $t(n,2k)$ and $t(n,2k-1)$, $k\geq 1$. In other words, the extended Clos network gives

$$t(n,k) = O(n^{1+1/\lfloor\frac{k+1}{2}\rfloor}). \tag{17}$$

Moreover, a SNB n-connector is certainly a rearrangeable n-connector, thus the result of Pippenger and Yao mentioned in Corollary 5.9 of the previous section implies that $t(n,k) \geq c(n,k) = \Omega(n^{1+\frac{1}{k}})$. Friedman (1988, [32]) gave the only other known lower bound on SNB n-connector of a given depth k:

$$t(n,k) = \Omega(n^{1+\frac{1}{k-1}}), \tag{18}$$

closing the gap when $k = 2, 3$. Concerning $t(n)$, Pippenger (1978, [69]) showed by a probabilistic argument that

$$t(n) \leq 90n \log_3 n. \tag{19}$$

improving a previous bound of $66 \log n$ by Bassalygo and Pinsker (1973, [14]). Shannon (1950, [80]) was the first to show that $c(n) = \Omega(n \log n)$. The best lower bound for $c(n)$ was shown in Theorem 5.7, which implies

$$t(n) = \frac{45}{7} n \log_6 n + O(n) \tag{20}$$

Open Problem 5.10. Close the gap between the lower bound $\Omega(n^{1+\frac{1}{k-1}})$ and upper bound $O(n^{1+1/\lfloor\frac{k+1}{2}\rfloor})$ of $t(n,k)$ when $k \geq 4$.

A SNB network is also a WSNB network, hence $w(n) \leq t(n)$ and $w(n,k) \leq t(n,k)$. A WSNB network is rearrangeable, hence $w(n) \geq c(n)$ and $w(n,k) \geq c(n,k)$. Thus, we already had

$$w(n,k) = \Omega(n^{1+\frac{1}{k}}). \tag{21}$$

Feldman, Friedman, and Pippenger (1988, [31]) showed that a WSNB generalized connector with a fixed depth k exists, whose size is $O(n^{1+1/k}(\log n)^{1-1/k})$, namely

$$w(n,k) = O(n^{1+\frac{1}{k}}(\log n)^{1-\frac{1}{k}}) \tag{22}$$

Open Problem 5.11. Close the gap between the lower bound $\Omega(n^{1+\frac{1}{k}})$ and upper bound $O(n^{1+\frac{1}{k}}(\log n)^{1-\frac{1}{k}})$ of $w(n,k)$.

In this section, we give an improved version of the result by Friedman, who showed relation (18).

Let $G = (I, O; V, E)$ be a SNB n-connector. Assume that V can be partitioned into stages $V_0 = I, V_1, \ldots, V_k = O$, by adding more edges if necessary, as doing so would not increase the lower bound. For any vertex $v \in V_i$, let $D_L(v)$ ($D_R(v)$) be the set of vertices in V_{i-1} (V_{i+1}) which are connected to v. Also, define $d_L(v) := |D_L(v)|$ and $d_R(v) := |D_R(v)|$.

Let us first gives a lower bound of $|E|$ when $k = 2$.

Lemma 5.12. *For any $i \in I$, $o \in O$, let $V_{io} := D_R(i) \cap D_L(o)$, then*

$$\sum_{v \in V_{io}} \left(\frac{1}{d_L(v)} + \frac{1}{d_R(v)} \right) \geq 1 + \frac{1}{n}$$

Proof. Assume $V_{io} = \{v_1, \dots, v_m\}$. Let σ and τ be two rearrangements of $\{1, \dots, m\}$ such that

$$d_L(v_{\sigma_1}) \leq d_L(v_{\sigma_2}) \leq \cdots \leq d_L(v_{\sigma_m})$$

and

$$d_R(v_{\tau_1}) \leq d_R(v_{\tau_2}) \leq \cdots \leq d_R(v_{\tau_m}).$$

We first claim that there is a $j \leq m$ such that either $d_L(v_{\sigma_j}) \leq j$ or $d_R(v_{\tau_j}) \leq j$. Assume for contradiction that $d_L(v_{\sigma_j}) \geq j+1$ or $d_R(v_{\tau_j}) \geq j+1$ for all $j = 1, \dots, m$. As the request (i, o) must be routed through one of $v_1, \dots, v_m$, if we could find a state of G not involving i and o which uses all vertices $v_1, \dots, v_m$, then we reach a contradiction. As $d_L(v_{\sigma_j}) \geq j+1$ or $d_R(v_{\tau_j}) \geq j+1$ for all $j = 1, \dots, m$, it is easy to find a matching from some subset $\{i_{\sigma_1}, \dots, i_{\sigma_m}\}$ of $I - \{i\}$ onto V_{io} and another matching from V_{io} onto some subset $\{o_{\sigma_1}, \dots, o_{\sigma_m}\}$ of $O - \{o\}$. The set of paths $P_j = i_{\sigma_j} \to v_{\sigma_j} \to o_{\sigma_j}$ use up all vertices in V_{io}, contradicting the fact that G is SNB.

Now, let j be a number $\leq m$ such that $d_L(v_{\sigma_j}) \leq j$ or $d_R(v_{\tau_j}) \leq j$. Notice that $1/d_L(v_i) \geq 1/n$ and $1/d_R(v_i) \geq 1/n$, for all $i = 1, \dots, n$. Thus,

$$\begin{aligned}
\sum_{v \in V_{io}} \left(\frac{1}{d_L(v)} + \frac{1}{d_R(v)} \right) &\geq \left(\frac{1}{d_L(v_{\sigma_1})} + \cdots + \frac{1}{d_L(v_{\sigma_j})} \right) + \frac{m-j}{n} + \\
&\quad \left(\frac{1}{d_R(v_{\tau_1})} + \cdots + \frac{1}{d_R(v_{\tau_j})} \right) + \frac{m-j}{n} \\
&\geq \frac{j}{d_L(v_{\sigma_j})} + \frac{j}{d_R(v_{\tau_j})} + \frac{2m-2j}{n} \\
&\geq 1 + \frac{j}{n} + \frac{2m-2j}{n} \\
&\geq 1 + \frac{1}{n} \qquad (23)
\end{aligned}$$

□

Theorem 5.13. *Let $G = (I, O; V, E)$ be a strictly non-blocking n-connector of depth 2, then G has at size at least $n^2 + n$. Moreover, there exists a strictly non-blocking n-connector of depth 2 and size exactly $n^2 + n$.*

Proof. Given the lemma above, the proof of this theorem is straight forward, as shown below.

$$\begin{aligned} n^2+n=n^2\left(1+\frac{1}{n}\right) &\leq \sum_{i\in I,o\in O}\sum_{v\in V_{io}}\left(\frac{1}{d_L(v)}+\frac{1}{d_R(v)}\right) \\ &= \sum_{v\in V_1}\left(\frac{1}{d_L(v)}+\frac{1}{d_R(v)}\right)\left|\{(i,o)\mid v\in V_{io}\}\right| \\ &= \sum_{v\in V_1}\left(\frac{1}{d_L(v)}+\frac{1}{d_R(v)}\right)d_L(v)d_R(v) \\ &= |E| \end{aligned}$$

We could construct a SNB n-connector of depth 2 and size n^2+n by setting $|V_0|=|V_1|=|V_2|=n$, connecting V_0 to V_1 by a perfect matching using n edges, and connecting every vertex in V_1 to every vertex in V_2 using n^2 more edges. The resulting graph is clearly a SNB n-connector. □

This line of reasoning could be extended to find a lower bound for $t(n,k)$ with $k\geq 3$.

Lemma 5.14. *For any pair $i\in I$ and $o\in O$, let $\mathcal{P}$ be the set of all paths from i to o, and let*

$$\begin{aligned} A_{io} &:= \{v\in V_1\cap V(\mathcal{P})\} \\ B_{io} &:= \{v\in V_{k-1}\cap V(\mathcal{P})\} \end{aligned}$$

Then,

$$\sum_{v\in A_{io}}\frac{1}{d_L(v)}+\sum_{v\in B_{io}}\frac{1}{d_R(v)}\geq 1+\frac{1}{n}.$$

Proof. Let $m=|A_{io}|$ and $m'=|B_{io}|$. Suppose $A_{io}=\{v_1,\ldots,v_m\}$ and $B_{io}=\{u_1,\ldots,u_{m'}\}$. Let σ and τ be two permutations on $\{1,\ldots,m\}$ and $\{1,\ldots,m'\}$, respectively, such that the arrays $\{d_L(v_{\sigma_i})\}$ and $\{d_R(u_{\tau_i})\}$ are weakly increasing. Similar to Lemma 5.12, we claim that there is either a $j\leq m$ such that $d_L(v_{\sigma_j})\leq j$, or a $j'\leq m'$ such that $d_R(v_{\tau_{j'}})\leq j'$. Assume otherwise, then there are distinct vertices $\{i_1,\ldots,i_m\}\subseteq I-\{i\}$ which could be matched one to one onto A_{io}, and distinct vertices $\{o_1,\ldots,o_{m'}\}\subseteq O-\{o\}$ which could be matched one to one onto B_{io}. Now, connect A_{io} to B_{io} by a maximal set of vertex disjoint paths $\mathcal{F}$. Make $\mathcal{F}$ a set of vertex disjoint paths from I to O by adjoining $\{i_1,\ldots,i_m\}$

and $\{o_1, \dots, o_{m'}\}$ as possible. Clearly, there does not exist a new path from i to o disjoint from $\mathcal{F}$. Without loss of generality, we assume there is a $j \leq m$ such that such that $d_L(v_{\sigma_j}) \leq j$. We have

$$\begin{aligned}
\sum_{v \in A_{io}} \frac{1}{d_L(v)} + \sum_{u \in B_{io}} \frac{1}{d_R(u)} &\geq \left(\frac{1}{d_L(v_{\sigma_1})} + \cdots + \frac{1}{d_L(v_{\sigma_j})}\right) + \frac{m'}{n} \\
&\geq \frac{j}{d_L(v_{\sigma_j})} + \frac{m'}{n} \\
&\geq 1 + \frac{1}{n}
\end{aligned} \tag{24}$$

□

Theorem 5.15. *Let $G = (I, O; V, E)$ be a strictly non-blocking n-connector of depth $k \geq 3$, then*

$$|E| \geq \frac{1}{8} n(n+2)^{\frac{1}{(k-1)}}$$

Proof. Let

$$\begin{aligned}
A &:= \{v \in V \mid d_L(v) \geq \frac{4|E|}{n}\} \\
B &:= \{v \in V \mid d_R(v) \geq \frac{4|E|}{n}\}, \text{ and} \\
V' &:= A \cup B
\end{aligned}$$

Since $\sum_v d_L(v) = \sum_v d_R(v) = |E|$, $|A| \leq \frac{n}{4}$ and $|B| \leq \frac{n}{4}$, which imply $|V'| \leq \frac{n}{2}$.

Let $\mathcal{P}$ be a maximal set of vertex disjoint paths from I to O such that each path $P \in \mathcal{P}$ hits at least one member of V'.

Let $\bar{V} := V - V' \cup V(\mathcal{P})$, then the induced subgraph of G on $\bar{V}$: $G|_{\bar{V}} = (\bar{I}, \bar{O}; \bar{V}, \bar{E})$ is strictly non-blocking because any path from I to O which is vertex disjoint from $\mathcal{P}$ does not hit V' due to the maximality of $\mathcal{P}$. Also note that $\bar{I}$ and $\bar{O}$ each has at least $\frac{n}{2}$ vertices, and that d_L and d_R of each $v \in \bar{V}$ are at most $\frac{4|E|}{n}$. We assume $I|_{\bar{V}}$ and $O|_{\bar{V}}$ both have size exactly $n/2$, removing some vertices in $\bar{I}$ and/or $\bar{O}$ if necessary. Let $\bar{V}_i = V_i \cap \bar{V}$, and for any pair $i \in \bar{I}$ and $o \in \bar{O}$, let

$$\begin{aligned}
\bar{A}_o &:= \text{set of vertices in } \bar{V}_1 \text{ which can reach } o \\
\bar{B}_i &:= \text{set of vertices in } \bar{V}_{k-1} \text{ reachable from } i
\end{aligned}$$

Then, clearly $|\bar{A}_o| \leq (4|E|/n)^{k-1}$ and $|\bar{B}_i| \leq (4|E|/n)^{k-1}$. We also define $\bar{A}_{io}$ and $\bar{B}_{io}$ in the sense of Lemma 5.14, then by Lemma 5.14

$$\begin{aligned}
\left(\frac{n}{2}\right)^2 (1+\frac{2}{n}) &\leq \sum_{i\in\bar{I},o\in\bar{O}} \left(\sum_{v\in\bar{A}_{io}} \frac{1}{d_L(v)} + \sum_{v\in\bar{B}_{io}} \frac{1}{d_R(v)} \right) \\
&= \sum_{o\in\bar{O}} \left(\sum_{v\in\bar{A}_o} (\#\ i\text{'s connected to } v) \frac{1}{d_L(v)} \right) + \\
&\quad \sum_{i\in\bar{I}} \left(\sum_{v\in\bar{B}_i} (\#\ o\text{'s connected to } v) \frac{1}{d_R(v)} \right) \\
&= \sum_{o\in\bar{O}} |\bar{A}_o| + \sum_{i\in\bar{I}} |\bar{B}_i| \\
&\leq n\,(4|E|/n)^{k-1}
\end{aligned}$$

This inequality and $k \geq 3$ imply

$$|E| \geq \frac{1}{4}\left(\frac{1}{4}\right)^{\frac{1}{k-1}} n(n+2)^{\frac{1}{k-1}} \geq \frac{1}{8} n(n+2)^{\frac{1}{k-1}}$$

□

Remark 5.16. The original theorem from Friedman gives

$$|E| \geq \frac{1}{32} n^{1+\frac{1}{(k-1)}}$$

There are room for improvement. Firstly the constant 4 in the definition of A and B might still be optimized. Secondly, we could do something better than just considering the second and the next-to-last stage of G.

5.3 Generalized connectors and generalized concentrators

Let $gc(n)$, $gt(n)$, $gw(n)$ and $g(n)$ be the minimum sizes of generalized n-connectors, generalized SNB n-connectors, generalized WSNB n-connectors, and generalized n-concentrators, respectively. Naturally, we also use $gc(n,k)$, $gt(n,k)$, $gw(n,k)$, and $g(n,k)$ to denote the minimum size of the graphs have depth-k.

As we have mentioned in the previous section, Pippenger and Valiant (1976, [74]) showed that

$$3n\log_3 n \leq c(n) \leq 6n\log_3 n + O(n)$$

For $n = 2^k$, Ofman (1965, [62]) implicitly showed

$$gc(n) \leq 10n \log_2 n + O(n)$$

While if $n = 3^k$, Thompson (1977, in a tech report at CMU) showed

$$gc(n) \leq 12n \log_3 n + O(n)$$

Pippenger (1978, [68]) connect the two functions by showing

$$gc(n) = c(n) + O(n)$$

In this proof, Pippenger needed a bound for generalized n-concentrators and showed that generalized n-concentrators with at most $120n$ edges exist. Fan Chung [24] improved this bound to $118.5n$, more precisely

$$g(n) \leq 118.5n + O(\log n) \tag{25}$$

Dolev, Dwork, Pippenger and Wigderson (1983, [29]) probabilistically showed

$$gc(n,k) = O((n \log n)^{1+\frac{1}{k}}) \tag{26}$$

and constructively proved

$$gc(n, 3j-2) = O(n^{1+\frac{1}{j}}) \tag{27}$$

Masson and Jordan (1972, [54]), and Nassimi and Sahni (1982, [60]) gave two different constructions which proves

$$gc(n,3) = O(n^{\frac{5}{3}}) \tag{28}$$

Kirkpatrick, Klawe and Pippenger (1985, [47]) explicitly constructed generalized n-connectors to show

$$gc(n,3) = O(n^{\frac{3}{2}}(\log n)^{\frac{1}{2}}) \tag{29}$$

and extended this to

$$gc(n, 2k-1) = O(n^{1+\frac{1}{k}}(\log n)^{\frac{k-1}{2}}) \tag{30}$$

The upper bound for $gc(2k-1)$ differs very little from that of the best explicit construction bound for $c(n, 2k-1)$ (see next section).

It is easy to see that $gc(n,k) \leq gw(n,k) \leq gt(n,k)$. Feldman, Friedman, and Pippenger (1988, [31]) showed that

$$gw(n,k) = O(n^{1+\frac{1}{k}}(\log n)^{1-\frac{1}{k}}) \tag{31}$$

which in effect gives the best upper bound so far for $gc(n,k)$ also. There is no known upper bound for $gt(n,k)$.

Open Problem 5.17. Find an upper bound for $gt(n, k)$.

Note also that the corresponding lower bounds of $c(n, k)$, $w(n, k)$ and $t(n, k)$ yield lower bounds for their generalized versions.

5.4 Explicit Constructions

Rearrangeable n-connectors with size $O(n \log n)$ were constructed by Beizer [15], and rediscovered by Beneš (1965, [16]), Joel (1968, [45]) and Waksman (1968, [87]). Pippenger (1978, [69]) slightly generalized these constructions to get the bound of $6n \log_3 n$. In the same paper, Pippenger also showed how the classical construction by Slepian, Duguid and LeCorre can be used to construct n-connectors with depth $(2j + 1)$ and size $O(n^{1+\frac{1}{j+1}})$. Another construction based on combinatorial designs by Richards and Hwang (1985, []) gives n-connectors of depth 2 and size $O(n^{5/3}$ (see also [31]). This construction can be used with the previous construction to construct n-connectors of depth $2j$ and size $O(n^{1+2/(3j-1)})$.

Clos (1953, [26]), Cantor (1971, [23]) and Pippenger (1978, [69]) showed that $t(n, 2k - 1) = O(n^{1+\frac{1}{k}})$ by explicit constructions. That $t(n) = O(n(\log n)^2)$ was shown by explicit constructions of Cantor [23] and its generalization by Pippenger [69].

Masson and Jordan (1972, [54]) constructed WSNB generalized n-connector of depth 3 and size $O(n^{5/3})$. Pippenger (1973, [66]) and Nassimi and Sahni (1982, [60]) constructed WSNB generalized connectors with depth $j^2 - 3j + 3$ and size $O(n^{1+2/j})$, implying the results by Masson and Jordan. Note that any construction of SNB network yields also a WSNB network.

Dolev, Dwork, Pippenger and Wigderson (1983, [29]) constructed generalized n-connectors of depth $(3k - 2)$ and size $O(n^{1+\frac{1}{k}})$. Kirkpatrick, Klawe and Pippenger (1985, [47]) explicitly constructed generalized n-connectors of depth 3 and size $O(n^{\frac{3}{2}}(\log n)^{\frac{1}{2}})$, and of depth $2k - 1$ and size $O(n^{1+\frac{1}{k}}(\log n)^{\frac{k-1}{2}})$.

Feldman, Friedman, and Pippenger (1988, [31]) constructed WSNB generalized n-connectors with depth 2 and size $O(n^{5/3})$, and with depth 3 and size $O(n^{11/7})$. Wigderson and Zuckerman (1999, []) constructed depth k WSNB generalized connectors of size $n^{1+\frac{1}{k}+o(1)}$, which is within a factor of $n^{o(1)}$ to the optimal bound of equation (22)).

6 Conclusions

In this chapter, we have surveyed studies on the complexity of switching networks, mostly on the tradeoff between the size and the depth of various types of switching networks. The graph models of different switching networks were investigated: expanders, concentrators, superconcentrators, expanders, rearrangeable, strictly non-blocking and wide-sense non-blocking connectors, plus their generalizations.

Researches on these graphs were thoughroughly collected, with more intuitive results presented. Many open questions were also specified, which hopefully will help practitioners new to this field identify research problems.

This research area is certainly very interesting, has deep connection to many different areas of Mathematics and Computer Science. Algebraic Graph Theory and Probabilistic Method are the two popular tools which are used to deal with questions arising from switching networks. Obviously, researches on switching networks have also enriched techniques and problems in the former two fields.

References

[1] A. V. AHO, J. E. HOPCROFT, AND J. D. ULLMAN, *The design and analysis of computer algorithms*, Addison-Wesley Publishing Co., Reading, Mass.-London-Amsterdam, 1975. Second printing, Addison-Wesley Series in Computer Science and Information Processing.

[2] M. AJTAI, *Recursive construction for 3-regular expanders*, Combinatorica, 14 (1994), pp. 379–416.

[3] M. AJTAI, J. KOMLÓS, AND E. SZEMERÉDI, *Sorting in* $c \log n$ *parallel steps*, Combinatorica, 3 (1983), pp. 1–19.

[4] M. AJTAI, J. KOMLÓS, AND E. SZEMERÉDI, *Generating expanders from two permutations*, in A tribute to Paul Erdős, Cambridge Univ. Press, Cambridge, 1990, pp. 1–12.

[5] N. ALON, *Eigenvalues, geometric expanders and sorting in rounds*, in Graph theory with applications to algorithms and computer science (Kalamazoo, Mich., 1984), Wiley, New York, 1985, pp. 15–24.

[6] N. ALON, *Eigenvalues and expanders*, Combinatorica, 6 (1986), pp. 83–96. Theory of computing (Singer Island, Fla., 1984).

[7] ——, *Eigenvalues, geometric expanders, sorting in rounds, and Ramsey theory*, Combinatorica, 6 (1986), pp. 207–219.

[8] N. ALON, Z. GALIL, AND V. D. MILMAN, *Better expanders and superconcentrators*, J. Algorithms, 8 (1987), pp. 337–347.

[9] N. ALON AND V. D. MILMAN, λ_1, *isoperimetric inequalities for graphs, and superconcentrators*, J. Combin. Theory Ser. B, 38 (1985), pp. 73–88.

[10] N. ALON AND P. PUDLÁK, *Superconcentrators of depths 2 and 3; odd levels help (rarely)*, J. Comput. System Sci., 48 (1994), pp. 194–202.

[11] N. ALON AND Y. ROICHMAN, *Random Cayley graphs and expanders*, Random Structures Algorithms, 5 (1994), pp. 271–284.

[12] D. ANGLUIN, *A note on a construction of Margulis*, Inform. Process. Lett., 8 (1979), pp. 17–19.

[13] L. A. BASSALYGO, *Asymptotically optimal switching circuits*, Problemy Peredachi Informatsii, 17 (1981), pp. 81–88.

[14] L. A. BASSALYGO AND M. S. PINSKER, *The complexity of an optimal non-blocking commutation scheme without reorganization*, Problemy Peredači Informacii, 9 (1973), pp. 84–87.

[15] B. BEIZER, *The analysis and synthesis of signal switching networks*, in Proc. Sympos. Math. Theory of Automata (New York, 1962), Polytechnic Press of Polytechnic Inst. of Brooklyn, Brooklyn, N.Y., 1963, pp. 563–576.

[16] V. E. BENEŠ, *Mathematical theory of connecting networks and telephone traffic*, Academic Press, New York, 1965. Mathematics in Science and Engineering, Vol. 17.

[17] F. BIEN, *Constructions of telephone networks by group representations*, Notices Amer. Math. Soc., 36 (1989), pp. 5–22.

[18] N. BIGGS, *Algebraic graph theory*, Cambridge University Press, Cambridge, second ed., 1993.

[19] M. BLUM, R. M. KARP, O. VORNBERGER, C. H. PAPADIMITRIOU, AND M. YANNAKAKIS, *The complexity of testing whether a graph is a superconcentrator*, Inform. Process. Lett., 13 (1981), pp. 164–167.

[20] L. M. BRÈGMAN, *Certain properties of nonnegative matrices and their permanents*, Dokl. Akad. Nauk SSSR, 211 (1973), pp. 27–30.

[21] A. E. BROUWER, A. M. COHEN, AND A. NEUMAIER, *Distance-regular graphs*, Springer-Verlag, Berlin, 1989.

[22] M. W. BUCK, *Expanders and diffusers*, SIAM J. Algebraic Discrete Methods, 7 (1986), pp. 282–304.

[23] D. G. CANTOR, *On non-blocking switching networks*, Networks, 1 (1971/72), pp. 367–377.

[24] F. R. K. CHUNG, *On concentrators, superconcentrators, generalizers, and nonblocking networks*, Bell System Tech. J., 58 (1979), pp. 1765–1777.

[25] F. R. K. CHUNG, *Spectral graph theory*, Published for the Conference Board of the Mathematical Sciences, Washington, DC, 1997.

[26] C. CLOS, *A study of non-blocking switching networks*, Bell System Tech. J., 32 (1953), pp. 406–424.

[27] D. M. CVETKOVIĆ, M. DOOB, AND H. SACHS, *Spectra of graphs*, Johann Ambrosius Barth, Heidelberg, third ed., 1995. Theory and applications.

[28] N. G. DE BRUIJN, P. ERDŐS, AND J. SPENCER, *Solution 350*, Nieuw Archief voor Wiskunde, (1974), pp. 94–109.

[29] D. DOLEV, C. DWORK, N. PIPPENGER, AND A. WIGDERSON, *Superconcentrators, generalizers and generalized connectors with limited depth (preliminary version)*, in Proceedings of the Fifteenth Annual ACM Symposium on Theory of Computing, Boston, Massachusetts, apr 1983, pp. 42–51.

[30] M. EICHLER, *Quaternäre quadratische Formen und die Riemannsche Vermutung für die Kongruenzzetafunktion*, Arch. Math., 5 (1954), pp. 355–366.

[31] P. FELDMAN, J. FRIEDMAN, AND N. PIPPENGER, *Wide-sense nonblocking networks*, SIAM J. Discrete Math., 1 (1988), pp. 158–173.

[32] J. FRIEDMAN, *A lower bound on strictly nonblocking networks*, Combinatorica, 8 (1988), pp. 185–188.

[33] J. FRIEDMAN, J. KAHN, AND E. SZEMERÉDI, *On the second eigenvalue in random regular graphs*, in Proceedings of the 21st ACM STOC, 1989, pp. 587–598.

[34] O. GABBER AND Z. GALIL, *Explicit constructions of linear size superconcentrators*, in 20th Annual Symposium on Foundations of Computer Science (San Juan, Puerto Rico, 1979), IEEE, New York, 1979, pp. 364–370.

[35] ——, *Explicit constructions of linear-sized superconcentrators*, J. Comput. System Sci., 22 (1981), pp. 407–420. Special issued dedicated to Michael Machtey.

[36] C. D. GODSIL, *Algebraic combinatorics*, Chapman & Hall, New York, 1993.

[37] R. A. HORN AND C. R. JOHNSON, *Matrix analysis*, Cambridge University Press, Cambridge, 1985.

[38] X. D. HU AND F. K. HWANG, *An improved upper bound for the subarray partial concentrators*, Discrete Appl. Math., 37/38 (1992), pp. 341–346.

[39] J. H. HUI, *Switching and traffic theory for integrated broadband networks*, Kluwer Academic Publishers, Boston/Dordrecht/London, 1990.

[40] F. K. HWANG, *The mathematical theory of nonblocking switching networks*, World Scientific Publishing Co. Inc., River Edge, NJ, 1998.

[41] F. K. HWANG AND G. W. RICHARDS, *The capacity of the subarray partial concentrators*, Discrete Appl. Math., 39 (1992), pp. 231–240.

[42] ——, *A two-stage network with dual partial concentrators*, Networks, 23 (1993), pp. 53–58.

[43] J.-I. IGUSA, *Fibre systems of Jacobian varieties. III. Fibre systems of elliptic curves*, Amer. J. Math., 81 (1959), pp. 453–476.

[44] S. JIMBO AND A. MARUOKA, *Expanders obtained from affine transformations*, Combinatorica, 7 (1987), pp. 343–355.

[45] A. E. JOEL, *On permutation switching networks*, Bell System Tech. J., 47 (1968), pp. 813–822.

[46] N. KAHALE, *Eigenvalues and expansion of regular graphs*, J. Assoc. Comput. Mach., 42 (1995), pp. 1091–1106.

[47] D. G. KIRKPATRICK, M. KLAWE, AND N. PIPPENGER, *Some graph-colouring theorems with applications to generalized connection networks*, SIAM J. Algebraic Discrete Methods, 6 (1985), pp. 576–582.

[48] D. E. KNUTH, *The art of computer programming. Volume 3*, Addison-Wesley Publishing Co., Reading, Mass.-London-Don Mills, Ont., 1973. Sorting and searching, Addison-Wesley Series in Computer Science and Information Processing.

[49] G. LEV AND L. G. VALIANT, *Size bounds for superconcentrators*, Theoret. Comput. Sci., 22 (1983), pp. 233–251.

[50] A. LUBOTZKY, *Cayley graphs: eigenvalues, expanders and random walks*, in Surveys in combinatorics, 1995 (Stirling), Cambridge Univ. Press, Cambridge, 1995, pp. 155–189.

[51] A. LUBOTZKY, R. PHILLIPS, AND P. SARNAK, *Ramanujan graphs*, Combinatorica, 8 (1988), pp. 261–277.

[52] G. A. MARGULIS, *Explicit constructions of expanders*, Problemy Peredači Informacii, 9 (1973), pp. 71–80.

[53] ——, *Explicit group-theoretic constructions of combinatorial schemes and their applications in the construction of expanders and concentrators*, Problemy Peredachi Informatsii, 24 (1988), pp. 51–60.

[54] G. M. MASSON AND B. W. JORDAN, JR., *Generalized multi-stage connection networks*, Networks, 2 (1972), pp. 191–209.

[55] R. MESHULAM, *A geometric construction of a superconcentrator of depth* 2, Theoret. Comput. Sci., 32 (1984), pp. 215–219.

[56] H. MINC, *Upper bounds for permanents of* $(0,\ 1)$*-matrices*, Bull. Amer. Math. Soc., 69 (1963), pp. 789–791.

[57] M. MORGENSTERN, *Existence and explicit constructions of* $q+1$ *regular Ramanujan graphs for every prime power* q, J. Combin. Theory Ser. B, 62 (1994), pp. 44–62.

[58] ——, *Ramanujan diagrams*, SIAM J. Discrete Math., 7 (1994), pp. 560–570.

[59] ——, *Natural bounded concentrators*, Combinatorica, 15 (1995), pp. 111–122.

[60] D. NASSIMI AND S. SAHNI, *Parallel permutation and sorting algorithms and a new generalized connection network*, J. Assoc. Comput. Mach., 29 (1982), pp. 642–667.

[61] A. NILLI, *On the second eigenvalue of a graph*, Discrete Math., 91 (1991), pp. 207–210.

[62] J. P. OFMAN, *A universal automaton*, Trudy Moskov. Mat. Obšč., 14 (1965), pp. 186–199.

[63] W. J. PAUL, R. E. TARJAN, AND J. R. CELONI, *Space bounds for a game on graphs*, Math. Systems Theory, 10 (1976/77), pp. 239–251.

[64] ——, *Correction to: "Space bounds for a game on graphs"*, Math. Systems Theory, 11 (1977/78), p. 85.

[65] M. S. PINSKER, *On the complexity of a concentrator*, in Proceedings of the 7th International Teletraffic Conference, Stockholm, June 1973, 1973, pp. 318/1–318/4.

[66] N. PIPPENGER, *The complexity of switching networks*, PhD thesis, Department of Electrical Engineering, Massachusetts Institute of Technology, Cambridge, Massachusetts, 1973.

[67] ——, *Superconcentrators*, SIAM J. Comput., 6 (1977), pp. 298–304.

[68] ——, *Generalized connectors*, SIAM J. Comput., 7 (1978), pp. 510–514.

[69] ——, *On rearrangeable and nonblocking switching networks*, J. Comput. System Sci., 17 (1978), pp. 145–162.

[70] ——, *A new lower bound for the number of switches in rearrangeable networks*, SIAM J. Algebraic Discrete Methods, 1 (1980), pp. 164–167.

[71] ——, *Superconcentrators of depth* 2, J. Comput. System Sci., 24 (1982), pp. 82–90.

[72] ——, *Communication networks*, in Handbook of theoretical computer science, Vol. A, Elsevier, Amsterdam, 1990, pp. 805–833.

[73] ——, *Self-routing superconcentrators*, J. Comput. System Sci., 52 (1996), pp. 53–60.

[74] N. PIPPENGER AND L. G. VALIANT, *Shifting graphs and their applications*, J. Assoc. Comput. Mach., 23 (1976), pp. 423–432.

[75] N. PIPPENGER AND A. C. C. YAO, *Rearrangeable networks with limited depth*, SIAM J. Algebraic Discrete Methods, 3 (1982), pp. 411–417.

[76] P. PUDLÁK, *Communication in bounded depth circuits*, Combinatorica, 14 (1994), pp. 203–216.

[77] J. RADHAKRISHNAN AND A. TA-SHMA, *Bounds for dispersers, extractors, and depth-two superconcentrators*, SIAM J. Discrete Math., 13 (2000), pp. 2–24 (electronic).

[78] H. ROBBINS, *A remark on Stirling's formula*, Amer. Math. Monthly, 62 (1955), pp. 26–29.

[79] Y. ROICHMAN, *Expansion properties of Cayley graphs of the alternating groups*, J. Combin. Theory Ser. A, 79 (1997), pp. 281–297.

[80] C. E. SHANNON, *Memory requirements in a telephone exchange*, Bell System Tech. J., 29 (1950), pp. 343–349.

[81] M. SIPSER AND D. A. SPIELMAN, *Expander codes*, IEEE Trans. Inform. Theory, 42 (1996), pp. 1710–1722. Codes and complexity.

[82] D. A. SPIELMAN, *Constructing error-correcting codes from expander graphs*, in Emerging applications of number theory (Minneapolis, MN, 1996), Springer, New York, 1999, pp. 591–600.

[83] R. M. TANNER, *Explicit concentrators from generalized N-gons*, SIAM J. Algebraic Discrete Methods, 5 (1984), pp. 287–293.

[84] L. G. VALIANT, *On non-linear lower bounds in computational complexity*, in Seventh Annual ACM Symposium on Theory of Computing (Albuquerque, N. M., 1975), Assoc. Comput. Mach., New York, 1975, pp. 45–53.

[85] ——, *Graph-theoretic properties in computational complexity*, J. Comput. System Sci., 13 (1976), pp. 278–285. Working papers presented at the ACM-SIGACT Symposium on the Theory of Computing (Albuquerque, N. M., 1975).

[86] J. H. VAN LINT, *Problem 350*, Nieuw Archief voor Wiskunde, (1973), p. 179.

[87] A. WAKSMAN, *A permutation network*, J. Assoc. Comput. Mach. 15 (1968), 159-163; corrigendum, ibid., 15 (1968), p. 340.

[88] A. WIGDERSON AND D. ZUCKERMAN, *Expanders that beat the eigenvalue bound: explicit construction and applications*, Combinatorica, 19 (1999), pp. 125–138.

MIX
Papier aus verantwortungsvollen Quellen
Paper from responsible sources
FSC® C105338

If you have any concerns about our products,
you can contact us on
ProductSafety@springernature.com

In case Publisher is established outside the EU,
the EU authorized representative is:
Springer Nature Customer Service Center GmbH
Europaplatz 3, 69115 Heidelberg, Germany

Printed by Libri Plureos GmbH
in Hamburg, Germany